UNEMPLOYMENT AND POLITICS

A *Study in English Social Policy*

1886–1914

BY

JOSÉ HARRIS

CLARENDON PRESS · OXFORD

Oxford University Press, Walton Street, Oxford OX2 6DP

London New York Toronto
Delhi Bombay Calcutta Madras Karachi
Kuala Lumpur Singapore Hong Kong Tokyo
Nairobi Dar es Salaam Cape Town
Melbourne Auckland

and associated companies in
Beirut Berlin Ibadan Mexico City Nicosia

Oxford is a trade mark of Oxford University Press

Published in the United States
by Oxford University Press, New York

British Library Cataloguing in Publication Data
Harris, José
Unemployment and politics.
1. Unemployment—Political aspects—
Great Britain   2. Unemployment—Great
Britain—History
I. Title
331.13'7941   HD5765.A6
ISBN 0–19–820072–2

Library of Congress Cataloging in Publication Data
Harris, José.
Unemployment and politics.
Reprint. Originally published: Oxford: Clarendon
Press, 1972. With corrections.
Revision of thesis—Oxford University.
Bibliography: p.
Includes index.
1. Unemployment—Great Britain—History. 2. Labor
policy—Great Britain—History. 3. Insurance, Unemploy-
ment—Great Britain—History. I. Title.
HD5765.A6H37 1984   331.13'7941   84-7937
ISBN 0–19–820072–2 (pbk.)

Printed in Great Britain by
Biddles Ltd, Guildford

To
My Father

ACKNOWLEDGEMENTS

MANY people have helped me in the preparation of this book. I am very grateful to the following for permission to use collections of manuscripts and private correspondence: Mr. Mark Bonham Carter, Mrs. Elizabeth Clay, Sir John Dilke, Sir William Gladstone, Sir Walter Moberly, Sir Steven Runciman, and the Marquess of Salisbury. I am grateful also to the General Secretary of the Fabian Society for use of the Fabian Society Papers at Nuffield College; to the Controller of H.M.S.O. for the use of Crown copyright material in the Public Record Office; and to the librarian of the British Library of Political Science for use of the Beveridge, Braithwaite, Giffen, Lansbury, Macdonald, and Passfield collections. Mrs. Diana Wills and the Provost and Fellows of Corpus Christi College, Oxford, kindly allowed me to refer to the papers of Sir Robert Ensor; and the Provost and Fellows of Oriel College, Oxford, gave me access to the papers of Dr. Lancelot Phelps.

I have also received invaluable assistance from the staffs of many libraries and from friends and colleagues. In particular I wish to thank Mr. C. Allen, of the British Library of Political Science; Mr. K. Mallaber (librarian of the Department of Trade and Industry); Mr. W. Pearson (librarian of the Department for the Environment); Mr. E. Whybrow (librarian of the Department of Health and Social Security); Dr. C. Hazlehurst; the late Professor Joslin; Dr. H. M. O'Connell; and Dr. Fred Reid. The Warden and Fellows of Nuffield College, Oxford, elected me to a research fellowship while my research was in progress; and I also received financial assistance from the Central Research Fund of London University.

My greatest debts are to Professor Richard Titmuss, for his generous advice and encouragement in supervising the doctoral thesis from which this book evolved; and to my husband, without whose persistent criticism it would never have been written.

May 1971   JOSE HARRIS

CONTENTS

INTRODUCTION I

I. SOME PROBLEMS OF THE LABOUR MARKET
  1886–1914 7
  Industrial Unemployment 9
  The Condition of the Unemployed 33
  Summary 47

II. UNEMPLOYMENT AND POLITICAL ACTION
  1886–1896 51
  Unemployment and the Eight-Hours Day 58
  The Campaign for Public Employment 73
  The Select Committees of 1895 and 1896 90

III. THE REGIMENTATION OF THE UNEMPLOYED 102
  Unemployment and Charity 102
  'Home Colonisation' and the Unemployed 115
  Social Salvation and the Unemployed 124
  Poor Law Labour Colonies 135

IV. UNEMPLOYMENT AND LOCAL ADMINISTRA-
  TION 1903–1908 145
  Experiments in London 1903–1905 150
  The Unemployed Workmen Act 1905 157

Machinery and Finance   165

Some Policy Alternatives 1905–1908   180

V. THE DEMAND FOR A NATIONAL POLICY   211

Liberalism and Unemployment   212

The Revival of the 'Right to Work'   235

The Royal Commission on the Poor Laws   245

Liberal Volte-face   264

VI. A SCIENTIFIC POLICY FOR THE UNEMPLOYED   273

The Organization of the Labour Market   278

Unemployment and National Insurance   295

Unemployment and National Development   334

VII. SUMMARY AND AFTERMATH   348

A Critique of Liberal Policies   349

The Origins and Principles of State Intervention   362

Neglected Remedies for Unemployment   366

APPENDICES

A. Social Policy and the Problem of Local Taxation   369

B. Unemployment Statistics before 1914   371

SOURCES AND SELECT BIBLIOGRAPHY   383

INDEX   405

LIST OF TABLES

TABLE 1 Estimate of Able-Bodied Males Relieved by the Poor Law in England and Wales, 1886–1912. 373

TABLE 2 Trade Union Unemployment recorded by the Board of Trade, 1870–1912. 374

TABLE 3 Charity Organisation Society: Cases of Unemployment relieved by London District Committees, 1886–1906. 375

TABLE 4 Salvation Army: Work for the Unemployed in the U.K., 1903–1912. 376

TABLE 5 Numbers Relieved by Distress Committees under the Unemployed Workmen Act, 1905/6 to 1913/14. 377

TABLE 6 Income and Expenditure under the Unemployed Workmen Act. 377

TABLE 7 Labour Exchange Registrations, 1911–1920. 378

TABLE 8 National Insurance Act, 1911; Unemployed Insurance Fund Account, Select Items of Income and Expenditure, 1912–1921. 379

TABLE 9 National Insurance Act, 1911; Relation between Contributions and Income in Working-Class Families, 1912. 380

TABLE 10 Income and Expenditure of the Development Fund, 1910–1931. 381

TABLE 11 Income and Expenditure of the Road Fund, 1910–1921. 382

ABBREVIATIONS USED IN REFERENCES AND FOOTNOTES

Add. MS.	Additional Manuscripts in the British Museum
CAB	Cabinet Papers
COS	Charity Organisation Society
*B.T.	Board of Trade Papers
*Ed.	Board of Education Papers
H. of C.	House of Commons Papers
*H.L.G.	Local Government Board Papers
*H.O.	Home Office Papers
*LAB	Ministry of Labour Papers
*PIN	Ministry of Pensions and National Insurance Papers
RC	Royal Commission
SC	Select Committee

(* Classification used by Public Record Office.)

INTRODUCTION

FOR fifty years after the Poor Law reform of 1834 unemployment as a serious theoretical and practical question was virtually ignored by English economic theorists and social reformers. It was excluded from the analysis of the trade cycle and neglected by the classical economists, most of whom believed in the logical impossibility of a general imbalance between supply and demand.[1] Its effects on the individual were largely overlooked by the mid-Victorian movement for classifying and quantifying social problems;[2] and measures for the prevention and relief of unemployment were in theory prohibited by utilitarian psychology, by fear of overpopulation, and by certain doctrines of Ricardian economics. It was believed by orthodox economic thinkers that gratuitous assistance to the unemployed would depress the level of wages, discourage labour mobility, and put a premium on reckless procreation; and that since the aggregate 'fund' for wages at any given time was inelastic, individual workmen were unemployed simply because they tried to sell their labour at too high a price. Moreover, attempts by public authorities to create 'artificial' employment were self-defeating, since they withdrew capital from private industry and thereby depleted the wages fund of workmen who were privately employed.[3] The truth of these doctrines was thought to have been demonstrated by the experience of the Old Poor Law in relieving able-bodied workmen, and by the failure of the Paris workshops in 1848.[4] Public employment and public relief for the unemployed were therefore

[1] S. G. Checkland, 'The Propagation of Ricardian Economics in England', *Economica*, N.S. 16, no. 61 (Feb. 1949), 40–52.

[2] This movement had concentrated predominantly on education, crime, and health (O. R. Macgregor, 'Social Research and Social Policy in the Nineteenth Century', *British Journal of Sociology*, 8, no. 2 (June 1957), 152).

[3] D. Ricardo, *Principles of Political Economy* (ed. Piero Sraffa 1951), pp. 133–4, 222.

[4] N. Masterman, *Chalmers on Charity* (1900), pp. 178–83; H. Clarence Bourne, *The Unemployed* (COS Occasional Paper No. 30, reprinted from *Macmillan's Magazine*, Dec. 1892).

regarded as both dangerous and futile; and lack of employment was seen either as a voluntary condition which workmen incurred wilfully, or as an inevitable occurrence which they should predict and provide for out of their earnings whilst employed.[1] The institutional embodiment of these doctrines was the New Poor Law, whose deterrent characteristics had been specifically devised to deal with the 'able-bodied' pauper, by driving him into the open labour market and forcing him to retain his independence when out of work.[2]

Nevertheless, these doctrines were never fully accepted or enforced, either inside or outside the administrative structure of the New Poor Law. There were some dissenters, even within the mainstream of classical political economy, from Ricardo's theory of public investment.[3] There were some who rejected the equation of unemployment with idleness and of destitution with moral delinquency.[4] There were many critics of the 'principles of 1834'; and many pragmatic adjustments were made in the administration of the Poor Law to meet the needs of different kinds of pauper, different localities, and 'cases of sudden and urgent necessity'.[5]

It was realized, moreover, by certain mid-nineteenth-century social reformers that prolonged unemployment might be not merely harmful to the welfare of the individual but economically wasteful and socially disruptive. Plug-drawing and machine-breaking were the direct outcome of trade depression;[6] and Nassau Senior found in 1841 that periods of

[1] e.g. W. Chambers, *Misexpenditure*, pp. 7–8.

[2] Cd. 4499/1909, *RC on the Poor Laws, Majority Report*, pp. 201–2.

[3] B. Corry, 'The Theory of the Economic Effects of Government Expenditure in English Classical Political Economy', *Economica*, N.S. 25, no. 97 (Feb. 1958), 34–48.

[4] e.g. Anon. *The Unemployed and the Proposed New Poor Law* (4 Jan. 1843); W. H. Beveridge, *Voluntary Action*, p. 7.

[5] On the series of orders issued by the central Poor Law authority modifying the treatment of able-bodied paupers see H. of C. 365/1895, *SC on Distress from Want of Employment*, Appendix 30, 'Powers and Duties of Guardians and their Officers as to Poor Relief', pp. 557–62. The most important modifications were the Outdoor Relief Regulation Order and the Outdoor Labour Test Order, which in many urban unions permitted guardians to give outdoor relief to the able-bodied, subject to the performance of a task of work.

[6] A. G. Rose, 'The Plug Riots of 1842 in Lancashire and Cheshire' (pamphlet reprinted from *Transactions of the Lancashire and Cheshire Antiquarian Society*, vol. 67, 1957), pp. 75–6, 85–6; G. D. H. Cole (ed.), *Chartist Portraits*, pp. 198–9.

enforced idleness tended to undermine a workman's skill and habits of regularity.[1] Unemployment was therefore a threat to the process of industrialization and to the growth and maintenance of technical competence and social discipline that this process entailed.

For these reasons even people who believed that unemployment was largely self-inflicted did not always adhere to policies consistent with this view; and there were therefore many practical departures from the prohibition on public works, notably the relief works authorized in Ireland in the 1840s and in Lancashire between 1863 and 1866.[2] The existence, moreover, throughout the nineteenth century, of charities, employment agencies, and thrift institutions which gave assistance to surplus workmen suggests that there were many practical deviations from the official policy of repression or indifference towards the unemployed.[3]

These spasmodic forms of assistance almost certainly helped to ward off widespread distress and social disturbance among the working class; but in almost every case unemployment was treated as a localized 'crisis' phenomenon rather than as a problem that was more or less endemic throughout the industrial labour market. It is, moreover, arguable that *ad hoc* remedies for unemployment positively hindered the development of a more constructive policy, by preventing 'free market' principles from being carried to their logical conclusion during the half-century after 1834. This uneasy compromise was irrevocably shattered, however, by the conjunction of certain new factors—economic, political, and administrative—in English society in the 1880s. The debate that grew up around the subject called in question many of

[1] H. of C. 296/1841, *Report of the Commission (under the Great Seal) for Inquiring into the Condition of the Unemployed Hand Loom Weavers in the United Kingdom*, pp. 21–2.

[2] *Public Works in Lancashire for the Relief of Distress Among the Unemployed Factory Hands, during the Cotton Famine, 1863–66* (1898). From the late 1860s onwards there were many proposals for a regular policy of 'useful' public works, modelled on the Lancashire experiment, which would not compete with private enterprise but would absorb the unemployed and prevent their deterioration during periods of depression; e.g. J. H. Stallard, *Pauperism, Charity and the Poor Laws*, 1868; G. Howell, *Waste Land and Prison Labour. A Pamphlet on the Cultivation of Waste Land by Convict and Unemployed Labour* (1877).

[3] Below, pp. 102–5.

the prevailing orthodoxies in political economy, political
theory, and social administration; and by the end of the
decade unemployment was seen by many writers on social
problems as the root of crime, vagrancy, and prostitution,
and as the 'sphinx of the age'.[1]

A symptom of the new-found interest in the problem of
unemployment was the actual conceptualization and definition of the term. Professor T. S. Ashton discovered casual
references to 'unemployment' as early as the 1840s;[2] but the
word was not introduced into the language of political economists until it was used by Alfred Marshall in 1888.[3] The first
formal definition of the unemployment was advanced by J. A.
Hobson in 1895;[4] and 'unemployment' was thereafter rapidly
incorporated into the popular vocabulary of social reform by
the Select Committees on Distress from Want of Employment in 1895 and 1896.[5]

Unemployment henceforward became the concern not
merely of philanthropists and Poor Law administrators, but
of politicians, 'efficiency' experts, and leaders of the organized
working class; and between 1886 and 1914 three new kinds
of policy were advanced by different groups of reformers,
which challenged the efficacy and sufficiency of existing
methods of relieving the unemployed. Firstly, there were
policies that aimed at segregating the unemployed per-

---

[1] H. V. Mills, *Poverty and the State*, p. 97; W. Clarke, 'Industrial Basis of Socialism', in *Fabian Essays in Socialism* (1931 ed.), p. 67.

[2] T. S. Ashton, 'The Relation of Economic History to Economic Theory',
Economica, N.S. 13 (May 1946), 86.

[3] Alfred Marshall, *Official Papers* (ed. J. M. Keynes), p. 92. This was probably
the origin of the reference cited by the Webbs in their *English Poor Law History*,
II. ii. 633, as the earliest known use of the word. The Webbs also stated that
'unemployed' was not used as a substantive noun until 1882; but in fact many of
the works which the Webbs themselves cited had used the word 'unemployed' as a
substantive during the previous fifty years; e.g. *Public Works in Lancashire* etc.,
p. 15.

[4] J. A. Hobson, 'The Meaning and Measure of Unemployment', *Contemporary
Review*, 67 (Mar. 1895), 415–32 and 'The Economic Cause of Unemployment',
ibid. (May 1895), 744–60.

[5] A Select Committee on 'distress from want of employment' was appointed on
13 Feb. 1895, and this rather cumbersome term was used during the first two
months of the Committee's inquiries. On 30 Apr. the Reverend William Tuckwell
referred to 'unemploy' using this term as a noun (H. of C. 363/1895, Q. 6948).
John Burns referred to 'un-employment' on 24 May (Q. 9683) and the Reverend
William Hunt, Secretary of the Social department of the Church Army, discussed
the economic meaning of 'unemployment' on 11 June 1895 (Q. 10189).

manently or temporarily from the normal labour market, either by employing them on artificial relief works, or by confining them in various types of labour colony, or by subjecting them to compulsory schemes of technical training. Secondly, there were policies that aimed to eliminate unemployment through revolutionary changes in the organization and control of industry and society. And thirdly, there were policies that aimed at minimizing the effects of unemployment without either removing the unemployed from the labour market or making major structural changes in the existing industrial system.

Of these three kinds of policy the first were tried and found wanting in a series of abortive experiments in the 1890s and early 1900s, and the second were not politically feasible at any time during this period. Between 1908 and 1914 reforming politicians therefore concentrated on remedies of a more traditional kind, which were designed to streamline the operation of the free market and to promote the self-sufficiency of the unemployed. Nevertheless, by 1914 fatalistic acceptance of the inevitability of the trade cycle and doctrinaire prejudice against the relief of unemployment seemed to have largely passed away. Unemployment had been transformed from a rather peripheral concern of the Poor Law guardians into a central problem of public administration. The Asquith government was in principle committed to a policy of counter-depressive public works. A department of the Board of Trade was responsible for insuring part of the labour force against irregular employment, penalizing employers who gave work on a casual basis, and finding vacant situations for unemployed workmen. Advanced liberals as well as socialists had endorsed the controversial doctrine of the 'right to work';[1] and both liberal and socialist reformers believed that they were moving towards a 'final solution for the problem of the unemployed'.[2]

This study is designed to examine, firstly, the new interpretations that were imposed by social scientists and social

---

[1] L. T. Hobhouse, *Liberalism* (1964 ed., first published 1911), pp. 83–4. See also below, pp. 244, 346.
[2] W. H. Beveridge in the *Morning Post*, 6 Nov. 1905 and 15 Nov. 1905. Fabian Society MSS., Sidney Webb to E. Pease, 12 Aug. 1908.

reformers upon the problem of unemployment and its social consequences between 1886 and 1914. Secondly, the formation and direction of political support for policies of relief and prevention. And, thirdly, the nature of remedial intervention in the problem by voluntary agencies, local authorities, and departments of the central government. Finally, it will be considered how far governmental action was a response to new ways of analysing the unemployment problem; how far it was provoked or constrained by political factors; and how far it was successful in reducing or preventing the incidence of unemployment and in relieving distress among the unemployed. A point that clearly emerges from this discussion and that should be stated in advance is that, although unemployment was increasingly recognized as an 'economic' question, the consideration of practical remedies involved few of the theoretical economic equations that characterized the debate on unemployment after the First World War. The majority of writers on the subject before 1914 did not identify unemployment with deflation, high interest rates,[1] adherence to the gold standard, or maintenance of the value of the pound; and measures for the relief of unemployment rarely came into collision with organized financial interests or with a conscious and articulate Treasury 'point of view'. Throughout this period the history of unemployment policy at all levels— voluntary and statutory, local and central—is therefore primarily concerned with problems of social administration; it is only rarely concerned with the regulation or elimination of unemployment by methods of economic, fiscal, or monetary control.[2]

---

[1] For most of this period high unemployment was assumed to coincide with *low* rates of interest (Cd. 4499/1909, *RC on the Poor Laws, Minority Report*, p. 1198). Beveridge's 'pulse of the nation' chart showed that for most of the latter part of the nineteenth century unemployment had fallen when the official bank rate was low (W. H. Beveridge, *Unemployment: A Problem of Industry* (1910 ed.), p. 44).

[2] Even the Webbs, who came closer than any other contemporary reformers to devising an 'economic' remedy for unemployment, were much more interested in creating an administrative science for the treatment of unemployed workmen analogous to the science of public health (Cd. 4499/1909, *RC on the Poor Laws, Minority Report*, p. 1179).

I

SOME PROBLEMS OF THE LABOUR
MARKET 1886–1914

During the 1880s unemployment was recognized by politicians and administrators for the first time for nearly half a century as one of the most harmful consequences of trade depression and as a chronic social problem among certain sections of the working class.[1] In popular political economy fear of 'over-production' tended to replace fear of overpopulation,[2] and in 1886 the Minority Report of the Royal Commission on the Depression of Trade and Industry drastically redefined the basic economic problem of the community as no longer the 'struggle for existence' but the struggle for work.[3] For the first time the causes and effects of unemployment were the subject of tentative theoretical analysis and of detailed empirical investigation; and many new explanations, both popular and scientific, were advanced for the so-called 'problem of the unemployed'. This chapter will be primarily concerned with the new definition imposed by economists and social scientists on different aspects of the problem of unemployment between 1886 and 1914. The study of defects in the labour market was, however, intimately connected with the search for administrative remedies; and much pioneering work in the quantification and analysis of unemployment was carried out by persons mainly concerned with the practical relief of the unemployed.

Two distinct types of social phenomenon came under con-

---

[1] *Report of the Mansion House Committee appointed March 1885 to Inquire into the Causes of Permanent Distress in London and the Best Means of Remedying the Same* (Dec. 1885), pp. 5–9.

[2] H. V. Mills, *Poverty and the State* (1886), pp. 105–11. The Royal Commission on the Depression of Trade and Industry endorsed the view of the classical economists that 'a general overproduction is of course impossible'; but it claimed that localized overproduction had occurred in many trades (C. 4893/1886, *RC on the Depression of Trade and Industry, Final Majority Report*, p. xvii, paras. 61–2).

[3] C. 4893/1886, *RC on the Depression of Trade and Industry, Minority Report*, p. iv, para. 57.

sideration; firstly, the incidence of industrial unemployment and, secondly, the social distress of the unemployed, which Beatrice Webb in an illuminating analogy compared to the twin spheres of public and private health.[1] The study of 'unemployment' and of the 'unemployed' posed two very different kinds of problem, both practical and conceptual. They were founded on information relating to two very different grades of workmen; information, moreover, that represented different regions, different occupations, and different types of industrial organization. The study of industrial unemployment was primarily based on the monthly returns made to the Board of Trade by highly skilled, benefit-paying trade unions, whose membership was particularly strong in the northern industrial towns.[2] The study of 'the unemployed' on the other hand was based on applications to the Poor Law, relief works, and private charities, which represented mainly unskilled and casual workmen in great commercial centres like London and Liverpool, where there was a rich charitable community but little organized self-help.[3] In both cases the statistical records of persons relieved referred to only a fraction of the total number out of work.[4] Attempts to supplement this information by conducting a 'census' of unemployed workmen proved in the 1880s to be a signal failure,[5] and throughout the

[1] Cd. 5066/1910, *RC on the Poor Laws, Minutes of Evidence*, Q. 78193.

[2] See Appendix B, p. 371.

[3] On the comparison frequently made between the independent, skilled unemployed workmen of the industrial north and the dependent casuals of London, see *Report of the Mansion House Committee, appointed March 1885 to Inquire into the Causes of Permanent Distress in London and the Best Means of Remedying the Same*, Dec. 1885, pp. 8–9; *Report of a Special Committee of the COS on the Best Means of Dealing with Exceptional Distress*, 1866, p. 9; Robert A. Woods, 'The Social Awakening in London', in *The Poor in Great Cities. Their Problems and What is Being Done to Solve Them* (1890), p. 30.   [4] See Appendix B, pp. 371–3.

[5] In 1887 a trial 'census' of the unemployed was conducted by Dr. William Ogle, The Registrar-General's Superintendent of Statistics (C. 5228/1887, *Tabulation of the Statements Made by Men Living in Certain Districts of London in March 1887*, pp. xi–xii, 4–9). The results were, however, grossly distorted because the workmen interviewed connected the survey with the prospect of public relief; and government statisticians thereafter maintained that it was impracticable to hold a 'census' of the unemployed (CAB 37/38/10, Memorandum by H. Llewellyn Smith on The Unemployed, 23 Jan. 1895, pp. 5–9). In this respect Great Britain lagged behind Germany, France, Denmark, and the U.S.A., all of which had incorporated the category of 'unemployed' into their official censuses of population by 1900 (*International Conference of Unemployment* (Paris 1910), Reports Nos. 2, 13, and 15; M. Lazard, *Le Chomâge et la profession* (1909), p. 354).

period under discussion statisticians could only indicate general trends in unemployment and in distress from unemployment.[1] They were unable to calculate precisely either the extent of unemployment, or the age, sex, and occupational composition of those who became unemployed.

INDUSTRIAL UNEMPLOYMENT

During the nineteenth century orthodox English economists had exhaustively analysed the dynamics of the trade cycle;[2] but want of employment they had virtually ignored, treating it merely as a subsidiary branch of wage theory, a prerequisite of labour mobility, and as an insignificant cause of working-class distress. With few exceptions this continued to be the case until 1914. In 1886 Professor Foxwell tentatively ascribed unemployment to the collapse of international prices caused by an inelastic supply of currency and inadequate facilities for credit; and as remedies he prescribed the suppression of speculation, the adoption of a bimetallic standard, and the issue and circulation of silver-based notes.[3] Foxwell's interpretation of the problem was echoed by many adherents of 'bimetallism' in the 1880s and 1890s,[4] but no other leading economic theorist made a detailed study of unemployment until 1913, when A. C. Pigou restated the classical doctrine that it was caused by the failure of wages to adjust to contractions of labour demand.[5] Alfred Marshall scarcely regarded it as a problem of political economy, believing that involun-

---

[1] See Appendix B, pp. 371–3.

[2] T. W. Hutchison, *A Review of Economic Doctrines 1870–1929*, pp. 344–73; J. Schumpeter, *Business Cycles*, i. 162–3.

[3] H. S. Foxwell, 'Irregularity of Employment and Fluctuations of Prices', in *The Claims of Labour* (1896), pp. 232–4, 240–2. Foxwell also proposed that 'public works and permanent improvements of all kinds should be reserved, as far as possible, for the years when prices are low' (ibid., p. 237).

[4] e.g. C. 5512/1888, *RC on Gold and Silver, Minority Report*, paras. 23–4; *Proceedings of the Bimetallic Conference held at Manchester*, 4 and 5 Apr. 1888, 15–16, 28, 31, 83–4, 101. Bimetallists argued that the international price fall during the 'Great Depression' had been caused partly by the failure of world gold supplies to keep pace with commercial expansion, and partly by the conversion of Germany, the U.S.A., and the Latin Union to the gold standard in the 1870s. They claimed that a gold-and-silver standard would increase currency and credit and thus revive trade (F. A. Walker, *International Bimetallism* (1895), pp. 155–217).

[5] A. C. Pigou, *Unemployment* (1913), p. 51.

tary idleness was being progressively diminished by the
expansion of stable service industries, and that residual unemployment was mainly concentrated among those who were
physically and morally incapable of work.[1]

This indifference to unemployment as a problem of economic theory was, however, challenged by writers outside the
orthodox economic school. Socialists ascribed unemployment
to the exclusion of workers from the profits of mechanization;[2]
protectionists to the subsidized competition of foreign producers;[3] and 'single-taxers' to monopolistic restrictions on
the use of land.[4] The 'underconsumptionist' school, headed
by J. M. Robertson and J. A. Hobson, blamed a top-heavy
income structure which gave rise to 'over-saving', to the
simultaneous underemployment of labour, land, and capital,
and to a consequent deficiency of consumer demand;[5] and
Hobson blamed also the shortage of capital in the home
market caused by the growth of British investment overseas.[6]
All these groups of writers were concerned at the failure of
domestic markets to keep pace with the growth of industrial
production, and all of them implicitly or explicitly queried the
classical doctrine that there could be no general maladjustment between supply and demand.

The detailed study of unemployment throughout this

---

[1] Alfred Marshall, *Economics of Industry* (1892 ed.), pp. 360–1; A. C. Pigou (ed.),
op. cit., pp. 446–7; Alfred Marshall to Percy Alden, 28 Jan. 1903. The problem
received no special treatment in Marshall's *Principles of Economics* (1890). But see
J. N. Wolfe, 'Marshall and the Trade Cycle', *Oxford Economic Papers*, N.S. 8
(1956), 90–101, where it is argued that Marshall anticipated the Keynesian school
in perceiving that unemployment in one industry generated unemployment elsewhere in the economy.

[2] Tom Mann, *What A Compulsory Eight Hours Working Day Means to the
Workers* (1886); Fred Hammill, *The Problem of the Unemployed* (1894).

[3] J. Crabb, *Bad Times—the Cause and Cure*, publ. by the National Fair Trade
League, Aug. 1885.

[4] Henry George, *Social Problems* (1884), pp. 123–32.

[5] J. A. Hobson, *The Evolution of Modern Capitalism* (1926 ed., first publ. 1894),
186–9; and *The Problem of the Unemployed. An Enquiry and an Economic Policy*
(1896), pp. 98–111; J. M. Robertson, *The Fallacy of Saving* (1892), pp. 95–7,
138–9.

[6] J. A. Hobson, *Imperialism: A Study* (1902 ed.), pp. 86–9. This was denied
by Robert Giffen, 'Notes on Imports versus Home Production, and Home versus
Foreign Investments', *Economic Journal*, 15 (Dec. 1905), 491; and by C. K.
Hobson, *The Export of Capital* (1913), pp. 220–1, who argued that foreign investment increased domestic employment by increasing overseas demand for British
goods.

period, however, was the work not of economic theorists but of a group of intellectual hybrids, who were concerned partly with general economic hypotheses, partly with sociological investigation, and partly with administrative reform. The most influential members of this group, both practically and theoretically, were Hubert Llewellyn Smith, Charles Booth, William Beveridge, and Sidney and Beatrice Webb, all of whom were interested not merely in the study of unemployment but in institutional remedies for dealing with the unemployed.[1] Their practical involvement in the question was reflected in the nature of their theoretical approach to unemployment, which they saw primarily as a problem of rationalizing the market for labour by administrative means; they were only peripherally concerned with the kind of macroeconomic analysis that later revolutionized the study of unemployment by relating it to public investment and consumer demand.

Two major problems arose over the definition of unemployment. Firstly, the term was not conceptualized until the mid 1890s;[2] and even then there was no general agreement about the circumstances to which it could be applied—whether, for

---

[1] Hubert Llewellyn Smith (1864–1945), statistician, civil servant, historian of the Great London Dock Strike, and promoter of technical education; first Commissioner for Labour, Board of Trade, 1893–1903; Controller-General of the Commercial, Labour and Statistical department 1903–7; Permanent Secretary 1907–19; seconded as Permanent Secretary of the Ministry of Munitions 1915–18; chief Economic Advisor to H.M. Government 1918–27 and Director of the New Survey of London Life and Labour 1928–35.

Charles Booth (1860–1916), Liverpool shipowner and pioneering social investigator; advocate of tariff reform and non-contributory Old Age pensions; a member of the Royal Commission on the Poor Laws 1905–8.

William Beveridge (1879–1963), sub-warden of Toynbee Hall 1903–5; leader-writer on the *Morning Post* 1905–8; civil servant in the Board of Trade, Ministry of Munitions, and Ministry of Food 1908–19; Director of L.S.E. 1919–37, and Master of University College, Oxford, 1937–44; Liberal M.P. for Berwick-on-Tweed 1944–5, and subsequently Liberal leader in the Lords; author of Cmd. 6404/1942, *Social Insurance and Allied Services*.

Sidney Webb (1859–1947), Fabian Socialist, Progressive member of L.C.C. until 1910; drafted Labour Party Constitution of 1918; President of the Board of Trade 1923–4; Baron Passfield 1929; with Beatrice Webb (née Potter 1858–1943) co-author of a *History of Trade Unionism* (1894), *Industrial Democracy* (1897), *English Local Government* (1906–29), and *Soviet Communism: A New Civilisation* (1935). Throughout this study Beatrice Webb is referred to by her married name, although her earliest studies of unemployment dated from before her marriage in 1892.   [2] Above, p. 4.

instance, it could be used to describe short-time working, or periods of 'leakage' between jobs, or idleness directly or indirectly caused by strikes and lockouts. Secondly, although unemployment was increasingly viewed as a pathological condition and a 'disease' of industry it was not at all clear what set of industrial conditions would correspond to a state of health. Sociologists who examined the unemployed in their homes had little doubt about the ideal situation they desired to find, which was as close an approximation as possible to middle-class norms of 'regularity', providence, and good household management; social distress was indeed often defined in terms of deviation from these norms. In the study of industry, however, the most desirable level of employment was not at all self-evident; nor whether this should be assessed with regard to maximum profit or maximum welfare or a compromise between the two. Sociologists and social reformers throughout the period were therefore confused about whether unemployment was an evil to be reduced or even eliminated; or whether it was a necessary by-product of industrial efficiency and economic progress, which should simply be met by increasing the skill, mobility, and self-sufficiency of the unemployed.

During the 1880s and 1890s a classification was gradually evolved of the different kinds of industrial situation in which a workman might be described as unemployed. A committee of the Fabian Society in 1886 first distinguished clearly between the three most obvious types of unemployment—seasonal, cyclical, and casual;[1] and in 1893 Llewellyn Smith, the author of the Board of Trade's unemployment index, devised a two-dimensional classification of different types of trade fluctuation and different types of unemployed workmen, which with slight variations was used as the standard framework of analysis for many years to come.[2]

'Fluctuations' were classified by Llewellyn Smith according to eight hypothetical causes. Firstly, there were three different kinds of seasonal fluctuation, depending on the

[1] *The Government Organisation of Unemployed Labour*, Report by a committee of the Fabian Society, 1886, pp. 3–4.

[2] C. 7182/1893, *Report on Agencies and Methods for Dealing with the Unemployed*, prepared for the Controller-General of the Commercial, Labour and Statistical Department of the Board of Trade by H. Llewellyn Smith.

weather and on the annual rhythm of foreign and domestic
trade. Secondly, there were 'cyclical' fluctuations, which
coincided with general commercial depressions. And, finally,
there were fluctuations determined by changes in fashion,
in the location of industry, in technical and managerial
processes, and by the onset of unpredictable disasters like
the Lancashire Cotton Famine of 1863–6.[1] As a cause of unemployment, Llewellyn Smith remarked that 'incompar-
ably the most serious kind of fluctuations are those I have
spoken of as cyclical oscillations', which were increasing in
frequency and intensity and occurred most severely in the
heavy 'instrumental trades' that set the pace for the rest of the
economy.[2] The 'unemployed' he divided into four groups;
firstly, those engaged for short periods, who were frequently
in transition from job to job; secondly, those subject to
various kinds of trade fluctuation; thirdly, those who were
'economically superfluous', because of an over-supply of
labour in their trades; and, fourthly, those who were unemployed because they were 'below the standard of efficiency
usual in their trades, or because their personal defects are
such that no one will employ them'.[3]

Llewellyn Smith admitted that this classification was rather
abstract and would not stand up to rigorous verification, since
the composition of the 'unemployed' was constantly changing
and definitions of efficiency and redundancy varied with the
level of wages and demand for labour.[4] Nevertheless, he suggested that it was possible to identify different types of unemployment in different trades and at different levels of
industrial organization. At one extreme were the unorganized
'casual' industries in which employment was chronically irregular and workmen 'might at any given moment be counted
with almost equal appropriateness as employed or unemployed'.[5] At the other extreme were the heavy manufacturing
industries, characterized by a high degree of technical skill

---

[1] C. 7182/1893, pp. 10–11.
[2] H. of C. 365/1895, *SC on Distress from Want of Employment, Minutes of Evidence*,
QQ. 4547, 4549, 4839.   [3] C. 7182/1893, pp. 9–10.
[4] C. 7182/1893, p. 10: '. . . the proportion of superfluous labour in any trade is
not an *absolute* but a *relative* quantity; i.e. it depends on the standard of efficiency
and remuneration current in that trade. . . .'
[5] C. 7182/1893, p. 8.

and of trade-union and managerial organization. Employment in these industries was normally stable and regular; but
they were liable to violent epidemics of unemployment when
as much as 30 per cent of their labour force might be thrown
out of work.[1] Fluctuations of this kind distinguished clearly
between the states of 'employment' and 'unemployment'; and
they exercised a selective influence on industry, by concentrating unemployment on the margin of least efficient men.[2]

Between these two extremes lay many variations in stability of employment, industrial organization, and technical
skill; and, as Llewellyn Smith himself pointed out, the impact
of fluctuations upon individual workmen was also influenced
by many other factors, such as the customary length of engagement in a district,[3] or the practice of spreading employment by working 'short time'.[4] Moreover, as an explanation
of the causes of unemployment throughout industry, an
analysis based on the frequency of trade fluctuation had severe
limitations. In particular, it underrated the importance of
local variations in the structure of the labour market and in
the relative ease of communication between potential employers and the unemployed. Communications were much
more efficient in an industry like shipbuilding, where employment was highly localized and controlled by a small
number of large firms, than in the docks, where employment
was geographically concentrated but divided among many
firms, or in the building industry, where over 60,000 employers were dispersed in several hundred labour markets
throughout the United Kingdom.[5]

[1] H. of C. 365/1895, *SC on Distress from Want of Employment, Minutes of Evidence*, Q. 4547; CAB 37/38/10; Memorandum on 'The Unemployed', by H.
Llewellyn Smith, 23 Jan. 1895, p. 10.

[2] H. of C. 365/1895, *SC on Distress from Want of Employment, Minutes of Evidence*, QQ. 4733–5, 4880.   [3] Ibid., Q. 4657.

[4] Ibid., QQ. 4540–1. Short-time working was the normal method of meeting
depressions in 'piece-rate' industries, e.g. mining and shoe-making. In the latter
industry, however, Llewellyn Smith observed that the transition to mass production
was involving a transition to 'time-work' and consequently more 'unemployment'
and less 'short time'.

[5] Beveridge MSS. (first deposit) B. 13, Typescript memorandum on 'Classification of Employers by the numbers employed in engineering, shipbuilding and
building'. On the fragmentation of the London market for builders' labourers see
E. Hobsbawm, 'The Nineteenth Century London Labour Market', in *London.
Aspects of Change*, edited by the Centre for Urban Studies, pp. 6, 8, 10.

Nevertheless, Llewellyn Smith's account of industrial unemployment imposed a certain form and clarity on an extremely diffuse and ill-defined problem. A similar kind of
analysis was adopted by Charles Booth, the architect of the
first major survey of metropolitan poverty, who saw the inter-
relationship between frequency of unemployment and the
internal organization of industry as part of a much wider process of social development. Booth was concerned not merely
with the description and quantification of trade depressions,
but with their evolutionary significance, seeing the 'strange
and monstrous strangulation of over-production' as a challenge
to human ingenuity and a catalyst of technical innovation.[1] He
believed that organization and mechanization tended to reduce unemployment in advanced industries by banishing
inefficient workmen into more archaic forms of production;[2]
and he implied that there was an element of historical inevitability, a necessary stage of industrial and human progress
involved in this state of affairs.[3] Booth was therefore most
interested in those sections of the unemployment problem
where organization was least advanced and where rational
human intervention could most effectively expedite the process of industrial evolution. Hence he focused attention on
the casual labour market, where the conjunction of primitive
organization with the waxing and waning of different kinds
of trade fluctuation was most severe.

The casual labour market was composed of many small
centres of employment, each geographically and organiza-
tionally separate from the rest. A casual labourer had by
definition no security of employment, being hired by the day
or by the hour for a specific task of work. Every morning he
had to compete with other casual workmen at one of the many
centres of employment for a limited number of jobs; and he
had little chance of finding work elsewhere if he failed at his
first place of call. Moreover, the casual labour market harboured a concealed surplus of labour, created partly by
superior labourers who were attracted into casual employment

[1] Charles Booth, *Life and Labour of the People in London*, ix (1897), 166.
[2] Ibid. i (1892), 153–4.
[3] Ibid. ix (1897), 344–5. This passage was actually written by Booth's chief
assistant, Ernest Aves.

in periods of inflated demand and partly by men who were too inefficient or too disreputable to secure regular work. Many employers encouraged this situation, because competition kept wages low in periods of recession and made possible rapid expansion during a boom. Consequently all casual labourers were more or less liable to endemic shortage of work, caused by time-lags between engagements, ignorance of work available at other centres of casual employment, and the almost random distribution of too few jobs among too many men. It was therefore impossible to distinguish clearly between employed and unemployed because, as Booth pointed out, 'we have to deal not with individuals out of work, but with a body of men some of whom are superfluous; though each individual may be doing a job of work the total number of the superfluous is the true measure of the unemployed.'[1]

In London, where there was no single large industry but many workshops, small factories, and daily markets, many trades had a fringe of casual labour; but the case that attracted most attention from social scientists was the docks, which, according to Booth acted as a kind of 'distress meter' to indicate the less obvious ebb and flow of unemployment in other industries.[2] Nowhere was the primordial struggle of men for work more savagely realized than among unskilled labourers in the Port of London. But, if dock employment was chaotic, it was a chaos that seemed incapable of natural evolution, since all the evils of disorganization that prevailed in the 1880s and 1890s had been described by Henry Mayhew forty years before.[3]

The long history of the casual system helps to explain the resistance of both employers and workmen to organizational improvement in the labour market.[4] In the early 1870s dock labourers[5] had enjoyed an age of comparative prosperity,

---

[1] Charles Booth, *Life and Labour of the People in London*, i (1892), 151.
[2] Ibid., p. 42.
[3] Henry Mayhew, *London Labour and the London Poor*, iii. 310–22.
[4] Charles Booth, *Life and Labour of the People in London*, vii (1896), 404.
[5] The term 'dock labourer' refers to unskilled labourers who handled goods on the quayside, as opposed to 'stevedores' who stowed and discharged cargoes on board ship, lightermen who ferried goods to and from ships in deep water, and specialist groups like grain and timber porters, who were more regular and highly paid in the Port of London. These terms were, however, subject to local variation; and on the north bank of the Thames dock labourers were also employed in the

when trade flourished, wage rates advanced, and steam-shipping had reduced their daily dependence on wind and weather.[1] But this prosperity was short-lived, since higher wages had attracted more workmen to the docks; while the competition of Tilbury and continental ports, combined with higher wage bills, had driven dock employers to economize on labour. Hence the introduction of 'more efficient manage-ment, labour-saving machinery and piece-work', which had regularized the employment of some workmen, but intensi-fied competition among the rest.[2]

This was the situation that Beatrice Webb discovered, when as an investigator for Booth's survey of London she visited the docks in 1887. She found that—apart from work-men who performed a specialist function, like stevedores and corn and timber porters—there were three distinct grades of dock employee, 'permanent' men, 'preference' men, and purely casual labourers. 'Permanent' men earning 20s. to 25s. a week were fairly secure in their employment.[3] But 'pref-erence' men, earning 15s. to 20s. a week, shared in the alterna-tion of overwork and unemployment. They had to live near the docks and always be on hand in order not to lose their prior claim to work. 'These men, together with the more con-stant of the casuals are . . . the real victims of irregular trade' wrote Beatrice Webb.

If they be employed by small contractors, unprincipled foremen or corrupt managers, they are liable to be thrust on one side for others who stand drink, or pay back a percentage of the rightful wage. Physically they suffer from the alternation of heavy work for long hours, and the unfed and uninteresting leisure of slack seasons; and the time during which they are 'out o' work' hangs heavily on their hands.[4]

They were subject to continuous competition from, and cor-ruption by, the purely casual class, reinforced by the refuse of other occupations. The latter class earned 12s. to 15s. a week

more highly skilled work of discharging cargoes, stevedores being used only for loading (Cd. 4391/1908, *Dock Labour in Relation to Poor Relief*, Report to the Presi-dent of the L.G.B. by Hon. Gerald Walsh, p. 4).

[1] B. Webb, 'The Docks', *Life and Labour of the People in London*, iv (ed. Charles Booth, 1893, reprinted from the *Nineteenth Century*, Sept. 1887), 13-14.

[2] Ibid., p. 14; *16th Annual Report of the L.G.B.* (1886-7), p. 57.

[3] B. Webb, loc. cit., pp. 24-7.   [4] Ibid., p. 26.

throughout the year; but this average concealed the wide fluctuations in actual earnings, which sabotaged 'thrift, temperance and good management'.[1] Moreover, dock employers claimed that 'after they have taken on the average number of hands they strike a quality of labour which is not worth a subsistence wage.'[2] In other words, casual labour in its lowest form was, like all sweated labour, basically inefficient. It was unprofitable alike to the employer and to the casual workman, who was forced to rely on supplementary forms of income, such as the earnings of wife and children, or on outdoor relief and charity.

Beatrice Webb's account of the situation of third-class dock labourers was in some respects ambiguous. She described their 'life and death' struggle for employment, and then implied that the casuals were a 'leisure' class.[3] She likened casual employment both to sweated labour and to 'a gigantic system of outdoor relief'.[4] She admired the primitive communism of casual life, but deplored its lack of trade-union organization.[5] These vantage-points were not entirely incompatible; but they were an indication both of the complexity of the problems of the docks and of the competing sets of hypotheses that Beatrice Webb brought to social problems at this stage in her career.[6] The diversity of life in the docks was confirmed by Llewellyn Smith's study of metropolitan immigration in the late 1880s which revealed that very few dock labourers had started their working lives on the quayside. He discovered *déclassé* shipwrights, soldiers, valets, and office workers among the casual class and was 'told by dock officials of the son of a general, a clergyman and a baronet, who at various times picked up a living in this way'.[7]

The accounts of Beatrice Webb and Llewellyn Smith were, however, soon rendered obsolete by the Great London Dock Strike of 1889, by which dockers obtained 6*d*. an hour and minimum four-hour engagements.[8] The strike led to the

---

[1] B. Webb, loc. cit., p. 27.   [2] Ibid., p. 30.
[3] Ibid., p. 31.   [4] Ibid., pp. 21, 31.   [5] Ibid., pp. 22, 32.
[6] B. Webb, *My Apprenticeship*, Chapter VI; S. R. Letwin, *The Pursuit of Certainty*, pp. 356–62.
[7] H. Llewellyn Smith, 'Influx of Population (East London)', *Life and Labour of the People in London*, iii. 89.
[8] Vaughan Nash and H. Llewellyn Smith, *The Story of the Dockers' Strike*, p. 152.

formation of two unskilled unions, and was hailed as significant evidence of the dockers' material and organizational
progress.[1] But insecurity of employment was actually enhanced by the strike, firstly because the Joint Committee of
the London and India Dock Companies increased its 'permanent' and 'preference' staffs, thus stabilizing employment for
a few but increasing competition among the rest. Secondly,
the dock companies relinquished much of their control over
unloading to the shipping companies, who did not recognize
the unions and whose employment was entirely casual.
Thirdly, the strike encouraged the further introduction of
machinery and 'American methods' of saving labour.[2] These
changes coincided with other changes in dock employment,
not connected with the strike. The increased draught of
ocean-going steamers meant that many of them could no
longer use the docks, and cargoes had to be stowed and discharged in deep water.[3] This meant more work for lightermen
and stevedores, less for the unskilled quay porters; it also
meant reduced profits for the dock companies, since lightermen and barge-owners were exempt from harbour dues.[4] The
dock companies therefore had a further incentive to economize on labour. Hence the progress towards organization
and decasualization was counteracted by the altered division
of labour between dock companies, barge-owners, and shipping lines, and by intensified competition among the 'residuum' of casual labour.[5]

The development of this new situation was described in

---

[1] Charles Booth, 'Inaugural Address', *Journal of the Royal Statistical Society*,
55 (Dec. 1892), 524–8. On the subsequent history of dock unionism see Cd. 4361/
1908, *Dock Labour in Relation to Poor Relief*, pp. 9–10, and E. Hobsbawm, 'National
Unions on the Waterside', in *Labouring Men*, pp. 204–30.

[2] H. of C. 111/1895, *SC on Distress from Want of Employment, Minutes of Evidence*, QQ. 2007–16, 2317–23.

[3] Cd. 1151/1902, *RC on the Port of London, Report*, pp. 24–9.

[4] Ibid., pp. 78–80. In 1902 it was estimated that 80 per cent of cargoes in London
were loaded directly by lightermen and 75 per cent were unloaded in the same way.

[5] Between 1897 and 1904 the average daily number employed by the dock companies in the Port of London fell from 15,384 to 12,988, while annual tonnage
cleared rose from 8,992,409 to 16,145,277 (Cd. 4391/1908, *Report to the President
of the L.G.B., on Dock Labour in Relation to Poor Relief* by Gerald Walsh, p. 11).
The census of 1901 recorded 19,710 dock labourers in London, but this was almost
certainly an underestimate of those who from time to time sought work in the
docks (ibid., p. 12).

Booth's presidential address to the Royal Statistical Society in 1892, when he gave a dock-by-dock analysis of 'constant', 'nearly constant', and 'irregular' employment. Booth showed that there was more or less regular employment in the Port of London for 14,500 to 15,000 men, out of a dock-labouring clientele of 22,000, of whom 10,000 were trade unionists.[1] The conclusions which Booth drew from this analysis were revealed in the remedies that he advocated before public inquiries into labour questions over the next few years. He proposed the complete decasualization of dock employment, by the transference of necessary workmen to permanent or preference lists, and the exclusion of a residuum of about 6,000 or over a quarter of the existing supply of dockers. In times of acute demand employers would draw, not upon a labour reserve, but upon the unemployed of other industries in which employment fluctuated inversely with that of the docks.[2] The sufferings of the excluded residuum would be a

step towards the cure of the evil in the end. Those for whom there is no longer a living must . . . be gradually absorbed into other industries, or, if the worst comes to the worst, they pass through the workhouse and finally die . . . it also has another influence, and that is of an educational character; it tends to subdue the repugnance to regular work . . . which I think goes hand in hand with the facilities for its indulgence. . . .[3]

Booth's analysis was challenged by Tom Mann,[4] who argued that it was the geographical dispersion of docks, quays, and wharves along 27 miles of waterfront that created isolated pockets of casual labour. Mann therefore proposed that a cutting should be built through the West India Dock, which would bypass $2\frac{1}{2}$ miles of river bend, concentrate dock facilities, and provide construction work for the London unemployed for several years to come. He urged that a single public authority should be set up, modelled on those already

[1] Charles Booth, 'Inaugural Address', loc. cit., pp. 528, 532, 548.

[2] Ibid., pp. 550–4.

[3] H. of C. 365/1895, *SC on Distress from Want of Employment, Minutes of Evidence*, Q. 10534.

[4] Tom Mann (1856–1941), founder of the Eight Hours League and one of the leaders of the Dock Strike of 1889; member of the Royal Commission on Labour 1891–4. See below, pp. 60–72.

established in Bristol, Liverpool, Glasgow, and several continental ports, to control all the docks and wharves, stabilize
trade in the port, and regularize employment.[1] Booth's
analysis was also criticized by Geoffrey Drage, the Secretary
of the Royal Commission on Labour, who pointed out that
Booth gave no adequate definition of the 'superfluous' and
objected to the implication that the term could be used to
describe all unemployed workmen.[2] Nevertheless, Booth's
account of the 'demoralised residuum' was widely accepted,
particularly since in winter after winter in the 1890s and
1900s, schemes for the relief or employment of temporarily
unemployed workmen were flooded out with applications
from the casually unemployed.[3] The analysis of the casual
labour problem that he propounded in the 1890s was essentially that elaborated and publicized by William Beveridge
as the theory of 'under-employment' between 1904 and
1909.[4]

Beveridge, unlike Booth and Llewellyn Smith, first came
to the problem of unemployment as a charitable volunteer,
with little prior knowledge of industrial and social conditions.[5] Both as a private individual and later as a civil servant,
he inquired into the problem with the specific aim of devising
a scheme of relief; and his initial concern was not with the
explanation or even the reduction of unemployment, but with
the 'preservation of efficiency' among workmen while they
were unemployed.[6] Nevertheless he soon came to the con-

---

[1] C. 7063–I/1893, *Minutes of Evidence taken before the Royal Commission on Labour sitting as a Whole*, Q. 2171. Mann's proposal for a single statutory public authority was echoed by the Royal Commission on the Port of London (Cd. 1151/1902, *Report*, pp. 111–24), and was implemented in 1908–9 by the creation of the Port of London Authority (D. Owen, *Ports of the United Kingdom*, pp. 51–60). A public harbour authority was in no sense a guarantee of decasualization, however, as was shown by the case of Liverpool where, in spite of the existence of the Mersey Docks and Harbour Board the ratio of men employed to tonnage cleared was much greater than in London. Moreover, the 'preference' system was actually declining in the 1890s and 1900s, owing to the resistance of the Dockers' Union (Eleanor Rathbone, *Report of an Inquiry into Conditions of Dock Labour at the Liverpool Docks*, 1904).   [2] G. Drage, *The Unemployed*, p. 145.

[3] *Report of a COS Special Committee on the Relief of Distress Due to Want of Employment*, 1904, p. 11.

[4] W. H. Beveridge, 'Labour Exchanges and the Unemployed', *Economic Journal*, 17 (Mar. 1907), 74.   [5] W. H. Beveridge, *Power and Influence*, p. 23.

[6] W. H. Beveridge, 'Unemployment in London: the Preservation of Efficiency', *Toynbee Record*, Dec. 1904, pp. 43–7.

clusion that an over-emphasis on the relief of individuals was obscuring the crux of the problem, which lay in faults in the organization of the market for labour.[1]

His account of industrial maladjustment at first adhered closely to Booth's, concentrating on the disguised surplus of workmen in the labour market. In a paper read to a conference at the London School of Economics early in 1906 he proposed that the State should subsidize organized thrift and regulate naval shipbuilding in order to soften the impact of depressions; and that the casual labour market should be subject to compulsory decasualization.[2] Once employment had been stabilized and organized in these ways, those who were still without work and in need of public assistance should be disfranchized, confined in detention centres, and deprived of the right to marry and bear children. This underprivileged remnant was, moreover, to consist not merely of those who were technically 'unemployable'. To those who 'may be born personally efficient, but in excess of the number for whom the country can provide, a clear choice will be offered; loss of independence by entering a public institution, emigration or immediate starvation.'[3]

This formula was strongly criticized by other members of the conference, notably by Hobson and by the statistician, Arthur Bowley. Hobson attacked the view that unemployment was exclusively a problem of labour rather than a 'problem of the simultaneous unemployment of all the factors of production', caused by the restriction on national consumption imposed by the unequal distribution of wealth.[4] Bowley objected that more evidence was needed about the actual incidence of unemployment and the character and identity of the unemployed before the cure for unemployment could be reduced to such a simple and inexorable process as Beveridge had described;[5] to which Beveridge retorted that 'the mere number of the unemployed does not mean anything' and that the collection of further statistical data would be irrelevant to the inherent logic of his decasualization scheme.[6]

---

[1] Cd. 5066/1910, *RC on the Poor Laws, Minutes of Evidence*, Q. 77832, para. 2.
[2] W. H. Beveridge, 'The Problem of the Unemployed', *Sociological Papers*, 3 (1906), 328–31.   [3] Ibid., p. 327.
[4] Ibid., pp. 332–4.   [5] Ibid., pp. 337–8.   [6] Ibid., p. 341.

Beveridge's analysis of the problem was, however, broadened by further empirical research, particularly into highly unionized as well as casual industries; and his scheme of reform was modified by the constraints of personal involvement in public administration.[1] This was apparent in the evidence that he collected for the Board of Trade in 1907-8, and published as *Unemployment: a Problem of Industry* in 1909.[2] In this work Beveridge examined and dismissed many of the rival interpretations of unemployment that were being advanced in the 1900s. Firstly, it was not caused by overpopulation, since all economists agreed that national wealth and national income had grown much faster than population over the previous hundred years. Certain industries might be overstocked, but this was a function of their internal structure rather than of an over-all imbalance in the supply and demand for labour;[3] and having come to this conclusion Beveridge had no need to pursue his earlier proposal for the penal repression of surplus workmen. Secondly, he dismissed 'monetary' explanations of the problem. Unemployment was not caused by an inelastic gold supply, because 'precious metals form but an insignificant part in the actual means of exchange', and because fluctuations in the gold supply showed no significant correlation with fluctuations in the percentage of unemployed.[4] Thirdly, unemployment could not be ascribed to under-consumption, since 'Mr. Hobson's thesis really explains little or nothing that cannot be explained as mere misdirection of productive energy.'[5] Nor could it be eliminated by investment in public works, since these had either to be profit-making and therefore liable to displace workmen in private industry; or else non-profit-making, in

[1] Below, pp. 200 ff.

[2] The first edition of this work was largely based on Beveridge's evidence to the Royal Commission on the Poor Laws, and the research papers that he composed for the Commission on behalf of the Board of Trade (Cd. 5066/1910, QQ. 77831–78379; Cd. 5068/1910, Appendix XXI (A), (B), (C), and (K). Draft copies of these memoranda survive in Beveridge MSS. (first deposit), Parcel 2, Folder D.

[3] Beveridge, *Unemployment* (1910 ed.), pp. 4–7.

[4] Ibid., p. 57. It should be noted, however, that monetary explanations of unemployment, and in particular 'instability of credit' played an important explanatory role in later editions of Beveridge's work (*Unemployment* (1930 ed.), pp. 326–33).

[5] Beveridge, *Unemployment* (1910 ed.), p. 59.

which case they would demoralize the workmen by accepting low standards of efficiency.[1]

Beveridge's notes to Llewellyn Smith on unemployment insurance in 1908–9 reveal that he was aware of the inter-dependence of contractions of employment and contractions in the level of consumer demand.[2] But he never fully explored this aspect of the unemployment problem before the 1940s.[3] Instead he identified three main types of unemployment, caused firstly by the decay of particular industries, secondly by the retention of a surplus of casual labour, and thirdly by various kinds of trade fluctuation. Of these he selected the first two as the most explicable and the most amenable to reforms of industrial organization.[4]

Strictly speaking, unemployment in obsolescent industries was caused, not by the actual process of decline, but by the failure of the free labour market to divert labourers from such industries and to redirect them to alternative employment.[5] This deficiency might, Beveridge suggested, be supplied by labour exchanges which could introduce workmen to poten-tial employers, rationalize the process of job selection, pro-mote mobility throughout the labour market, and minimize the employment 'leakage' incurred by workmen in transition from district to district and from job to job. Industries that engaged their workmen on a casual basis were retarded rather than declining; but the same process of rationalization through labour exchanges could be used to identify and re-

---

[1] Beveridge, *Unemployment* (1910 ed.), p. 195.

[2] Beveridge MSS. (first deposit), Parcel 2, Folder A, 'Questions and Answers', submitted by H. Llewellyn Smith to W. H. Beveridge, Oct. 1908, Q. 7: 'Different trades are to a very considerable extent dependent upon one another. Each finds its market very largely in the purchasing power of men in other trades. Labour in every trade is depressed by unemployed men from other trades seeking work on desperation terms.'

[3] In 1944 Beveridge claimed that his proposals of 1909 were 'not contradictory but complementary' to the policies subsequently advocated by the Keynesian school (W. H. Beveridge, *Full Employment in a Free Society*, p. 106). But he admitted that he and most other writers before 1914 had greatly underestimated the effect upon unemployment of deficiencies in consumer demand (ibid., p. 91.). On the difficulty with which Beveridge eventually accepted the Keynesian analysis, see Harold Wilson, *Beveridge Memorial Lecture*, delivered to the Institute of Statisti-cians, 18 Nov. 1966, pp. 2–3.

[4] Beveridge, *Unemployment* (1910 ed.), pp. 13–14.

[5] Ibid., pp. 114–16.

deploy the invisible surplus of workmen among the casually employed.[1]

By 1907 Beveridge had seriously modified the rather artificial account of the casual labour problem that had been advanced by Booth in the early 1890s. It was not, he realized, a problem virtually sealed off from the labour market as a whole, since even highly sophisticated industries like shipbuilding harboured pockets of casual employment. Hence, although 'under-employment' was caused primarily by geographical dispersion and labour congestion in certain trades, it also occurred in other cases because employers found it convenient to give employment on a casual basis or because workmen voluntarily chose a casual way of life.[2] Moreover, it was not in practice possible to apply the clearly defined causal distinctions to different instances of unemployment that had been suggested by Llewellyn Smith. Instead, Beveridge pointed out that

a riverside labourer in Wapping during February 1908 might be suffering at one and the same time from chronic irregularity of employment, from seasonal depression of his trade, from exceptional or cyclical depression of trade generally, from the permanent shifting of work lower down the river and from his own deficiencies of character or education.[3]

Beveridge conceded that in the last resort the problem could not be solved without an understanding of the causes of cyclical fluctuation. But he assumed that these causes were still mysterious and that fluctuations 'probably cannot be eliminated without an entire reconstruction of the industrial order'.[4] Since he was not concerned with promoting an economic revolution,[5] but simply with enhancing the inherent logic of the free market system, Beveridge therefore concluded that within the context of practical politics fluctuations were incurable. Like Booth he believed that they were also a precondition of general economic advance. 'Trade fluctuation is indeed at times most obviously and directly the means

[1] Ibid., pp. 201–11.   [2] Ibid., pp. 96–101.
[3] Ibid., p. 3.   [4] Ibid., p. 67.
[5] For Beveridge's views on the politics of social welfare see Beveridge MSS., I, b. 356, 'Unemployment in Utopia', address by W. H. Beveridge to L.S.E. Students' Union (dated 1905 but from its contents clearly given in 1907/8).

by which the standard of comfort is driven upwards ... each
wave leaves wages higher or prices lower and productivity
higher than did the wave before.'[1]

Finally, he agreed with Booth that woven into all the other
causes of unemployment was the 'character' and 'capacity' of
the worker. Beveridge thought it was unlikely that personal
characteristics had any effect on the total volume of employment; but within a competitive system, personal deficiencies
of skill, physique, or morality would determine the order of
dismissal, and the accuracy of this selective process would be
improved by the rationalization of the market.[2] Even so,
Beveridge emphasized the relativity of the concept of the
'unemployable', which depended largely on the prevailing
level of wages and the work available. A man could only be
defined as unemployable within the context of his particular
trade, and 'the best carpenter in the world is unemployable as
a compositor'.[3] At the bottom of the industrial system, however, there lurked a class of parasites who

cannot appropriately be described as men out of work because they
are never in work ... each of these is ... as definitely diseased as are
the inmates of hospitals, asylums and infirmaries and should be classed
with them. Just as some suffer from distorted bodies and others from
distorted intellects, so these suffer from a distortion of judgment, an
abnormal estimate of values, which makes them, unlike the vast
majority of their fellows, prefer the pains of being a criminal or vagrant
to the pains of being a workman.[4]

Between 1905 and 1909 the problems of the labour market
were exhaustively considered by the Royal Commission on
the Poor Laws, which revealed that, in spite of a long-term
decline in able-bodied pauperism, the impact of fluctuations
was becoming more frequent and the pressure for work in the
lower ranks of industry more severe.[5] Discussion of the
practical recommendations of this Commission must be deferred until a later chapter; but, in discussing innovations in
social policy, both the Majority and Minority Reports of the

---

[1] Beveridge, *Unemployment* (1910 ed.), p. 64.   [2] Ibid., pp. 134, 138–9.
[3] Ibid., p. 135.   [4] Ibid., p. 134.
[5] Cd. 4499/1909, *RC on the Poor Laws, Majority Report*, pp. 360–3. All references,
except where otherwise stated, are to the folio edition of the Royal Commission's
Majority and Minority Reports.

Commission made a specific contribution to the study of industrial unemployment and to the analysis of cyclical depressions of trade.

The commissioners who signed the Majority Report accepted the classical doctrine that there could be no over-all deficiency in the demand for labour;[1] but they remarked that the introduction of machinery and other labour-saving devices frequently led to local and temporary maladjustment in supply and demand.[2] They suggested, moreover, that the whole of commerce and industry was based on the anticipation of a consumer demand so capricious as to be virtually unpredictable;[3] and they pointed out that the contraction of employment in a single industry had depressive repercussions on all other branches of trade.

Some workers are thrown out of work. Having no wages, they buy less from the shops. The shops with unsold goods on their shelves and diminished takings in their tills, cannot give the usual orders to merchants and manufacturers who supply them. . . . This contraction in demand in course of time affects shipping and all transit trades and the persons in them curtail their purchases. . . . At every step some shop or other . . . is affected and spreads the contagion back to those who fill its shelves. And so the dislocation spreads; until the thoughtful observer wonders how industry is ever to get a start out of the depression and stagnation.[4]

The Minority Report, drafted by Sidney and Beatrice Webb, did not consider the long-term causes of industrial fluctuation; but it proposed that the volume of employment should be stabilized and increased by the concentration of public expenditure into periods when the private demand for labour was slack and interest rates were low.[5] This recom-

[1] Bryce MSS. Box E. 28, Lancelot Phelps to James Bryce, 29 Mar. 1909. Phelps was chairman of the committee that drafted the Majority Report (below, p. 260).

[2] Cd. 4499/1909, *RC on the Poor Laws, Majority Report*, p. 345, paras. 222–3.

[3] Ibid., pp. 329–30, paras. 147–50.

[4] Ibid., p. 331, para. 151. This part of the Majority Report was composed by William Smart, Professor of Political Economy at Glasgow University (Phelps MSS., Lancelot Phelps to Lord George Hamilton, 27 July 1908).

[5] Cd. 4499/1909, *RC on the Poor Laws, Minority Report*, pp. 1195–8. A similar proposal had been advanced by the Fabian Society as early as 1886 (*The Government Organisation of Unemployed Labour. Report by a Committee of the Fabian Society*, 4 June 1886, p. 15).

mendation has often been seen as the germ of later devices for regulating the level of unemployment by monetary controls;[1] and it was an implicit rejection of the doctrine of an automatic adjustment between supply and demand.

The Webbs themselves, however, were less concerned in the 1900s with the economic regulation of unemployment than with devising a new administrative science for the treatment of the unemployed.[2] They rejected as inadequate all existing categorizations of unemployment according to frequency of fluctuations, personal character, or industrial skill; and instead they proposed a new classification of unemployed workmen, based on the regularity of the situations in which they had been previously engaged. This consisted, firstly, of workmen normally in permanent employment; secondly, of those who shifted continuously from job to job; thirdly, of casual labourers; and, fourthly, of those who had been 'ousted' or had 'wilfully withdrawn themselves' from the ranks of the employed.[3] The similarity between this classification and that previously adopted by Booth, Llewellyn Smith, and Beveridge tends to disguise the basic antagonism between their views of the unemployment problem and that of the Webbs. Like Booth the Webbs were evolutionists; and like Beveridge they were most immediately concerned in the 1900s with the reduction of unemployment among casual workmen and the 'under-employed'. But they had no respect for the 'evolutionary' function of trade depressions, and they believed that unemployment at all levels of industry was wasteful, unnecessary, and personally destructive to the individual unemployed.[4] They regarded the 'free market', not as basically efficient, but as crude and irrational;[5] and they looked forward to a time when all types of unemployment would be eliminated and all aspects of the movement and employment of labour subjected to rational administrative control.[6]

---

[1] R. Skidelsky, *Politicians and the Slump*, pp. 33–5; T. Hutchinson, *A Review of Economic Doctrines*, pp. 414–17.

[2] Cd. 4499/1909, *Minority Report*, p. 1179.

[3] Ibid., pp. 1131–2.   [4] Ibid., p. 1177.

[5] S. and B. Webb, *Industrial Democracy* (1897), ii. 654–74.

[6] Cd. 4499/1909, *Minority Report*, p. 1215. On the details of the Webbs' practical proposals, see below, pp. 258–9.

The impact of fluctuations and the disorganization of the labour market were the two central themes pursued by social scientists who examined the causes of industrial unemployment between 1886 and 1914. There were, however, several subsidiary themes, of which the three most important were, firstly, the pressure of various kinds of immigration; secondly, the premature employment of juvenile labour; and thirdly, deficiencies in technical education and industrial skill.

During the 1880s and the 1890s local administrators and trade unionists frequently ascribed the shortage of employment in commercial and industrial centres to the competition of alien immigrants[1]—particularly Jewish refugees from East Europe, who were accused of displacing British workmen by working for long hours at sub-standard rates.[2] Official inquiries into the impact of alien immigration, however, could find no conclusive evidence to prove that this was the case.[3] Robert Giffen, the head of the Commercial, Labour and Statistical department of the Board of Trade, claimed in 1891 that Jewish immigrants had actually increased employment by opening up previously undeveloped branches of industry; and he argued that any increase of competition for employment must necessarily be balanced by increased consumer demand.[4] Llewellyn Smith, who investigated the problem for

[1] M.H. 19/203, 'Majority Report of the Committee of Metropolitan Poor Law Guardians on the Immigration of Foreign Poor', 13 Apr. 1891; *Report of the Bradford T.U.C.* 1888, p. 41; *Report of the Liverpool T.U.C.*, 1890, p. 47; *Report of the Glasgow T.U.C.*, 1892, pp. 29, 69; *Report of the Norwich T.U.C.*, 1894, p. 59; *Report of the Cardiff T.U.C.*, 1895, p. 45.

[2] 120,000 Jewish immigrants settled in England between 1870 and 1914, mainly in London, Leeds, Manchester, and other commercial centres (Lloyd P. Gartner, *The Jewish Immigrant in England 1870-1914*, p. 30). Fluctuations in emigration and immigration closely coincided, the former rising and falling slightly in advance of the latter (CAB 37/30/30, Board of Trade Report on 'Emigration and Immigration for 1890', circulated to the Cabinet, 2 June 1891, pp. 2-3); but in no year during this period did immigration exceed emigration, even when immigration reached a peak in the mid 1900s (B. R. Mitchell and P. Deane, *Abstract of British Historical Statistics*, p. 50, Table B).

[3] H. of C. 311/1889, *SC on Emigration and Immigration (Foreigners) Report*, pp. ix–xi; M.H. 19/203, Summary of Replies from East End Vestries, 4 Apr. 1891; H.L.G. 77, ff. 116–18, Summary of Replies Received from the Guardians of the Metropolitan Unions and of certain Provincial Unions as to the Number of Destitute Aliens, especially Russian and Polish Jews, relieved by them during the Year 1893.

[4] CAB 37/30/31, 'Immigration of Foreigners', Memorandum by Robert Giffen, 6 June 1891, pp. 11–13.

the Royal Commission on Alien Immigration in 1903, could
find no statistical evidence that immigrants caused 'displacement';[1] and he showed that wages were rising among English
workmen in trades where aliens were most commonly employed.[2]

A much more serious industrial problem was the influence
upon urban labour markets of immigration from the countryside. At the beginning of the 'Great Depression' it was often
assumed that the urban unemployed were largely composed
of redundant agricultural labourers who had travelled to the
cities in search of work.[3] This pattern of explanation was,
however, modified by the discovery that agricultural labourers
were leaving the countryside faster than was warranted by
decline of cultivation.[4] Moreover, Llewellyn Smith showed
in 1888 that far from joining the ranks of the urban unemployed, rural immigrants were preferred by employers for
their superior health and physique.[5] His analysis of the 'case
papers' of unemployed waterside labourers who applied to
a Mansion House relief committee in 1892–3 revealed that
less than 1 per cent of those relieved originated from rural
areas.[6] This was endorsed by another Board of Trade investigator, Arthur Wilson Fox, who discovered that country-
born workmen in large towns were very rarely unemployed.[7]

[1] Cd. 1742/1903, *RC on Alien Immigration, Minutes of Evidence*, Q. 22656;
Cd. 1741–I/1903, *RC on Alien Immigration, Appendix to Minutes of Evidence*,
pp. 18–19, Tables XIV–XV.

[2] Cd. 1742/1903, QQ. 22494–5; Cd. 1741–I/1903, pp. 20–6, Tables XVI–XVIII.
See also Gerald Balfour MSS., P.R.O. 30/60/45, Gerald Balfour to 'Bob', 12 Dec.
1903. Balfour, at this time President of the Board of Trade, argued that aliens
increased the level of consumption more than they competed in production; he
conceded, however, that fear of the 'competition of aliens in the labour market'
was mainly responsible for the popular demand for restrictive legislation.

[3] Alsager Hill, 'Unemployed Labour. What Means are Practicable for Checking
the Aggregation and Deterioration of Unemployed Labour in Large Towns?',
Transactions of the National Association for the Promotion of Social Science, 1875,
pp. 657–8.

[4] C. 3309/1882, *RC on Agriculture, Final Report*, pp. 26–7.

[5] H. Llewellyn Smith, 'Influx of Population (East London)', *Life and Labour
of the People of London*, iii. 96–7.

[6] C. 7182/1893, *Agencies and Methods for Dealing with the Unemployed*, p. 244.

[7] H. of C. 376/1906, *SC on the Housing of the Working Classes Acts Amendment
Bill*, Appendix 23, pp. 456–60. Wilson Fox found that the proportion of 'town-
born' workmen among the inmates of Salvation Army and Church Army homes
and shelters and among applicants to distress committees varied from 86 per cent
to 94 per cent.

It was therefore concluded that rural immigrants exacerbated the congestion of urban labour markets, not by failing to gain employment, but by displacing the least efficient town-bred workmen and forcing them downwards into 'less regular and worse paid occupations'.[1]

Labourers in unskilled occupations were also threatened by competition from low-paid 'juvenile' workers;[2] but social scientists who investigated the problem of premature juvenile employment were less concerned by the direct displacement of adult workmen than by the perpetuation of an untrained and potentially 'casual' class. Charles Booth in the 1890s described the process whereby school-leavers, instead of seeking skilled training, tended to drift into relatively well-paid, 'blind-alley' situations, from which they were dismissed without skill or prospects when they became eligible for adult wages. Hence 'the seed is sown of a future crop of unemployed adult labour'.[3] This problem was believed to be increasing in the 1890s and 1900s, particularly in large commercial centres where there were few highly skilled industries and many openings for school-leavers as errand-boys and messengers.[4] Beveridge maintained that the growth of an unskilled juvenile labour supply in the 1890s was an important source of increased competition in the adult labour market after 1900;[5] and the Royal Commission on the Poor Laws

[1] Cd. 4499/1909, *RC on the Poor Laws, Majority Report*, pp. 351–2, paras. 254–7.

[2] Cd. 5068/1910, *RC on the Poor Laws, Minutes of Evidence*, Q. 96218, para. 2 (vi), b. The term 'juvenile' was normally used to describe persons of both sexes over 12 years and under 18 years of age.

[3] Charles Booth, *Life and Labour of the People of London*, ix (1897), 393.

[4] Cd. 5068/1910, *RC on the Poor Laws, Minutes of Evidence*, Q. 93031, Answer VII, p. 183.

[5] W. H. Beveridge, 'Population and Unemployment', address to Section F of the British Association for the Advancement of Science, Dec. 1923; reprinted in *Essays in the Economics of Capitalism and Socialism* (ed. R. L. Smyth), pp. 260–2. Beveridge ascribed this process to the decline in infant mortality after 1870, which swelled the rising generation of manual labourers who were due to enter the market in the 1880s. The introduction of primary education postponed the impact of this new generation until the 1890s; with the result that an 'earthquake wave' of surplus unskilled adult labourers hit the labour market and depressed real wages and employment after the turn of the century. Beveridge's reasoning is not entirely convincing, however, since juvenile workers declined from 12·70 per cent to 11·06 per cent of the labour-force between 1891 and 1901; and a percentage increase in juvenile employment occurred only in the building, furniture, and 'precious metals' industries and in the armed forces and public administration (Cd. 4758/1909, Report

reported that one of the most potent causes of casual employment was not simply the disorganization of the labour market, but the pressure of a large and increasing supply of urban labourers who were mentally and physically unfitted for any other kind of work.[1]

A solution advanced by many social scientists and reformers for the problem of juvenile labour and for the unemployment problem generally was the improvement of technical training and the advancement of industrial skill.[2] In the 1890s voluntary Apprenticeship and Skilled Employment Committees were established in London, Liverpool, and other large towns to divert school-leavers from 'blind-alley' occupations.[3] But traditional forms of apprenticeship were being rendered increasingly obsolete by the advance of mass production;[4] and reformers therefore turned to new forms of technical education to supply the deficiency of industrial skill and to improve a workman's prospects of regular employment.[5]

The arguments that equated technical training with the reduction of unemployment were, however, rather conjectural, because the relationship between levels of skill and likelihood of unemployment was not at all clear.[6] It was virtually impossible to determine whether unemployment was more or less prevalent among skilled or unskilled workmen and among

of the Consultative Committee of the Board of Education on *Attendance, Compulsory or Otherwise, at Continuation Schools*, vol. i, Appendix C, pp. 265–7).

[1] Cd. 4499/1909, *RC on the Poor Laws, Majority Report*, pp. 325–8, paras. 136–42.

[2] E. M. Hogg, *Quintin Hogg. A Biography* (2nd ed., 1904), p. 213; M. K. Bradby and F. H. Durham, 'Apprenticeship in Relation to the Unemployed', in *Methods of Social Advance* (ed. C. S. Loch, 1904), pp. 118–30.

[3] G. Williams, *Recruitment to Skilled Trades*, pp. 6–7.

[4] N. Adler and R. H. Tawney, *Boy and Girl Labour* (publ. by the Women's Industrial Council, 1909).

[5] G. Drage, *The Unemployed* (1894), p. 183; Cd. 4757/1909, *Attendance, Compulsory or Otherwise, at Continuation Schools*, vol. i, Memorandum by R. H. Tawney, p. 300.

[6] Sidney Webb who was an enthusiastic advocate of technical training, both for its own sake and for the unemployed, nevertheless denied that it would improve their chances of employment—he supported it 'solely as an "occupation and deterrent"' (CAB 37/98/40, 'The Poor Law Commission', memorandum circulated by H. H. Asquith, 2 Mar. 1909, p. 21). B. S. Rowntree and B. Lasker, *Unemployment. A Social Study* (1911), pp. 3–4, found, however, that failure to reach the seventh standard of elementary education was twice as common among unemployed juveniles as in the whole juvenile work force.

more or less efficient workmen within the same level of skill, because of the highly imperfect and strictly incomparable statistical evidence relating to different types of unemployed. It was often assumed that, because the wages of skilled workmen were higher than for unskilled, the demand for their services was greater and that they were therefore less liable to become unemployed.[1] But Llewellyn Smith suggested that this was not necessarily the case and that highly skilled workmen experienced extremes of unemployment unknown in unskilled trades.[2] Moreover, skilled workmen were less mobile between different occupations than the unskilled;[3] and when out of work they had less incentive to 'make shift' with casual labouring jobs or to accept employment below their normal wage.[4] What was undeniable, however, was that the high wages and communal organization of skilled workmen enabled them to provide for themselves in ways that were rarely available to unskilled or casual labourers. But this was often confused with the actual prevention of unemployment; and many advocates of technical training appeared to assume that the increase of skill and industrial efficiency would not merely promote working-class independence, but would eventually reduce the numbers of the unemployed.[5]

THE CONDITION OF THE UNEMPLOYED

As Llewellyn Smith pointed out to a Select Committee of the House of Commons in 1895, unemployment was not the same thing as distress from unemployment,[6] although the two problems were self-evidently related and public discussion of the incidence of distress was often based on statistics that referred to workmen maintained by their trade unions when

[1] J. A. Hobson, *The Problem of the Unemployed: An Enquiry and an Economic Policy* (1896), p. 20.

[2] CAB 37/38/2, 'Memorandum on A Recent Estimate of the Number of the Unemployed', by H. Llewellyn Smith, 8 Jan. 1895, pp. 2–3.

[3] On the mobility of casual workmen between different trades, see B. F. C. Costelloe, 'The Housing Problem', *Transactions of the Manchester Statistical Society*, 1898–1900, p. 42.

[4] C. 7221/1894, *RC on Labour, Fifth and Final Report*, para. 221.

[5] Cd. 4757/1909, *Attendance, Compulsory or Otherwise, at Continuation Schools*, vol. i, Report, pp. 42–3, 176, 219.

[6] H. of C. 365/1895, *SC on Distress from Want of Employment*, Q. 4677.

unemployed.[1] Detailed information about the social and domestic situation of unemployed workmen was, however, biased towards those who sought relief in distress, since those who made adequate provision for loss of income and those who suffered distress but did not seek assistance tended to escape both sociological and administrative scrutiny.[2] Just as statistical evidence about the recurrence of industrial unemployment was derived from the experience of a small and privileged group of highly skilled artisans, so case studies of the social condition of the unemployed related mainly to 'improvident' and 'inferior' workmen by no means representative of the working class as a whole. Moreover, the fact that social investigation was closely linked with charitable action meant that it was workmen in London and other great cities about whom most was revealed and on whom generalizations were primarily based—thereby reinforcing the bias of the evidence towards casual workmen and the unskilled unemployed.

Nevertheless, many different kinds of data both quantitative and qualitative were accumulated in the investigation and relief of unemployed distress. This data could be analysed in various ways; but three of its aspects were most significant in the development of a sociological framework for social policy. These were, firstly, the impact of unemployment upon individual standards of living. Secondly, the relation between unemployment and other agents of social distress—notably sickness, homelessness, and overcrowding. And, thirdly, the habits and characteristics of the victims of unemployment, or what contemporary social investigators rather loosely referred to as the 'character' of the unemployed.

The most immediate social effect of unemployment was loss of income; and, in so far as it affected individuals, unemployment was first and foremost a problem of poverty.

[1] See Llewellyn Smith's critique of Keir Hardie's assumption that because 10 per cent of members of trade unions making returns to the Board of Trade were unemployed in Jan. 1893, 1,300,000 persons were unemployed in the labour force as a whole, of whom all were in distress (CAB 37/38/2, 'Memorandum on A Recent Estimate of the Number of the Unemployed', 8 Jan. 1895, pp. 1–5).

[2] *Special Committee of the COS on the Relief of Distress Due to Want of Employment, 1904, Minutes of Evidence*, Q. 328.

This seems a truism; but it was only gradually acknowledged by persons who thought that unemployment was a voluntary condition or that it was a predictable hazard for which workmen should provide out of their wages.[1] Victorian social administrators had been increasingly concerned with distress caused by factors outside the individual's control, such as epidemics and bad sanitation. But for most of the nineteenth century inadequacy of income as such, whether caused by low wages, irregular employment, or even by sickness, was not considered a legitimate object of state interference, beyond the conditional subsistence guaranteed by poor relief.[2]

In the mid 1880s, however, social investigators began to observe that irregular employment could have a chronically depressive effect upon the living standards of a whole community;[3] and Charles Booth in 1895 redefined the 'unemployed' as 'those whose periods of unemployment are excessive, with the result that their earnings fall below the needs of life'.[4] Nevertheless, it was difficult to calculate precisely the contribution of unemployment to working-class poverty, because most wage estimates referred to average rates or earnings rather than to the actual earnings of workmen who were either employed or unemployed.[5] Moreover, analyses of household budgets rarely took sufficient account of variations of income through time. Booth himself emphasized that a year was the shortest unit of time over which the social effects of unemployment could be adequately calculated;[6] but his own investigation of domestic budgets was limited to five-week periods.[7] A more prolonged survey con-

---

[1] e.g. Helen Bosanquet, 'Wages and Housekeeping' in *Methods of Social Advance* (ed. C. S. Loch, 1904), pp. 136–8.

[2] T. H. Marshall, *Social Policy in the Twentieth Century* (1967 ed.), pp. 14–15.

[3] *Report of the Mansion House Committee, appointed March 1885 to Inquire into the Causes of Permanent Distress in London and the Best Means of Remedying the Same* (publ. 1886), pp. 8–9.

[4] H. of C. 365/1895, *SC on Distress from Want of Employment, Minutes of Evidence*, Q. 10519.

[5] G. H. Wood, 'Some Statistics Relating to Working Class Progress since 1860', *Journal of the Royal Statistical Society*, 62 (Dec. 1899), 640.

[6] Charles Booth, *Life and Labour of the People in London*, i (1892), 151.

[7] Ibid., pp. 136–8. Henry Mayhew fifty years earlier had produced figures more relevant to the study of irregular employment by recording fluctuations in the income of a timber-porter over a period of seventy-two weeks (Henry Mayhew, *London Labour and the London Poor*, iii. 306–7).

ducted by the Economic Club in 1891–4 was too narrowly based to be really significant[1] and none of the other records of family income and expenditure compiled by social scientists during this period covered more than a few consecutive weeks.[2]

From this limited evidence, however, two conclusions were drawn. Firstly, it was found in London that irregular employment was by far the most important single cause of poverty among the working class. In an intensive study of 4,000 families known to School Board visitors in the late 1880s, Booth found that 43 per cent of poverty among those whom he classed as 'C' and 'D' and 51 per cent among those whom he classed as 'A' and 'B' was caused by casual and irregular work.[3] He therefore concluded that 'it is fairly certain that the standard of comfort is fixed by the regularity rather than by the rate of pay'.[4] This was confirmed in 1906–7 by a nation-wide inquiry conducted by Rose Squire and Arthur Steel-Maitland, who discovered that

the chief industrial causes of pauperism were as follows: first, casual and irregular employment. Secondly, bad housing. Third, unhealthy trades and insanitary, injurious and exhausting conditions of employment. Fourth, low wages, that is, earnings habitually below what is required for healthy subsistence.[5]

It was endorsed also by Beveridge, who condemned the Great London Dock Strike as a 'tragedy of misdirected enthusiasm' for having subordinated security of employment to a higher scale of pay.[6]

[1] The Economic Club, *Family Budgets: Being the Income and Expenses of Twenty-eight British Households, 1891–1894* (publ. 1896).

[2] e.g. S. B. Rowntree and B. Lasker, *Unemployment. A Social Study* (1911), Chapter VII; M. S. Pember Reeves, *Round About a Pound a Week* (1914), pp. 73–93.

[3] Charles Booth, *Life and Labour of the People in London*, i. 147.

'Class A'—'the lowest class of occasional labourers, loafers and semi-criminals'.
'Class B'—those with 'casual earnings'.
'Class C'—those with 'intermittent earnings'.
'Class D'—those with 'small regular earnings'.

Booth's results should, however, be compared with Rowntree's study of York in 1899, where 5·14 per cent of poverty was ascribed to irregular employment (S. B. Rowntree, *Poverty: A Study of Town Life*, p. 121). This was due partly to the smaller proportion of casual labour in York, partly to the fact that Rowntree's investigation was carried out at the peak of a trade cycle.

[4] Charles Booth, *Life and Labour of the People in London*, ix. 328.

[5] Rose Squire, *Thirty Years in Public Service*, p. 124.

[6] Beveridge, *Unemployment* (1910 ed.), p. 207.

Secondly, it was found that the economic repercussions of unemployment were largely determined by the age-structure and size of families. The position of an unemployed single man was likely to be less serious than that of a man with wife and children to support. The domestic situation of a family man of the kind most vulnerable to irregular employment was described to a commission of inquiry in Liverpool in 1894:

We will take the case of a man with four children; the minimum rent will be 5s. Then there are lights, 1s. at the very lowest; there is fire, 1s. 6d.; there is food . . . which will cost 3s. a week for each member of the family . . . if you tot up those items you will see that they amount to 25s. 6d. a week without anything else . . . if you take into consideration that the average wage of a dock labourer does not exceed £1 a week, it seems . . . inexplicable how they live at all. . . .[1]

This budget included no provision for necessary items such as clothing, quite apart from 'luxuries' like alcohol and tobacco. It laid down a more stringent minimum expenditure for a four-child family than Rowntree's classic study of York in 1899,[2] and it allowed no margin for contingencies like shortage of work. The family unit was a potential source of relief as well as hardship, however;[3] and the same family several years later, when all the children were earning, might have been comparatively prosperous, even if all its members were irregularly employed. Social investigators frequently found that children worked to support unemployed parents and wives to support unemployed husbands;[4] and contemporary budgetary analysis revealed that, at certain stages in its life-cycle, the family was often a microcosm of institutional thrift, a 'natural, mutual, benefit and insurance society by whose *vis mediatrix* so many of the ills of the body politic are dispelled'.[5]

---

[1] *Commission of Inquiry into the Subject of the Unemployed in the City of Liverpool, 1894, Minutes of Evidence*, QQ. 595–6.

[2] S. B. Rowntree, op. cit., p. 110. Rowntree's scale would have prescribed a minimum income for a four-child family of 26s. a week.

[3] C. S. Loch, *La Lutte pour le travail et les inemployés*, paper read at the Congress of the Institut International de Sociologie, July 1906, pp. 7, 12.

[4] Macdonald MSS. ii, ff. 31–2, M. J. Bell to Margaret Macdonald, 10 Nov. 1904; ibid., f. 45, Herbert Day to Margaret Macdonald, Nov. 1904; H. of C. 290 Ind./1907, *SC on Homework, Minutes of Evidence*, Q. 2323.

[5] Henry Higgs, 'Workmen's Budgets', *Journal of the Royal Statistical Society*, 56 (June 1893), 270; Lady Bell, *At the Works* (1911 ed.), pp. 163–71.

The connection between unemployment and other social problems was highly complex; a compound of industrial, demographic, physiological, and administrative factors. The coincidence in many urban centres of irregular employment, domestic overcrowding, rural immigration, and the technical and physical inferiority of many unemployed workmen gradually provoked a new sociology of town life, in which irregular employment was seen as part of a spectrum of endemic social distress.[1] The ramifications of 'permanent distress' were clearly outlined by a Mansion House committee of 1885, which described a vicious circle of social and economic problems which was to become increasingly familiar to social investigators in the metropolis over the next twenty years. An over-supply of labour, swollen by immigration, competing for average wages of 12s. a week; depression of trade in general and the eclipse of shipbuilding and sugar-refining in particular; indiscriminate charity and inconsistent Poor Law administration, attracting a concentration of distress in certain areas, with the consequent evils of high rents and overcrowding; the whole pattern being linked together by a rather discordant explanatory reference to the 'character' of the local unemployed.[2]

This picture was corroborated by the Royal Commission on the Housing of the Working Classes, which reported in the same year that 'the precarious element in the struggle for employment' was one of the chief causes of overcrowding in Central London and other large towns.[3] Since the 1860s the supply of accommodation in the centre of London had been steadily diminished by slum clearance and commercial and municipal development;[4] but casual labourers could not easily migrate to the new working-class suburbs, since the nature of the casual system meant that they had to live within

[1] F. W. Lawrence, 'The Housing Problem', in *The Heart of the Empire* (1902 ed.), pp. 53–110; Cd. 2175/1904, *Report of the Interdepartmental Committee on Physical Deterioration*, paras. 79–105 on the 'Urbanisation of the People'.

[2] *Report of the Mansion House Committee, appointed March 1885, to Inquire into the Causes of Permanent Distress in London and the Best Means of Remedying the Same* (publ. 1886), pp. 11–12.

[3] C. 4402/1885, *RC on the Housing of the Working Classes, First Report*, p. 18.

[4] Ibid., pp. 20–3; J. N. Tarn, 'The Peabody Donation Fund: the Role of a Housing Society in the Nineteenth Century', *Victorian Studies*, 10, no. 1 (Sept. 1966), 15–20, 31.

striking distance of all possible sources of work.[1] Similarly, costermongers and small traders had to live where there was a working-class clientele for their wares, since 'the poor form their own markets, and there is the same difficulty of moving a market as there is of moving an industry, and both these factors increase the pressure of overcrowding'.[2] Even when casual workmen sought work elsewhere, they tended to leave their wives and children behind in congested areas, where credit was easily available and food was cheap.[3] When unemployment forced families through loss of income to leave their dwellings it was often in exchange for more expensive accommodation where they were allowed longer arrears of rent.[4] Alternatively they might enter common lodging-houses or take to the streets; and eye-witness accounts of the problem of urban vagrancy suggested that, in periods of depression, the so-called 'homeless' were often virtually identical with the casual class of unemployed.[5] Hence, the over-supply of labour and the under-supply of housing reinforced each other; and, like other aspects of the casual labour problem, the connection between unemployment and overcrowding was invested by contemporary social scientists with an 'evolutionary' significance in the process of social change.[6]

[1] C. 4402/1885, RC on the Housing of the Working Classes, First Report, p. 18. Moreover, although rents in the suburbs were cheaper, few casual workmen could afford the daily cost of travel (B. F. C. Costelloe, 'The Housing Problem', Transactions of the Manchester Statistical Society, 1898–1900, p. 48). And of the railway companies operating in London, only the Great Eastern made any serious attempt to provide cheap workmen's trains (P. Hall, 'The Development of Communications', in J. T. Coppock and H. C. Prince, Greater London, pp. 65–6).

[2] C. 4402/1885, p. 18.

[3] Ibid., pp. 15, 18.

[4] Beveridge MSS. (first deposit), Parcel 2, Folder A, Memorandum on 'Unemployment Insurance', by H. Llewellyn Smith, Apr. 1909. It was argued that one of the secondary benefits of unemployment insurance would be that it enabled unemployed workmen to remain in cheap accommodation.

[5] Bennet Burleigh, 'The Unemployed', Contemporary Review, 52 (Dec. 1887), 770–80; Report of a Special Committee of the COS on the Homeless Poor of London, June 1891, p. xx.

On 17 Feb. 1905 an inquiry carried out on behalf of Sir Shirley Murphy, the L.C.C.'s Medical Officer of Health, found 5,958 persons in London either homeless or resident in common lodging-houses and casual wards (Cd. 2852/1906, Departmental Committee on Vagrancy, Report, para. 64). These persons were classed by the police as 'vagrants', but it is not at all clear how far this term was synonymous with the unemployed.

[6] John Simon, English Sanitary Institutions (1890), pp. 438, 447–8.

The relationship between unemployment and physical disability was only partially investigated before 1914. The influence of sickness in causing unemployment was largely self-evident. A skilled workman who fell ill might be retained by his employer, though he was unlikely to be paid in such circumstances; but for a casual labourer sickness was synonymous with unemployment since absence from his place of hire for whatever reason meant no work for the day. Similarly, it was recognized that superior health and physique improved a workman's personal chances of employment, this being one of the main reasons for the displacement of urban by rural labourers;[1] and the physical standard of volunteers for military service, a 'last resort' for workmen who could find no other occupation, fluctuated inversely with depressions of trade.[2] There were, moreover, many firms that refused to employ workmen suffering from various kinds of physical disability;[3] and a Home Office factory inspector stated in 1904 that 'occupations had a profoundly selective effect', whereby physically superior workmen tended to gravitate upwards to highly skilled and secure employment, while the 'weaker vessels' were pressed downwards to the bottom of the industrial scale.[4] In 1910 the juvenile employment committees established by the Board of Trade found that many school-leavers had difficulty in finding employment because of deficiencies in health and physique. 'Perhaps the commonest bar to employment for boys and girls', reported one of the Board's officials,

... is the fact that the applicant is undersized or lacks general strength. Deafness, stammering and lameness are frequent causes of rejection of applicants or of early dismissal. . . . Deafness or bad eyesight are

---

[1] C. 7421/1894, *RC on Labour, Fifth and Final Report*, para. 235.

[2] Cd. 2175/1904, *Interdepartmental Committee on Physical Deterioration, Report*, para. 19.

[3] e.g. the Post Office rarely employed men with defective vision, even when corrected by spectacles (LAB 2/210/LE701, 'The Influence of Physical Fitness on Employment', memorandum by Frederick Keeling (file dated 22 Dec. 1909, but from its contents clearly written after the opening of labour exchanges in Feb. 1910)).

[4] Cd. 2175/1905, *Interdepartmental Committee on Physical Deterioration, Report*, para. 142. This inspector also endorsed the complaint of trade unionists that the Employers' Liability Act of 1897 was causing employers to dismiss workmen with physical disabilities (*Report of the Bath T.U.C.*, 1907, pp. 139–40).

frequently the cause of actions or mistakes on the part of boys and girls which are assumed to arise either from deliberate insubordination or stupidity. . . . Mental deficiency, or slowness, which may almost be classed as such, often prevents a child from obtaining employment.[1]

The opposite sequence of causation—the influence of unemployment in actually generating disease or disability—was only tentatively explored during this period.[2] Public inquiries in the 1900s bore witness to a gradual disillusionment with the conviction of Victorian sanitary reformers that destitution and disease would be abolished by measures of public health;[3] and out of this came a realization that loss or irregularity of income might be causes as well as consequences of physical decline.[4] The relief-works movement found that many unemployed workmen were not 'shy' of hard work, but—through lack of exercise and malnutrition—physically incapable of it; and after a period of relief employment they often improved in health and physique if in no other respect.[5] There was, moreover, evidence to suggest that the effects of unemployment were psychological as well as physical;[6] and C. S. Loch, the Secretary of the Charity Organisation Society, suggested that unemployment could have the same effect as prolonged hospitalization—the unemployed were unable to readjust themselves to a competitive environment and grad-

[1] LAB 2/210/LE 701, 'The Influence of Physical Fitness on Employment'.
[2] S. B. Rowntree, *The Way to Industrial Peace and the Problem of Unemployment* (1914), pp. 136–44.
[3] Cd. 4499/1909, *RC on the Poor Laws, Majority Report*, pp. 50–1.
[4] S. Rowntree and B. Lasker, op. cit., pp. 226–8. John Burns in 1895 claimed that much of the 'sick benefit' paid by trade unions, should be counted as 'unemployment benefit', since it was paid in respect of sickness arising from want of work (H. of C. 365/1895, *SC on Distress from Want of Employment, Minutes of Evidence*, QQ. 4958–61).
[5] Beveridge MSS., Col. B, vol. iv, item 35, Extracts from the 2nd Report of the Central Unemployed Body 1906–7, with comments by the Chairman, Russell Wakefield, para. 1.
[6] C. B. Hawkins, *Norwich: A Social Study* (1910), p. 63, reported that men grew 'twenty years older in five years' through anxiety about loss of work. See also Charles Booth, *Life and Labour of the People in London*, ix. 331, statement of an operative brushmaker. 'When I thought it likely that I should be thrown out of employment, it seemed to paralyze me completely. . . . I used to sit at home brooding over it until the blow fell . . . the fear of being turned off is the worst thing in a working-man's life, and more or less acutely it is always, in the case of the vast majority, present in his mind.'

ually became incapable of regular work.[1] Moreover, medical judgements on the unemployed were often closely linked with moral;[2] and in the 1890s and 1900s it was often suggested that the unemployed as a class were a cross-section of the 'unfit'.[3] 'On the whole, the casual worker or unemployed person is of a lower mental type than those in regular employment' recorded the General Relieving Officer of Bethnal Green in 1910.

> Those classes include many persons (many more than is generally admitted) whose mental condition is so weak as not merely to make them unemployable, but also to prevent their supporting themselves, except by the aid of the rates . . . the mental weakness is not that condition which we call lunacy, it is much more subtle than that, it is rational without being intelligent, it is a failure to grasp the essential facts of life.[4]

Both Beveridge and the Webbs believed that a shift from moral and personal to industrial and environmental explanations of unemployment was one of the chief characteristics of the new analysis of the subject that developed at the end of the nineteenth century. Unemployment, they maintained, was seen no longer as a condition of voluntary idleness but as a 'disease of industry' and a product of deficiencies in labour organization.[5] This account of the reformulation of the problem is, however, in certain respects misleading, since students of society in the mid nineteenth century were by no means uniformly convinced of the personal delinquency of the unemployed.[6] Moreover, as the contemporary writings of

---

[1] C. S. Loch to the Editor of *The Times*, 31 Dec. 1903, quoted in the *Charity Organisation Review*, N.S. 15 (Jan.–June 1904), 48.

[2] e.g. Alexander Scott, 'Physical Fitness as a Cause of Unemployment', and subsequent discussion, *Proceedings of the National Conference on the Prevention of Destitution*, 30 May–2 June 1911, pp. 446–81.

[3] Charles Booth, *Life and Labour of the People in London*, i. 149–50.

[4] E. J. Lidbetter, 'Some Examples of Poor Law Eugenics', *Eugenics Review*, 2, no. 3 (Nov. 1910), 223.

[5] W. H. Beveridge, *Insurance for All and Everything* (1924), pp. 4–5. S. and B. Webb, *English Poor Law History*, II. ii. 637–8.

[6] See e.g. H. of C. 296/1841, *Report of the Commission (under the Great Seal) for Inquiring into the Condition of the Unemployed Hand-Loom Weavers in the United Kingdom (chairman Nassau Senior)*, p. 18; C. P. Bosanquet, *London: Some Account of its Growth, Charitable Agencies and Wants* (1868), pp. 146–8; Thomas Mackay (ed.), *The Autobiography of Samuel Smiles* (1905), p. 114.

Beveridge and the Webbs themselves prove, the new generation of unemployment theorists was in some respects no
less censorious than its predecessors;[1] and the growth of
a scientific analysis of unemployment was paralleled by the
growth of a harsher and more pessimistic attitude towards its
victims, which was directed primarily against those who failed
to support themselves but extended also to all who became
unemployed.

The reasons for this attitude were complex and to a certain
extent conjectural. Firstly, the accumulation of evidence
about the material progress of the working class as a whole
reflected adversely upon those who failed to share in this
prosperity;[2] and the advance of working-class thrift institutions tended to accentuate the gulf between regular and irregular, provident and improvident workmen[3]—a gulf that was
believed to be widening between the 1880s and the 1900s.[4]

Secondly, it has been pointed out that social investigation
tended to concentrate on unemployed workmen who sought
relief rather than on those who maintained their independence; and inquiries that exposed the sufferings of the 'very
poor' also dispelled illusions about them.[5] 'The large towns
of England are unhappily full of a class of low, loafing, tipsy
people, very different from the élite of the artisan and labouring classes, though shading gradually into them' wrote a
Poor Law inspector in 1887; '. . . the class described forms
unhappily one of the largest factors in the sum total of our
population. . . . How to deal with it is perhaps the most important social question of the day.'[6]

---

[1] Above, p. 26. Beatrice Webb's belief that unemployment was as much a personal
as an industrial problem persisted throughout her life. In 1942 she was highly
critical of the proposals for Unemployment Insurance in the Beveridge Report—
'which if *carried out* (which I think unlikely)—will increase the catastrophic mass
unemployment, which could happen here as in the U.S.A. The better you treat
the unemployed in the way of means, without service, the worse the evil becomes;
because it is pleasanter to do nothing than to work at low wages and in bad conditions . . .' (Fabian Society MSS., Box 3, B. Webb to Reginald Pott, 14 Dec.
1942).   [2] Cd. 4499/1909, *RC on the Poor Laws, Majority Report*, p. 52.
[3] *15th Annual Report of the L.G.B.* (1885–6), p. 38.
[4] F. Tillyard, 'Three Birmingham Relief Funds—1885, 1886, and 1905',
Economic Journal, 15 (Dec. 1905), 506; Cd. 4499/1909, *RC on the Poor Laws,
Majority Report*, pp. 49–50.
[5] Charles Booth, *Life and Labour of the People of London*, i. 177.
[6] *16th Annual Report of the L.G.B.* (1886–7), p. 83.

This class was increasingly identified with casual and irregular workmen. Beatrice Webb in 1887 condemned casual employment for generating not merely economic insecurity but immoral habits of life;[1] and the moral feebleness of irregular workmen was emphasized by a Mansion House Committee which organized relief work in 1887–8. This Committee investigated the circumstances of 456 applicants, of whom scarcely any belonged to a savings institution, although nearly all were engaged in chronically irregular trades; two-thirds were casual or unskilled labourers, and 68 per cent were under forty years old, thereby undermining the common assumption that unemployment was primarily a problem of advancing age. It was found that 36 per cent were so lacking in character and initiative that they could be given only temporary assistance; and 30 per cent were beyond help, displaying 'an incapacity for steady work, which is but pauperism under another name'.[2]

The bulk of the unemployed workmen who applied to Mansion House relief committees in the 1880s and 1890s in fact defied the reforming assumptions of charitable investigators; they were incurably urbanized, indifferent to self-improvement, and resistant to occupational change.[3] Similar inquiries in Liverpool in the early 1890s endorsed this account of the chronically underemployed, thus dispelling the common illusion that the casuals of London were more degenerate than those elsewhere;[4] and in 1895–6 a survey conducted by the Toynbee Trust found no significant variation in the character and condition of life of casuals in London and in eight provincial towns.[5] The picture constructed by the Toynbee Trust inquiry was, however, very different from that of a corrupt and feckless, but versatile and volatile class, portrayed in Beatrice Webb's account of the London docks. 'We

---

[1] B. Webb, 'The Docks', loc. cit., pp. 31–2.

[2] *Mansion House Conference on the Condition of the Unemployed*, 1887–8, Report of the Reference Committee, pp. 2–3.

[3] H. V. Toynbee, 'A Winter's Experiment', *Macmillan's Magazine*, Nov. 1893–Apr. 1894, p. 56.

[4] *Report of the Commission of Inquiry into the subject of the Unemployed in the City of Liverpool*, 1894, pp. xv–xviii.

[5] *Report of an Inquiry into the Condition of the Unemployed*, carried out under the Toynbee Trust (winter 1895–6) by Arthur Woodworth, p. 25.

find that the most striking feature in the returns is stolidity'
reported Arthur Woodworth who designed the survey. 'Opposed to the popular notion of a restless, shifting population
ready to turn its hand to anything, the results show dull,
apathetic men whose passive resistance to all outside influences constitutes their most hopeless feature.'[1] Of the cases
examined a majority were town-born, had no savings, and
lived off their wives and children when out of work.[2] The
inquiry also discovered a tendency for the unemployed to
over-class themselves—for an 'odd-job whitewasher' to describe himself as a painter and decorator; and their ambitions
were almost exclusively confined to regaining work in their
previous occupations.[3] Inquiries of this kind, which covered
a total of 141 families, were too selective and too circumscribed to supply conclusive evidence about the habits of
irregular workmen; but they helped to reinforce existing
prejudice about the inferior 'character' of the unemployed.[4]

A third and perhaps the most crucial element in the hardening of attitudes towards irregular workmen was the fact that
environmental explanations of distress from unemployment
were in some respects more severe and more pessimistic than
those that blamed the moral turpitude of the unemployed.
The point of view that ascribed social distress simply to
delinquency of character always held out the hope that the
character—and hence the whole way of life—of the individual
might be reformed; this was one of the mainsprings of
evangelical involvement in social problems.[5] But the idea that
distress and even character itself were determined by environment held out little prospect of improvement for the
existing generation of destitute poor. 'They . . . have fallen
into a state of habitual dependence from which it should be
the chief aim of the charitable to rescue their children', reported the Mansion House committee of 1887–8, comment-

[1] Ibid., p. 54.
[2] Ibid., pp. 26–7.
[3] Ibid., p. 31.
[4] e.g. *Charity Organisation Review*, N.S. 1 (1897), 235.
[5] *COS Special Committee on the Relief of Distress Due to Want of Employment,
Nov. 1904, Minutes of Evidence*, Q. 583, Statement of Colonel David Lamb of the
Salvation Army.

ing on the situation of the most irregularly employed.[1] Within this context, irregular workmen were seen by many of the new generation of social scientists as not merely recalcitrant but degenerate, their failure to work no longer as a misdirection of free will but a symptom of inferior moral and physical capacity. 'The problem of the unemployed' became 'not so much a matter of finding work, but of dealing with the waste products of our nineteenth century civilisation',[2] and to the old-fashioned categories of 'deserving' and 'undeserving' the social jargon of the 1890s therefore added the sub-division of 'unemployable'.[3]

Moreover, in the 1900s the environmentalists were challenged by the 'eugenic' school of social theorists, who believed that social dependency was not merely incurable but hereditary and that social reform without selective breeding would secure the preservation of the 'unfit'.[4] The Eugenics Education Society,[5] founded in 1909, published detailed pauper genealogies to show how poor relief had fostered a class of industrial parasites who were rarely if ever employed. 'There exists this hereditary race of persons, capable of work, but refusing to do it, either continuously or at intervals; and, when they work, spend their money in drink or debauchery' reported a committee on the Eugenic Aspect of Poor Law Reform in 1910.[6] This committee could produce no definite

---

[1] *Mansion House Conference on the Condition of the Unemployed*, 1887–8, Report of the Reference Committee, p. 2.

[2] Frederick Thoresby, 'How to Deal with the Unemployed', *Westminster Review*, 165 (Jan. 1906), 36.

[3] S. and B. Webb, *Industrial Democracy* (1897), ii. 784–5. The Webbs divided the unemployable into three groups: (*a*) children, the aged, and child-bearing women, (*b*) the mentally and physically sick, and (*c*) 'men and women who, without suffering from apparent disease of body or mind are incapable of steady or continuous application, or who are so deficient in strength, speed or skill that they are incapable, in the industrial order in which they find themselves, of producing their maintenance at any occupation whatsoever.'

[4] B. Semmel, *Imperialism and Social Reform*, pp. 44–52. The eugenists, headed by Sir Francis Galton and Karl Pearson, were a small minority among English social theorists; but their fears and prejudices were widely pervasive in the 1900s even among people who did not share their principles (e.g. Cd. 2175/1904, *Report of the Interdepartmental Committee on Physical Deterioration*, para. 249; Arthur Shadwell, *Industrial Efficiency* (1909 ed., first publ. 1905), pp. 663–4).

[5] Members included Galton, Frederick Harrison, Havelock Ellis, Sir Edward Brabrook, Lord Lytton, Arnold White, and Lady Ottoline Morrell.

[6] 'Report of the Committee appointed to consider the Eugenic Aspect of Poor

evidence of 'hereditary weakness' among the 'able-bodied unemployed'; but it concluded that

the unemployed as a whole seem unable to work at the standard required by industry, and it appears from the experience of distress committees that the men employed in specially provided work are unable to meet the normal day's requirements. Some kinds of preconception see the cause of this in the previously unfed condition of the men, or in the degeneration they had suffered through being a considerable time out of work. But with all this there is a sub-conscious conviction that unemployment as a whole represents inferior capacity, and that while undoubtedly the higher ranges of unemployment may be redeemed by judicious administrative assistance, beyond these higher ranges there is not much ground for hope.[1]

SUMMARY

In spite of the growth of empirical investigation, the evidence relating to all aspects of the problem of unemployment was highly imperfect for most of the period under discussion; and more reliable information only became available as a result of and not as a prelude to administrative reforms. In the interim the discussion of unemployment and the theories on which reforms were based, were derived partly from very limited statistical data and partly from highly conjectural preconceptions about the nature of the labour market and the characteristics of the unemployed.[2]

These preconceptions were strikingly apparent in the distinction made by the mainstream of contemporary social scientists between different kinds of unemployment and different types of unemployed. Thus the irregular employment of casual workmen was seen as an archaic survival from a pre-industrial era. By failing to eliminate surplus workmen, casual employment was a positive hindrance to industrial efficiency and labour mobility; and it depressed standards of

---

Law Reform, Section III. Investigation into Pauper Family Histories', *Eugenics Review*, 2, no. 3 (Nov. 1910), 193.

[1] 'Report of the Committee Appointed to consider the Eugenic Aspect of Poor Law Reform, Section I. The Eugenic Principle in Poor Law Administration', ibid., pp. 173–4.

[2] For further discussion of this point see J. Brown, 'Charles Booth and Labour Colonies 1889–1905', *Economic History Review*, 2nd series, 21, no. 2 (1968), 349–60.

remuneration and discouraged technological change. Cyclical fluctuations among highly skilled workmen were seen on the other hand as both a symptom and a catalyst of economic advance. It was believed that they were a spur to technical and organizational innovation, and that they tended to weed out inefficient workmen by distinguishing clearly between employed and unemployed.

No less ambivalent was much of the commentary during this period on the social condition of the unemployed. Distress from unemployment was seen as the product of a disorganized labour market; but, at the same time, since organization was used as an index of industrial quality, it was often inferred that casual and irregular workmen were inferior social specimens—a view that was confirmed by the evidence of 'improvidence' and lack of 'regularity' in the casual labourer's way of life.[1] Moreover, in much contemporary literature on the subject the lowest stratum of casual labourers was seen as not merely inefficient or improvident but as a degenerate class, doomed to obsolescence like some primitive tribe; and since the rationalization of industry would eventually deprive them of all economic status the casuals, like the savage, would have to adapt themselves or die. Casual labourers were therefore unfavourably compared with skilled and organized workmen, who subscribed to benefit institutions and maintained their independence when unemployed. But even the latter were subject to unfavourable moral judgement, since it was assumed that the most provident were the least likely to experience unemployment[2] and that in the upper reaches of industry workmen were engaged and dismissed in direct accordance with their personal reliability and industrial skill.[3]

The distinction between unemployment among the higher and lower grades of industry was reflected in the frequent assumption of contemporary reformers that, whereas the aim of social policy among skilled workmen should be merely to mitigate the effects of unemployment, its aim among casual

---

[1] COS Special Committee on the Relief of Distress Due to Want of Employment, 1904, Minutes of Evidence, QQ. 1488–9.

[2] H. of C. 365/1895, SC on Distress from Want of Employment, Minutes of Evidence, Q. 10519.

[3] Beveridge, Unemployment (1910 ed.), pp. 138–9.

and unskilled workmen should be the gradual elimination of the unemployed.[1] This dualistic analysis of the problem was, however, modified by certain other factors. Firstly, the distinction between the self-supporting unemployed of the labour aristocracy on the one hand and socially dependent, chronically irregular workmen on the other was increasingly called in question by the political leaders of the working class; and from the mid 1880s onwards the attention of public authorities was continually drawn to a large and heterogeneous class of persons, difficult to contain within the bounds of either classical or Darwinist economic and social theory, who although on the verge of destitution struggled to maintain their independence when out of work.[2] It was difficult to calculate the extent of such concealed destitution or to establish the identity of such persons because they rarely asked for poor relief or charity;[3] and Poor Law officials maintained that it was no part of their function to seek out distressed persons who did not actually apply for public support.[4] By the mid 1890s, however, it was clear that in certain areas the resources of this class were being strained to their limits;[5] and witnesses before the Select Committee on Distress from Want of Employment in 1895 suggested that it was with this class that innovations in social policy should be specially concerned.[6]

Secondly, personal criticism of the unemployed was not incompatible with a pattern of explanation that linked unemployment on the one hand with sickness, overcrowding, and environmental squalor; and on the other hand with a highly imperfect labour market and erratic consumer demand. It was increasingly recognized that the influence of unemployment was by no means confined to periods when a workman was out of work, since it could permanently damage his physique and efficiency and its anticipation and after-effects

---

[1] A. C. Pigou, *Unemployment* (1913), pp. 201–3.   [2] Below, p. 107.

[3] *COS Special Committee on the Best Means of Dealing with Exceptional Distress, 1886, Minutes of Evidence,* Q. 1645; *16th Annual Report of the L.G.B.* (1886–7), pp. 73–4.

[4] H. of C. 111/1895, *SC on Distress from Want of Employment, Minutes of Evidence,* QQ. 886–7.   [5] Below, p. 89.

[6] H. of C. 365/1895, *SC on Distress from Want of Employment, Minutes of Evidence,* Q. 5393.

continually depressed his 'standard of life'.[1] Even in families whose average income over a period of time was theoretically quite adequate, it disrupted household management and sabotaged the process of rational long-term budgeting on which the Victorian ideal of self-help was based.[2] It was realized, moreover, that the consequences of unemployment were not confined to unemployed individuals but that the living standards of a whole community might be depressed by occasional or chronic shortage of work.[3] In spite of its preconceptions and tendency towards oversimplification, social investigation therefore revealed that not one but many social and economic questions were contained in the problem of unemployment. Subsequent chapters will consider the impact of new kinds of social and economic analysis upon the political discussion of the problem and upon the formation of administrative remedies for the dislocation of the labour market and the distress of the unemployed.

[1] Cd. 5066/1910, *RC on the Poor Laws, Minutes of Evidence*, Q. 78461, para. 10 Statement by Walter Long on the cumulative effects of unemployment; B. S. Rowntree and B. Lasker, op. cit., p. 305.

[2] G. H. Wood, 'Trade Union Expenditure on Unemployed Benefits since 1860', *Journal of the Royal Statistical Society*, 62 (Mar. 1900), 92; Cd. 5068/1910, *RC on the Poor Laws, Minutes of Evidence*, Q. 93031, Answer VIII, p. 186.

[3] *16th Annual Report of the L.G.B.* (1886–7), p. 82; *Report of the Mansion House Committee, appointed March 1885 to Inquire into the Causes of Permanent Distress in London and the Best Means of Remedying the Same* (publ. 1886), pp. 8–9.

II

UNEMPLOYMENT AND POLITICAL ACTION 1886–1896

BETWEEN 1880 and 1890 the uneasy synthesis of Poor Law, thrift, and charity which had relieved distress from want of employment since the 1830s broke down. Political attention was shifted from the commercial problem of 'depression' to the social problem of the 'unemployed'.[1] The free-trade system was called in question, and its critics denounced 'the very tantalising gilded mockery . . . of dangling a so called cheap and large loaf before the eyes of the people if they have no settled work . . .'.[2] Socialists and social reformers attacked certain anomalies in the organization of industry—the conjunction of overwork with unemployment, rural depopulation with urban congestion, widespread poverty with apparent 'over-production'[3]—and began to demand work rather than poor relief or charitable assistance for the destitute unemployed.

This new movement must be considered against a background of economic, administrative, and political change. Historical assessments of Great Britain's economic situation between the mid 1870s and mid 1890s vary widely, but most authorities now agree that the 'Great Depression' was by no means so uniformly disastrous as it appeared to many contemporaries;[4] and that, in spite of a high decennial average

---

[1] C. 4893/1886, *RC on the Depression of Trade and Industry, Minority Report*, p. lv, paras. 56–7.

[2] Salisbury MSS., Class M, Box 22, Morley Alderson (Secretary of the Shop Hours Labour League) to Lord Salisbury, 6 Feb. 1886.

[3] e.g. H. V. Mills, *Poverty and the State*, pp. 105–14.

[4] For contemporary accounts of the causes and effects of the depression see C. 4893/1886, *RC on the Depression of Trade and Industry, Majority and Minority Reports*. Both reports blamed the increase of protected foreign competition rather than deficiencies in British industry; and both claimed that fund-holders who invested abroad were prospering to the detriment of the domestic economy.

For summaries of the conflicting modern interpretations of the depression see Charles Wilson, 'Economy and Society in Late Victorian Britain', *Economic History Review*, 2nd series, 18, no. 1 (Aug. 1965), 183–98; and D. H. Aldcroft

of unemployment, the 1880s was a period of rising real wages and therefore of rising prosperity for a majority of the working class.[1] Nevertheless, in several ways the depression helped to precipitate the 'problem of the unemployed'. Firstly, the conjunction of high real wages with widespread unemployment emphasized the economic and social distinction between those who were regularly and those who were irregularly employed;[2] and for the first time the 'unemployed' were seen as a distinct class or group, whose problems were different from those of the working class as a whole. Secondly, even though the purchasing power of wages was rising the prolonged depression tended to undermine the capacity of both organized and unorganized workmen to retain their independence whilst unemployed. Contemporary writers noted with approval that the 'pauperisation of the unemployed' was now 'limited' or 'deferred' by institutionalized thrift.[3] But, nevertheless, personal and mutual savings which might be sufficient for short periods of unemployment were likely to be far less adequate during several successive years of depression, such as occurred in certain industries and localities between 1884 and 1888. During those years the funds of many trade unions were severely strained by continuous 'out-of-work' payments;[4] and it was observed that the resources of the whole 'casual' population of London were being undermined by chronic shortage of work.[5]

This erosion of private resources among certain sections of the working class occurred at a time when the Local Govern-

(ed.), *The Development of British Industry and Foreign Competition*, pp. 11–36. The controversy centres chiefly on (a) the extent of industrial stagnation; (b) whether decline should be dated from the 1890s, 1870s, or 1850s; and (c) whether it was primarily caused by entrepreneurial failure, lack of technical innovation, falling *per capita* output, or a relative slackening of capital accumulation.

[1] B. R. Mitchell and Phyllis Deane, *Abstract of British Historical Statistics*, p. 344.

[2] C. 4893/1886, *RC on the Depression of Trade and Industry, Final Report*, p. xli, note appended by C. M. Palmer; *16th Annual Report of the L.G.B.* (1886–7), pp. 73–4, 76.

[3] C. 4893/1886, *RC on the Depression of Trade and Industry, Minority Report*, p. xlix, para. 41.

[4] *16th Annual Report of the L.G.B.* (1886–7), p. 82; *Report of the Swansea T.U.C.*, 1887, pp. 18–21; D. C. Cummings, *History of the United Society of Boilermakers and Iron and Steel Shipbuilders*, pp. 116–17.

[5] *Report of the Mansion House Committee, appointed March 1885 to Inquire into the Causes of Permanent Distress in London and the Best Means of Remedying the Same*, pp. 8–9.

ment Board inspectorate was trying to impose greater strin-
gency and greater uniformity on the administration of poor
relief.[1] Deterrent 'test workhouses' for indoor able-bodied
paupers were established in some areas;[2] and they were accompanied by a movement for the abolition of outdoor relief,
even in urban unions not covered by the Outdoor Relief
Prohibition Order of 1844.[3] This policy was only partially
successful, because guardians were often more susceptible
to local than central pressures; but, nevertheless, in certain
London unions where there was close co-operation between
guardians and organized charity outdoor relief had been
virtually abolished in the early 1890s;[4] and the national
incidence of 'able-bodied' outdoor pauperism declined steeply
between 1871 and 1891 in spite of the recurrent depression
of trade.[5]

Unemployed workmen accounted, however, for only a
small fraction of both indoor and outdoor pauperism;[6] and
more directly relevant to the problem of unemployment was
the adoption of a harsher policy towards 'casual' paupers, who
used the workhouse when travelling as a temporary place of
call. In the 1850s the Poor Law Board had directed local
guardians to provide separate accommodation for casuals and
to discriminate between habitual vagrants and persons genuinely seeking work.[7] After 1870, however, the Local Government Board virtually closed the casual wards to genuine unemployed workmen, by giving workhouse masters powers of
compulsory detention over all casual paupers and ordering

[1] Cd. 4499/1909, *RC on the Poor Laws, Majority Report*, pp. 146–7.
[2] Seven test workhouses were in operation at different periods, the last surviving at
Belmont in Fulham until the abolition of the guardians in 1929 (S. and B. Webb,
English Poor Law History, II. i. 378–83; and ii, pp. 972–6). These workhouses
were an exposition of the classic theory of the Poor Law, that its aim should be the
elimination rather than maintenance of pauperism (W. S. Jevons, *Methods of
Social Reform and Other Papers*, pp. 191–2). Their virtual emptiness was proof of
their effectiveness; but ratepayers objected to expenditure on empty workhouses,
and they were usually converted to 'general mixed workhouses' as soon as other
kinds of indoor pauper outgrew their accommodation (S. and B. Webb, op. cit. II.
i. 389).
[3] On the local variations in guardians' powers to relieve the unemployed see
H. of C. 365/1895, *SC on Distress from Want of Employment*, Appendix 30,
pp. 557–62.   [4] *Charity Organisation Review*, 11 (Dec. 1895), 481.
[5] Cd. 5077/1911, *RC on the Poor Laws, Statistics Relating to England and Wales*,
pp. 22–5.
[6] Appendix B, Table 1, p. 373.   [7] S. and B. Webb, op. cit. II. i. 402–6.

that nobody should be discharged from a casual ward before 9 a.m.—thereby effectively spoiling a workman's chances of finding employment for that day.[1] These measures were a logical outcome of a policy that the leaders of 'organized charity' had been urging upon the central Poor Law authority since the 1860s—a policy based on the premiss that, while the undeserving poor should be dealt with by a repressive Poor Law, the deserving should be preserved from pauperism and helped by intelligent casework to retain their independence.[2] Charitable associations in the 1880s were willing and indeed anxious to extend their control over the welfare of the poor. But their financial and administrative resources were inadequate to relieve more than a fraction of the distress caused by prolonged unemployment; and the practice of charitable 'casework' was in any case an inherently unsuitable method of relieving the unemployed.[3]

Hence in the mid 1880s economic and administrative factors combined to create a vacuum in the relief of unemployed workmen that no existing institution could adequately fill. This vacuum coincided with a period in which political attention was increasingly focused on 'social questions';[4] and, in certain quarters the unemployed themselves were seen as a potentially significant political force. 'Each succeeding winter brings up afresh the great question "what to do with the unemployed"', wrote Friedrich Engels in the autumn of 1886. 'But while the number of the unemployed keeps swelling from year to year there is nobody to answer that question; and we can almost calculate the moment when the unemployed, losing patience, will take their fate into their own hands.'[5]

---

[1] By the Pauper Inmates Discharge Act of 1871 and the Casual Poor Act of 1882. These Acts were partially relaxed by an L.G.B. circular of 1885, which gave guardians discretion to reduce the period of detention in respect of casual paupers 'who, it is believed, are really desirous of obtaining work'. But this circular does not appear to have been effective and was reissued in Nov. 1887 (*17th Annual Report of the L.G.B.* (1887–8), p. lvii).

[2] A. F. Young and E. T. Ashton, *British Social Work in the Nineteenth Century*, p. 96. S. and B. Webb, op. cit. II. i. 457–9.

[3] Below, Chapter 3, pp. 106–7.

[4] S. and B. Webb, op. cit. II. ii. 644.

[5] Karl Marx, *Capital* (transl. by S. Moore and E. Aveling, 1886 ed.), introduction by F. Engels, i. 6. See also H. H. Champion, *The Facts about the Unemployed. An Appeal and a Warning, by One of the Middle Class* (1886), pp. 15–16.

For a time in the mid 1880s it seemed that Engels's pro-
phecy of 'unemployed' direct action might well prove correct.
In 1884 the militant Social Democratic Federation began
to organize protest marches among irregular workmen in the
East End of London;[1] and in 1885 an 'East End Sugar
Workers' Committee' was established by agents of the
National Fair Trade League to arouse support for protection
among the London unemployed.[2] In the winter of 1885–6
unemployed demonstrations occurred in many provincial
cities;[3] and on 8 February 1886 the unemployed momentarily
terrorized the property-owners of London when two rival
protest meetings, summoned by socialists and protectionists,
gathered in Trafalgar Square and ran amok in the East End.[4]
In the autumn of 1886 the unemployed demonstrated at the
Lord Mayor's show,[5] and the Warden of Toynbee Hall,
Samuel Barnett, informed the government that unemployed
discontent was growing as trade revived.[6] In the summer and
autumn of 1887 the S.D.F. held weekly protest meetings
among unemployed and homeless families squatting in Trafalgar Square;[7] and on 13 November 1887 the unemployed
swelled the ranks of a giant civil liberties demonstration,
when mounted guardsmen from Whitehall Palace were summoned to disperse the crowd.[8]

[1] H. M. Hyndman, *Record of an Adventurous Life*, p. 370; G. Lansbury,
Looking Backwards—And Forwards, pp. 188–90.

[2] This committee received encouragement and financial assistance from Conservative leaders early in 1886 (Salisbury MSS., Class M, Box 22, Thomas Kelly to Lord
Salisbury, 6 Feb. 1886; Class E, Thomas Kelly to Lord Salisbury, 3 Jan. 1891).
The National Fair Trade League had been established in 1881 to press for reciprocal
tariffs on foreign manufactures (B. H. Brown, *The Tariff Reform Movement in
Great Britain 1881–1895*, pp. 17–28). On the 'fair trade' solution for unemployment see J. Crabb, *Bad Times—the Cause and Cure*, Aug. 1885.

[3] Demonstrations were recorded in Northampton, Birmingham, Manchester,
Grimsby, Hastings, Nottingham, Sheffield, Leicester, and Yarmouth (*Pall Mall
Gazette*, 16 Jan.–8 Mar. 1886).

[4] *Pall Mall Gazette*, 9 Feb. 1886, p. 8; H. Lee and E. Archbold, *Social Democracy
in Great Britain*, pp. 111–13; Add. MS. 46308, John Burns' Notes for an Autobiography, ff. 38–41. After the riot H. M. Hyndman, John Burns, Jack Williams,
and H. H. Champion, all members of the S.D.F., were tried and acquitted on a
charge of conspiracy (W. Kent, *John Burns, Labour's Lost Leader*, pp. 25–6).

[5] *Unemployed of London*, Address by the General Council of the S.D.F., 6 Oct. 1886.

[6] Salisbury MSS., Class E, C. T. Ritchie to Lord Salisbury, 4 Nov. 1886.

[7] Bennet Burleigh, 'The Unemployed', *Contemporary Review*, 52 (Dec. 1887),
770–80.

[8] This demonstration had been summoned to protest against the imprisonment

These demonstrations, particularly the riots of 8 November
1886 and 13 November 1887, helped to force the discussion
of unemployment to the forefront of political life;[1] but they
had little direct influence on the formation of social policies
for the relief of the unemployed.[2] They were seen by both
Liberal and Conservative governments as a problem of public
order rather than of social distress—as the responsibility of
the Home Office rather than of the Local Government Board.[3]
They revealed not the strength but the weakness of unconstitutional pressure; and tactically they were self-defeating,
since much of the West End was closed to public meetings
after November 1887 and not reopened until 1892.[4]

The 'unemployed' in the 1880s were in any case singularly
ill fitted for direct political action; they were a very heterogeneous group, whose composition was constantly changing and whose most chronically distressed members were
notoriously resistant to any kind of change. Even as an
electoral force they were almost certainly negligible, since,
in spite of successive extensions of the franchise, many casual
workmen were unregistered or disqualified or ineligible for
either the 'household' or the 'lodger' vote.[5] Nevertheless, in

of the Irish M.P., William O'Brien, by the Federation of London Working Men's
Clubs (*Hansard*, 3rd series, vol. 322, col. 1944). The meeting had been prohibited
by the Metropolitan police, and ninety-seven persons were convicted of unlawful
assembly, including John Burns and Robert Cunninghame-Graham, the Scots
radical M.P., who were gaoled for six weeks (ibid., col. 1913).

[1] *Hansard*, 3rd series, vol. 302, cols. 710–15, 1911–13; vol. 303, cols. 102–3,
444–5, 643–97.
[2] After the Trafalgar Square riot of 8 Feb. 1886 Joseph Chamberlain temporarily
relaxed the conditions for the granting of outdoor relief in London (*Pall Mall
Gazette*, 10 Feb. 1886, p. 1). But the riot was not responsible for the issue of the
famous Chamberlain circular which encouraged local authorities to start relief
works (below, pp. 75–6).
[3] Parliamentary criticism of public authorities after the riot of Nov. 1886
was focused almost entirely on the police and Home Office rather than the guardians
and the L.G.B. (*Hansard*, 3rd series, vol. 302, cols. 562–3, 571, 593–603; vol. 303,
cols. 741–9).
[4] Trafalgar Square (Regulation of Meetings) Act, 1888. Below, p. 81.
[5] The emergence of unemployment in the 1880s as a problem requiring political
attention has sometimes been ascribed to the extension of the franchise among the
working class (B. G. Gilbert, 'Winston Churchill versus the Webbs: the Origins
of British Unemployment Insurance', *American Historical Review*, 71, no. 3
(Apr. 1966), 847 and footnote). But there is little evidence to support this view,
particularly since the urban workman had gained no new rights under the Representation of the People Act of 1884. Moreover, by 1911 about 40 per cent of the

the mid 1880s certain groups emerged which challenged
political indifference to unemployment and, in certain cases,
claimed to speak on behalf of the unemployed. Radical
members of the Liberal party argued that a side-effect of land
reform would be the reduction of unemployment;[1] while Conservative backbenchers claimed that unemployment could be
abolished by the exclusion of aliens or by 'fair trade'.[2] Outside
Parliament the Social Democrats, whilst seeking to arouse the
unemployed to revolutionary consciousness, called also for
the 'palliatives' of an eight-hours day and public works.[3] The
Fabian Society, torn between a classical and a socialist analysis
of unemployment, proposed that public investment should
be used to counteract depression in the private sector,[4] and
that surplus labour should be employed in municipal workshops and county council farms.[5] Skilled trade unionists,
threatened by increased competition from unskilled and unorganized workmen,[6] for the first time turned their attention

adult male population were still unregistered (N. Blewett, 'The Franchise in the
United Kingdom 1885–1918', *Past and Present*, 32 (Dec. 1965), 31). Little information is available about unregistered voters; but it is probable that a disproportionately high percentage of casual and irregular workmen were unregistered,
either through failure to fulfil the residence qualifications or through receipt of
poor relief. In 1888 Salisbury ascribed a by-election defeat to 'the votes of those
who had been dispersed at Trafalgar Square' (CAB 41/21/3, Lord Salisbury to
the Queen, 12 Feb. 1888); but this surmise seems unlikely, since many of those
evicted from Trafalgar Square were homeless and thus almost certainly ineligible
to vote.

[1] *Hansard*, 3rd series, vol. 316, cols. 1501–24; vol. 326, cols. 452–4.

[2] On the exclusion of destitute immigrants see *Hansard*, 3rd series, vol. 311,
col. 1724; vol. 312, col. 1777; vol. 315, cols. 514–19. On 'reciprocal free trade',
ibid., vol. 318, cols. 1726–7; vol. 319, col. 488.

[3] *State Organisation of Unemployed Labour. As An Alternative to the Harmful
Scheme of State-Aided Emigration*, Nov. 1883; *Manifesto of the S.D.F. after the
West End Riots of 8 Feb. 1886*, 15 Feb. 1886.

[4] *The Government Organisation of Unemployed Labour*, Report by Hubert
Bland, F. S. Hughes, Frank Podmore, J. G. Stapleton, and Sidney Webb, 4 June
1886, p. 15. This committee tentatively endorsed the 'right to work' (p. 13),
and proposed that the old-fashioned 'navvy' should be replaced by a semi-military
corps of publicly employed engineering labourers (pp. 19–20). But at the same time
it echoed the fear of orthodox economists that public employment could only be
financed by depleting the resources available for private investment (p. 12).

[5] Annie Besant, 'Industry Under Socialism', in *Fabian Essays in Socialism*
(1931 ed., first publ. 1889), pp. 143–4.

[6] *Report of the Bristol T.U.C.*, 1878, p. 14. For much of the nineteenth century
skilled craft unions had protected themselves from competition by limiting the
ratio of apprentices to journeymen and thereby limiting recruitment to skilled

to the problem of unemployment outside the trade-union movement and endorsed the demands for land reform and the exclusion of aliens, for public works and the eight-hours day.[1]

Many of these remedies figured prominently in the political debate on unemployment for the next thirty years; but the two policies that in the late 1880s and early 1890s attracted most active political support were, firstly, the limitation of the working day and, secondly, the provision of public employment for the unemployed. The campaign for the eight-hours day and the campaign for public works will here be considered in some detail, since they helped to unite the various constituents of the unemployed movement and illustrate both central and local government responses to organized political pressure on behalf of the unemployed.

UNEMPLOYMENT AND THE EIGHT-HOURS DAY

The 'eight hours movement' had a variety of origins and covered a multiplicity of policies and motives.[2] Firstly, it was part of the socialist demand for a new social order, in which workers reaped the benefit of industrial growth. Secondly, it was a logical development of trade-union pressure for shorter hours and higher wages; and in this context it was the first major practical issue since the 1840s to raise the question of whether the organized labour movement should pursue its aims through parliamentary legislation or through the tradi-

---

trades (Ellic Howe and H. E. Waite, *The London Society of Compositors*, pp. 66–83; J. B. Jefferys, *The Story of the Engineers*, p. 102). But the advance of mass production, which was accelerated by the depression, tended in certain industries to downgrade the skill of the artisan and to make him more easily replaceable by semi-skilled or unskilled labour (E. Hobsbawm, 'The Labour Aristocracy in 19th Century Britain', in *Democracy and the Labour Movement* (ed. J. Saville), pp. 211–12; A. E. Duffy, 'New Unionism in Britain 1889–1890: A Re-appraisal', *Economic History Review*, 2nd series, 14 (1961–2), 311–12).

[1] *Report of the Southport T.U.C.*, 1885, pp. 35, 39, 44; *Report of the Bradford T.U.C.*, 1888, p. 41.

[2] An often-cited influence on the 'eight hours movement' was the economic historian J. Thorold Rogers, whose *Six Centuries of Work and Wages*, claiming that artisans had enjoyed an eight-hours day in the fourteenth and fifteenth centuries, was published in 1884. See R. A. Hadfield and H. de B. Gibbins, *A Shorter Working Day*, p. 27; S. Webb and H. Cox, *The Eight Hours Day*, p. 14.

tional process of collective bargaining and strikes.[1] Thirdly, the juxtaposition of long hours with high unemployment gave rise to the belief that a general limitation of the working day would automatically lead to the absorption of the unemployed.[2] In this third sense the demand for the eight-hours day was a reformulation of the old discredited theory of the 'work fund'—that available employment at any given time was a fixed quantity and that men who laboured too hard or too long were depriving others of work.[3] It was reinforced by fears of overproduction and of redundancy caused by mechanization;[4] and the limitation of hours seemed a rational way both of limiting output and of 'sharing work' between employed and unemployed.

The connection between the restriction of hours and reduction of unemployment was first pressed upon trade unionists by Adam Weiler, a German Marxist delegate to the Trades Union Congress in 1878;[5] and in 1883 the T.U.C. passed a resolution calling for a statutory eight-hours day in government employment as a means of absorbing the unemployed.[6] At the Southport conference of the T.U.C. in 1885 James Threlfall, secretary of the newly founded Labour Electoral Association, drew attention to the depressive effect of unemployment upon the level of consumption; and as remedies he proposed the limitation of the working day and the cultivation of 'deserted' land.[7] In 1887 conflicting resolutions were laid before the Congress, one supporting legislation and the other combination to obtain the eight-hours day.[8] The advantages of legislation were outlined by a Welsh delegate, W. Bevan, who warned trade

[1] The history of the restriction of hours contained precedents for both legislation and collective bargaining. In the 1840s and 1850s the ten-hours day in factories had been secured by statute for women and children, and indirectly for adult male workmen; in the 1860s and 1870s many trade unions had obtained a nine-hours day by negotiation and strikes.

[2] *Report of the Hull T.U.C.*, 1886, pp. 15, 21.

[3] C. 7684–II/1895, *RC on the Aged Poor, Minutes of Evidence*, Q. 10278, Statement of Alfred Marshall.

[4] James Leatham, *An Eight Hours Day with Ten Hours Pay. How to Get it and How to Keep it* (1890), pp. 1–2.

[5] *Report of the Bristol T.U.C.*, 1878, pp. 32–3.

[6] *Report of the Nottingham T.U.C.*, 1883, p. 47.

[7] *Report of the Southport T.U.C.*, 1885, pp. 15–16.

[8] *Report of the Swansea T.U.C.*, 1887, pp. 34–7.

unionists that prolonged strike action would ruin their finances and gain nothing for the 'vast hordes' of unorganized unemployed. He urged instead that the unions should use the power of the vote, and that no parliamentary candidate who did not agree to promote an Eight Hours Bill should get trade-union support.[1]

Nevertheless, not all trade unionists favoured the limitation of hours; and many unions, including the long-established craft unions which dominated the T.U.C.'s Parliamentary Committee, argued that the trade-union movement should not compromise its independence by seeking legislation and that the eight-hours day should be obtained by sectoral negotiation between employers and employed.[2] Moreover, many unionists, especially in trades that countered unemployment with 'short-time' arrangements, were not particularly concerned with the conjectural connection between restriction of hours and relief of the unemployed.[3]

These attitudes were, however, challenged by the socialists, who were increasingly influential in the trade-union movement after 1886. The leader of the Social Democratic Federation, H. M. Hyndman, was sceptical about the practical results of an eight-hours day and believed that the sharing of work under the private enterprise system would merely depress the wages of those already employed.[4] But, nevertheless, an eight-hours day in public employment had been part of the S.D.F.'s short-term programme for reducing unemployment since 1883;[5] and in 1886 an Eight Hours League, committed to a policy of limiting the working day by legislation, was founded by Tom Mann, a member of the Battersea

[1] Report of the Swansea T.U.C., 1887, pp. 18–21.

[2] Theodore Llewellyn Davies, Notes on the Trade Union Congress 1890, f. 9.

[3] This was the case in the mining industry, where the arguments in favour of shorter hours centred mainly on the dangers of work at the coal-face. The miners' campaign is considered here only in so far as it demonstrated the general arguments for and against eight-hours legislation; for a more detailed discussion see B. McCormick and J. E. Williams, 'The Miners and the Eight-Hour Day, 1863–1910', Economic History Review, 2nd series, 12, no. 2 (Dec. 1959), 222–37.

[4] On Hyndman's adherence to Lassalle's 'iron law of wages' see C. Tsuzuki, Hyndman and British Socialism, p. 55. On his scepticism of all palliatives for unemployment see H. M. Hyndman, Commercial Crises of the Nineteenth Century, pp. 163–74.

[5] The State Organisation of Unemployed Labour, As An Alternative to the Harmful Scheme of State-Aided Emigration, 1883.

branch of the S.D.F.[1] The creation of the League was important, not merely because it gave the 'eight hours movement'
a formal organization, but because one of Mann's avowed
objectives was to establish co-operation between socialist and
trade-union proponents of the eight-hours day.[2] In June
1886 Mann published a pamphlet which outlined a social
and economic policy for the movement and analysed the
relationship between unemployment, under-consumption,
and the length of the working day. He estimated that of the
7,000,000 adult industrial workers in the United Kingdom,
nominally working nine hours a day, 900,000 were unemployed. If, therefore, the working day were reduced to eight
hours, 750,000 extra workers would have to be employed to
maintain the existing level of output; and 'remembering that
these 750,000 would immediately begin to buy more food,
clothing and general comforts, this . . . would give an impetus
to trade, and so add greatly to the comfort of the entire community. . . .' Mann did not envisage that the stimulus given
to employment would be permanent, since labourers would
again be displaced by 'more efficient machinery and advancement of scientific knowledge'; but several years' respite from
overwork and unemployment would enable the workers to
study and understand their situation and thence to proceed
to a revolutionary goal.[3]

In the late 1880s and early 1890s, the legislative eight-
hours day temporarily eclipsed all other remedies for unemployment in all sections of the labour movement. John
Burns, who had at first disparaged the movement, was soon
converted and joined Tom Mann in organizing support for
an eight-hours day among the London unions.[4] Even H. M.
Hyndman campaigned for a statutory restriction of hours,
claiming that it was a 'valuable palliative to our industrial

[1] The League grew out of the Battersea Progressive Society, which adopted an
eight-hours policy in Apr. 1886. It was formally established in London and
Newcastle in Oct. 1886 (Dona Torr, *Tom Mann and His Times*, pp. 214–15;
C. Tsuzuki, op. cit., p. 81).

[2] Tom Mann, *Memoirs*, p. 62.

[3] Tom Mann, *What a Compulsory Eight Hour Working Day Means to the Workers*
(1886).

[4] Tom Mann, *Memoirs*, pp. 61, 80–1. Burns had, however, included the eight-
hours day in his electoral campaign at Nottingham in 1885 (W. Kent, *John Burns:
Labour's Lost Leader*, p. 20).

anarchy', which would 'create an enormous home trade more
valuable than any foreign trade'.[1] At the same time the demand for an Eight Hours Bill was annually debated at the
Trades Union Congress; and it is clear that the major diver-
gence of trade unionist opinion on the use of legislation, which
the Webbs dated from the emergence of 'New Unionism' in
1889, had been brewing over such a major item of policy as
the eight-hours question for at least two years before.[2]
Furthermore, the split did not occur along socialist and non-
socialist, nor even along 'old' and 'new' unionist lines.[3] The
laissez-faire approach of the Parliamentary Committee was
outflanked not merely by socialists and unskilled workmen,
but by cautious Liberal trade unionists to whom parliamen-

[1] *Report of a Debate on the Eight Hours Movement between H. M. Hyndman and
Charles Bradlaugh*, 23 July 1890, pp. 14–16. Bradlaugh's arguments represented
precisely the old-fashioned radical view of state responsibility for the welfare of
labour; he believed in the shortest working day compatible with a maximum of
profit-making, and that the limitation of hours should be negotiated privately
in each industry. An Eight Hours Act would, he claimed, inevitably lead to the
lowering of wages and the ruin of British industry by foreign competition (C.
Bradlaugh, *The Eight Hours Movement*, pamphlet reprinted from the *New Review*,
1889).

[2] S. and B. Webb, *The History of Trade Unionism* (1920 ed.), p. 418. The Webbs
equated New Unionism with the organization of unskilled labour, the abandonment of *laissez-faire*, and the search for social reform by legislation as well as
combination—the movement being embodied in the great expansion of unskilled
unionism during and after the London Dock Strike of 1889. Recent authorities
have modified this interpretation, showing that the Webbs underestimated the
extent of unskilled unionism before 1889, that the 'upsurge' of union membership
in 1889–91 was ephemeral and that the contrast between the aims of Old and New
Unionism was exaggerated (A. E. Duffy, 'New Unionism in Britain 1889–90:
A Re-Appraisal', *Economic History Review*, 2nd series, 14 (1961–2), 306–17; H.
Clegg, A. Fox, and A. Thompson, *A History of British Trade Unionism*, i. 52–4, 96).
In fairness to the Webbs, however, it should be noted that, in spite of their pre-
occupation with the significance of the events of 1889, they themselves recorded
many of the earlier symptoms of change (*The History of Trade Unionism*, pp. 358–
421).

[3] See, for example, *Report of the Hull T.U.C.*, 1886, pp. 19–26, speech of Fred
Maddison, the editor of the *Railway Review*. Maddison was in no sense a socialist;
as M.P. for Burnley in 1906–10 he was one of the most conservative members of
the Lib.-Lab. group and in 1908 was a vehement opponent of the Labour party's
campaign for the right to work (H. Clegg, A. Fox, and A. Thompson, op. cit.,
p. 400). In 1892 John Burns told Joseph Chamberlain that 'acceptance of "legal
eight hours day" does not . . . prove and is not indicative of socialist tendencies';
but he thought that 'conversion to the legal eight hours day, where previously
hostile, leads the way to municipal socialism, and often far on the road to collectivism' (Add. MS. 46290, ff. 323–4, John Burns to Joseph Chamberlain, 20 Sept.
1892).

tary interference was distasteful but the prospect of prolonged industrial warfare infinitely worse.[1]

Resistance to legislation was in fact largely concentrated in particular industries and regions, its most consistent opponents being the Lancashire cotton-spinners, whose hours and conditions were already regulated by law,[2] and the miners of Northumberland and Durham, who worked for only seven-hour shifts and feared that an Eight Hours Act might increase rather than decrease their working day.[3] Elsewhere, however, the miners favoured legislation; and in 1888 the militant Miners' Federation of Great Britain was founded, covering all coalfields except Northumberland and Durham, to promote legislation for a minimum wage and a maximum eight-hours day.[4] In 1887 a conference of all London trade societies at the Bricklayers' Hall in Southwark declared itself in favour of eight-hours legislation;[5] and similar progress was made in Newcastle and Birmingham.[6] The results of two plebiscites, published by the Parliamentary Committee in 1888 and 1889, suggested that a majority of organized workmen, while favouring the eight-hours day, were still opposed to legislation.[7] But it was observed at the Congress of 1889 that

on few questions has public opinion made such rapid advance as on

---

[1] *Report of the Dundee T.U.C.*, 1889, p. 16.

[2] Ibid., p. 53. Many textile trade unionists had, however, been converted to eight-hours legislation by the end of 1892 (William Mather, *Labour and the Hours of Labour* (Nov. 1892), pp. 2–3).

[3] William Whitefield, *The Miners' Eight Hours Bill* (1891).

[4] H. Clegg, A. Fox, and A. F. Thompson, op. cit. i. 98–105.

[5] Tom Mann, *Memoirs*, p. 62.

[6] Tom Mann, *The Eight Hours Movement* (1889), pp. 2–3.

[7] The results of both plebiscites were thoroughly ambiguous. Only fifty trade unions and nine trades councils responded to the questionnaire circulated by the Parliamentary Committee in 1888. In societies voting by member, 24,351 unionists favoured, and 7,304 opposed the eight-hours day; 17,267 wanted to obtain it by legislation and 7,395 by combination. Of societies voting *en bloc*, 40 favoured and 11 opposed the eight-hours day; 7 wanted to obtain it by legislation and 7 by combination; but the unions that supported an Eight Hours Bill had a much smaller membership than those that opposed (*Report of the Bradford T.U.C.*, 1888, pp. 29–30). In the following year 39,656 unionists in societies voting by member favoured the eight-hours day, and 67,390 were opposed. Twenty-nine societies voting *en bloc* favoured the eight-hours day, and seven were opposed (*Report of the Dundee T.U.C.*, 1889, pp. 52–3). On the efforts of the Parliamentary Committee to influence the voting see James Bartley, *The Eight Hours Movement. The Points of the Parliamentary Committee of the T.U.C. in the Circular issued by them to Trade Unions of the United Kingdom*, n.d.

that of Parliamentary interference with the working hours of adult males. Within two years our unions have almost entirely changed front on this point, and the general body of our members are now in advance of their leaders.[1]

In 1890 the T.U.C. for the first time gave a majority verdict in favour of legislation, and instructed its Parliamentary Committee to press for a universal eight-hours day.[2]

In May 1889 Tom Mann published a further pamphlet, reviewing the progress and confounding the opponents of the movement for the eight-hours day. Against those who claimed that the limitation of hours would depress wages, he argued that it was not short hours but the competition of surplus labour that kept wages low. To those who objected that industry would be ruined by the increased wages bill, he replied that the consequent increase of working-class purchasing power would greatly enlarge the market for manufactures and thus make possible cheaper production and economies of scale.[3] In the same year the Fabian Society produced a draft Eight Hours Bill, the first of many Fabian exercises in legislative model-building.[4] This Bill proposed that the eight-hours day should be made compulsory in mines, railways, and government employment; and that all contracts for the hire of labour should be assumed to contain an eight-hours clause unless specific provision was made to the contrary. Local sanitary authorities would be empowered to extend the compulsory provisions of the Bill by local by-laws; and an eight-hours day could be introduced by administrative orders in any trade where a majority of the employees so desired.[5] The Fabians recommended that the Bill should be laid before all parliamentary candidates as a test of their willingness to support eight-hours legislation.[6] 'I am getting all the leaders of the advanced Liberal wing in London to accept the bill as

[1] *Report of the Dundee T.U.C.*, 1889, p. 16.
[2] *Report of the Liverpool T.U.C.*, 1890, pp. 48–53.
[3] Tom Mann, *The Eight Hours Movement* (1889), pp. 5–7.
[4] Fabian Tract No. 9, *An Eight Hours Bill in the form of an Amendment to the Factory Acts, with Further Provision for the Improvement of the Conditions of Labour*, drafted Nov. 1889, published May 1890.
[5] For a further discussion of the different ways of obtaining the eight-hours day see Fabian Tract No. 48, *Eight Hours by Law: A Practical Solution*, Dec. 1893.
[6] Add. MS. 46287, ff. 311–16, G. B. Shaw to John Burns, 12 Aug. 1892.

a good draft basis', wrote Sidney Webb to John Burns in November 1889, '. . . it is the provincial Liberals who are the difficulty . . .'[1]

Meanwhile, the movement was reinforced by international pressure for the eight-hours day. Realizing the potency of the employers' argument that shorter hours would ruin industry in the face of international competition, labour leaders in Britain, France, Germany, and America began to press for international action;[2] and in July 1889 the International Socialist Workers' Congress in Paris passed a resolution calling for an international demonstration in favour of the eight-hours day.[3] In January 1890 the Bloomsbury Socialist Society and the Gasworkers' and General Labourers' Union agreed to promote a joint May Day demonstration on behalf of the 'legal' eight-hours day; and a co-ordinating Central Committee was established by Edward Aveling and Eleanore Marx.[4] In April the hitherto conservative London Trades Council was persuaded by Tom Mann to hold a similar demonstration in favour of introducing an 'eight hour working day', though not necessarily by legislative means;[5] and on 4 May 1890 three separate eight-hours demonstrations, organized by the Central Committee, the London Trades Council, and the S.D.F., met in Hyde Park, forming one of the largest gatherings of working men that London had ever seen.[6] In July a Legal Eight Hours and International Labour League, representing the London trade unions and radical and socialist clubs was formed in Vauxhall. This League organized lec-

[1] Add. MS. 46287, ff. 258–9, Sidney Webb to John Burns, 14 Nov. 1889. Maltman Barry, the ex-Marxist Conservative agent, optimistically claimed on 12 Aug. 1890 that 90 per cent of prospective Liberal candidates and 70 per cent of prospective Conservative candidates would be prepared to take an eight-hours pledge (Maltman Barry, *The Labour Day*, p. 47, published as a pamphlet by the Aberdeen Trades Council, 1890). But see A. M. McBriar, *Fabian Socialism and English Politics*, p. 244.

[2] Tom Mann, *The Eight Hours Movement*, p. 4.

[3] James Joll, *The Second International 1889–1914*, p. 43; *Report of the Dundee T.U.C.*, 1889, p. 47.

[4] 'The Legal Eight Hours Demonstration in London. A Brief History of the Movement. Portraits and Biographies of some of its Promoters', reprinted from the *Workman's Times*, 1 May 1891.

[5] Dona Torr, 'Tom Mann and His Times, 1890–92', *Our History* (Pamphlets Nos. 26–7), Summer–Autumn 1962, p. 9.

[6] Ibid., pp. 10–11.

tures and published eight-hours propaganda throughout the winter of 1890–1.[1]

Political opposition to eight-hours legislation was, however, formidable.[2] In February 1890 the leader of the Opposition, Mr. Gladstone, told a miners' delegation that it was an intolerable infringement of personal liberty to prevent a man from working for as long as he wished to work;[3] and the Conservative Home Secretary, Henry Matthews, could 'hold out no hope that the Government will support any legislation which has for its objects to impose restrictions upon the freedom of adult males in the disposal and management of their own labour'.[4] This was endorsed by Salisbury and Balfour in an interview with the organizers of the May Day demonstrations on 16 June.[5] Gladstone, who grudgingly received a further deputation from the London Trades Council, refused to consider the eight-hours question until after the settlement of Irish Home Rule;[6] and in November 1890 the Commons did not grant a second reading to a bill introduced by Robert Cunninghame-Graham for a universal maximum eight-hours day.[7]

Nevertheless, during the next two years, the 'legal' eight-hours movement made considerable headway among Liberals in the House of Commons. This was largely because in the election of 1892 support for an Eight Hours Bill was made a condition of labour support in many industrial constituencies;[8] and, in spite of the over-all Liberal victory, several Liberal opponents of legislation were returned with reduced majority or actually lost their seats.[9] Moreover, in 1892 the

[1] 'The Legal Eight Hours Demonstration etc.', reprinted from the *Workman's Times*, 1 May 1891.

[2] The Conservative cabinet refused 'even to discuss propositions for diminishing production, or for regulating the labour of adult males' (CAB 41/21/27, Lord Salisbury to the Queen, 26 May 1889. CAB 41/21/36, Lord Salisbury to the Queen, 15 Feb. 1890).

[3] *Report of Deputations from Representatives of Miners of the United Kingdom to Henry Matthews, Lord Dunraven and W. E. Gladstone on the Miners (Eight Hours) Bill*, 17 and 18 Feb. 1890, p. 72.   [4] Ibid., p. 86.

[5] H. M. Hyndman, *Gladstone and the Eight-Hours Law* (1890), p. 5.

[6] Ibid., pp. 5–6.   [7] *Hansard*, 3rd series, vol. 349, col. 113.

[8] William Mather, *Labour and the Hours of Labour* (Nov. 1892), pp. 1–2, 12–13.

[9] Henry Broadhurst, the ex-secretary of the T.U.C.'s Parliamentary Committee, lost his seat at West Nottingham and John Morley was relegated to second place in the poll at Newcastle, almost certainly through their opposition to the Eight

Progressive majority on the London County Council—which was closely identified with the parliamentary Liberals—set up a Works Department, which was committed to the eight-hours day, the payment of standard rates and the replacement of 'contract' by 'direct' labour;[1] and the L.C.C.'s programme was widely canvassed as a precedent for the national government, especially by John Burns and by the radical M.P. for Tower Hamlets, Sydney Buxton.[2] It was the London radicals who had originally persuaded the L.C.C. to turn itself into a model employer;[3] but it was the Fabians who systematized the economic arguments for the reform of public employment. In 1891 Sidney Webb, in collaboration with the Cambridge economist Harold Cox, published a book on the eight-hours question which firmly maintained that employment would be increased by the limitation of hours and that neither wages nor *per capita* output need fall.[4] Indeed they cited the statistical results of previous reductions in hours to show that they had been followed by higher wage rates;[5] and they published testimonials from Liberal business-men who during the previous three years had introduced the eight-hours day into their factories without any permanent fall in wages or loss of productive power.[6]

The very success of these experiments served, however, to undermine the connection between the eight-hours day and the reduction of unemployment; and Sidney Webb's conclusions on this point were out of accord with the evidence that he adduced. The belief that shorter hours would increase employment depended on the assumption that, if hours were reduced, individual daily output would necessarily fall; in order, therefore, to maintain the previous level of production

Hours Bill (*Hansard*, 4th series, vol. 7, col. 240; D. A. Hamer, *John Morley. Liberal Intellectual in Politics*, pp. 256–62, 275–7).

[1] Gwilym Gibbon and Reginald Bell, *History of the London County Council 1889–1939*, pp. 234–6; Sidney Webb, *The Economic Heresies of the L.C.C.*, address to the Economic Section of the British Association for the Advancement of Science, 13 Aug. 1894 (pamphlet edition, 1894).

[2] George Dew, *Government and Municipal Contracts Fair Wages Movement. A Brief History* (May 1896), pp. 8–14.

[3] A. M. McBriar, op. cit., pp. 191–3, 198.

[4] Sidney Webb and Harold Cox, *The Eight Hours Day*, pp. 121–2.

[5] Ibid., pp. 94–102.

[6] Ibid., pp. 254–64.

additional workmen would have to be employed.[1] This was
not borne out, however, by the experience of three Liberal
industrialists—William Mather, William Allan, and Mark
Beaufoy[2]—who provisionally adopted an eight-hours day
between 1889 and 1893. Both William Mather and William
Allan reduced the working day in their engineering factories
in response to electoral pressure;[3] and both were influenced
by the prospect of absorbing the unemployed. 'We accept
in good faith the assurance made to us that your object is to
afford more employment to your fellow workmen' wrote
William Mather in an explanatory circular to his employees
on 18 February 1893.[4] William Allan, giving evidence before
the Royal Commission on Labour in December 1892 denounced the 'economic absurdity' of overtime when so many
workmen were unemployed; and he claimed that the 'tendency of overtime working, like piece-work, is to create a
superabundance of employment for one section and scarcity
for another.'[5] Both Mather and Allan therefore expected that
the reduction of hours would reduce productivity, and when
introducing the eight-hours system they reduced wages accordingly. But both found that production under the eight-
hours system was as great as before and that they had no need
to engage extra hands. In both firms daily wages were restored
to their former level; and William Allan claimed that all
employers who adopted the eight-hours system would 'be in
pocket by the change'.[6] This was confirmed by Mark Beaufoy, the jam and vinegar manufacturer, who found that under

[1] Sidney Webb and Harold Cox, *The Eight Hours Day*, p. 107.

[2] William Mather was the proprietor of the Salford Ironworks and Liberal
M.P. for Salford South 1885–6, for Gorton 1889–95, and for Rossendale, Lancs.,
1900–4; an advocate of land reform and technical and progressive education.
William Allan, the owner of the Scotia Engine Works of Sunderland, was Liberal
M.P. for Gateshead 1893–1903. Mark Beaufoy, a jam and vinegar manufacturer,
was Liberal M.P. for Kennington 1889–95.

[3] William Mather, who had previously opposed an Eight Hours Bill, except
on a basis of 'local trade option', had his majority seriously reduced at the General
Election of 1892 (W. Mather, *Labour and the Hours of Labour*, Nov. 1892, p. 1).
William Allan, who captured a seat from a Unionist, ascribed his success to his
positive support for the eight-hours day (*Hansard*, 4th series, vol. 7, cols. 368–9).

[4] *Report by William Mather, M.P., on the Result of a Year's Experiment with the
Eight Hours' Day at Salford Ironworks*, p. 15.

[5] C. 7063–I/1893, *RC on Labour (Sitting as a Whole), Minutes of Evidence*,
Q. 6839.

[6] R. Hadfield and H. de B. Gibbins, *A Shorter Working Day*, p. 145.

the eight-hours system 'we did more business than in any year I can remember, but not one hour of overtime was worked'.[1]

The results of these experiments were not entirely conclusive, since the characteristics of one industry were not always comparable with those of another; and, where output per man was inelastic, as in the gas and transport industries, the reduction of hours necessarily resulted in the employment of extra workmen.[2] Nevertheless, the conclusions drawn by Beaufoy, Mather, and Allan appear to have had a decisive influence on the Liberal party's attitude to the eight-hours movement, particularly in factories under direct government control. Since the mid 1880s trade unionists had complained that one of the industries in which the conjunction of long hours with irregular employment was most notorious was the manufacture of armaments in government arsenals.[3] The Government was accused of violating the 'progressive' principle that public authorities should lay down norms of industrial conduct for private employers; and after the election of 1892 John Burns, now M.P. for Battersea, continually pressed the new Liberal government for the introduction of an eight-hours day in factories controlled by the Admiralty and War Office.[4] A series of trade-union deputations lobbied the ministers of both departments, supporting the eight-hours day with the mutually exclusive arguments that it would increase output by improving 'physiological' efficiency and at the same time reduce the number of unemployed.[5]

By 1893 one hundred and fifty-seven parliamentary Liberals were known to be in favour of eight-hours legislation;[6] but Gladstone, now Prime Minister for the fourth time, was still doggedly opposed to any statutory limitation of the

[1] S. Webb and H. Cox, op. cit., p. 263. The eight-hours day was also successfully introduced by Arnold Hills, the proprietor of the Thames Ironworks and by the chemical firm of Brunner-Mond (*The Eight Hours Day. A Ton of Practice* (A.S.E. pamphlet), 1897; R. Hadfield and H. de B. Gibbins, op. cit., p. 141).

[2] S. Webb and H. Cox, op. cit., p. 107.

[3] *Report of the Hull T.U.C.*, 1886, pp. 14–15.

[4] *Daily Chronicle*, 6 Jan. 1894.

[5] Add. MS. 46282, ff. 8–22, 'An Eight Hours Day for Government Employees', Report of a deputation led by John Burns and received by Campbell-Bannerman, 6 July 1893.

[6] Fabian Society MSS., Box 20, Lists of Liberal, Liberal Unionist and Irish Nationalist M.P.s pledged to support an Eight Hours Bill.

adult working day.[1] The Secretary for War, Campbell-
Bannerman, however, inquired into the results of the experiment conducted by William Mather, and became convinced
of the economic and political expediency of the 'forty-eight
hour week'.[2] An eight-hours day was introduced into a
branch of the Woolwich Arsenal in the autumn of 1893; and
it was found that 'as much is turned out as under the longer
hours system'. Campbell-Bannerman therefore decided to
extend the eight-hours system to all War Office factories early
in 1894. Justifying the measure to the Cabinet, he argued
that the eight-hours day brought both 'gain to the employer'
and 'moral and physical benefits to the men'; and that, if
neglected by the Liberals, it would certainly be introduced by
the next Conservative government.[3] The whole Liberal cabinet, with the exception of Gladstone and Morley, appear to
have endorsed his views.[4] A few months later the eight-hours
day was extended to the Admiralty, and in 1895 to certain
branches of the G.P.O.[5] Three years later Campbell-Bannerman described the success of the experiment to a *Daily News*
reporter: 'Up to the time that I left office all the reports went
to show conclusively that the production was as great, if not
greater, under the eight hours system, as it had been under
the nine, and I have not the slightest reason to believe that this
is not the case today.'[6]

The conversion of the Liberal party to the principle of the
eight-hours day in public employment was a concession that
the labour movement could not afford to turn down. But the
grounds of this conversion completely undermined the arguments of those who believed that the problem of unemployment could be solved by the eight-hours day. Between 1886
and 1893 this argument was the spearhead of the short-term
policies advanced by all sections of the unemployed move-

[1] *Some Farrer Memorials, Selections from the letters of Lord Farrer, 1819–99*,
pp. 92–3.

[2] J. A. Spender, *The Life of the Right Honourable Sir Henry Campbell-Bannerman*,
i. 142.

[3] Add. MS. 41233, ff. 176–7, Memorandum by 'H. C. B.', 1 Jan. 1894.

[4] Add. MS. 41233, ff. 178–9, Cabinet ministers' comments on the eight-hours
scheme, Jan. 1894. Gladstone, however, came round before his retirement to 'local
option' for a miners' eight-hours day (A. M. McBriar, op. cit., p. 245).

[5] *The Eight Hours Day: A Ton of Practice*, pp. 4–6.

[6] Quoted in ibid., p. 2.

ment—trade unionists, London radicals, Fabians, and the
S.D.F. But to retain this argument Burns, Mann, Hardie,
and other proponents of eight-hours legislation would have
had to forfeit the opposite argument—that shorter hours led
to greater efficiency and increased productivity, which made
the employment of extra workmen unnecessary. To have
denied this argument would have been to jeopardize the
success of the whole movement; and this the leaders of the
movement were not prepared to do.

This change of emphasis in the eight-hours movement was
soon apparent in the reasoning of labour leaders about unemployment. The dilemma was clearly defined and pointed
out to John Burns by the Secretary of the Arsenal branch of
the South Side Labour Protection League in August 1892.[1]
Earlier in the year Tom Mann had published a third report
on the reduction of hours, which was a classic statement of the
short-term alternatives and long-term aims of the eight-hours
movement. He had modified his demand for a universal eight-
hours day, and now supported permissive legislation, which
would be enforceable by local authorities in particular trades
at the request of a majority of local workmen. Special provision should also be made for the redistribution of work among
casual, unorganized, and low-paid workers in order to raise
their purchasing power and thereby promote the wealth of
the community as a whole.[2] Mann's evidence to the Royal
Commission on Labour at the end of 1892, however, revealed his growing doubts about the efficacy of the eight-
hours day as a means of reducing unemployment. He still
maintained that the limitation of hours in certain industries
would help to counteract fluctuations. But his main argument
for the eight-hours day had shifted from the immediate absorption of the unemployed to a long-term increase in the
working-class standard of living; and his main prescription
for unemployment was no longer the regulation of hours in
private industry but the extension of public ownership and
public works.[3]

[1] Add. MS. 46290, ff. 278–9, Arthur Harris to John Burns, 31 Aug. 1892.
[2] Tom Mann, *The Eight Hour Day. How to Get it by Trade and Local Option*,
1892.
[3] C. 7063–I/1893, *RC on Labour (Sitting as a Whole), Minutes of Evidence*
QQ. 2538–40.

The shift in Mann's opinion was echoed by Sidney Webb, who told the Royal Commission that an Eight Hours Act would probably increase employment in public transport, but that 'when you come to manufacturing industries I am unable to make any assumption about the absorption of the unemployed'.[1] Robert Giffen, the head of the Board of Trade's Commercial department thought that a compulsory eight-hours day might be introduced in certain industries on grounds of health or safety;[2] but he was doubtful whether a reduction of hours could be imposed without a reduction in wages, which—far from increasing employment—would have a depressive effect on consumer demand.[3] The views of trade-union leaders were equally diverse.[4] Many believed that shorter hours must necessarily spread employment over a larger number of workmen; but their members were reluctant to abandon overtime, and were apparently more afraid of wage cuts than of unemployment or excessive hours.[5]

By June 1894, when the Majority and Minority Reports of the Royal Commission on Labour were published, the practical results of the eight-hours experiments were well known.[6] The Majority Report outlined both the short and long-term arguments that a reduction of hours would increase employment;[7] but it advised against legislative interference with the adult working day, even on a basis of trade option.[8] The Minority Report, drafted by Sidney Webb, strongly supported the introduction of an eight-hours day through

[1] C. 7063-I/1893, RC on Labour (Sitting as a Whole), Minutes of Evidence, QQ. 4762-3, 4758.

[2] Ibid., QQ. 7038-42.

[3] Ibid., Q. 7028.

[4] C. 6795-II/1892, Digest of the Evidence taken before Group B of the RC on Labour, ii. 91-4.

[5] In 1894 an A.S.E. member wrote to John Burns, deploring the reluctance of skilled engineers to sacrifice overtime for the sake of the unemployed: '. . . many of them have become so used to it that they consider they are on short time when working the ordinary hours, and yet they will probably join the May Day demonstration and shout about eight hours . . . members are willing to work the clock twice round sooner than make an honest effort to get some of the men off the books' (Add. MS. 46287, f. 6, Samuel Robinson to John Burns, 27 Jan. 1894).

[6] See J. Stephen Jeans, The Eight Hours Day in the British Engineering Industry. An Examination and Criticism of Recent Experiments, pp. 28-30; John Rae, Eight Hours for Work, pp. 179-213. Both were published in 1894.

[7] C. 7421/1894, RC on Labour, Fifth and Final Report, paras. 174-6.

[8] Ibid., paras. 319-29.

the extension of existing factory legislation. But it was defended chiefly on the grounds that it would promote health,
efficiency, combination, and self-help among the workers.[1]
The reduction of unemployment, which two years earlier
had been one of the main avowed objectives of the eight-hours
movement, was now virtually ignored.

This change was even more apparent in the proceedings of
the Select Committee on Distress from Want of Employment in 1895, when only one out of more than forty witnesses claimed that unemployment could be significantly
reduced by the introduction of an eight-hours day.[2] Advocates of the eight-hours day continued to press for the limitation of hours for its own sake. It was still seen as a means of
regularizing casual employment and of promoting working-
class consumer demand. Among trade unionists the limitation of hours and abolition of overtime were still periodically
revived as a means of reducing unemployment.[3] But as a
serious contribution to unemployment policy, the old static
'work-fund' notion of the relationship between hours and
employment was destroyed by the experience of the early
1890s;[4] and, within the labour movement, the eight-hours
day was replaced by other and more complex policies for
relieving the unemployed.

THE CAMPAIGN FOR PUBLIC EMPLOYMENT

Pressure for the eight-hours day as a remedy for unemployment stemmed mainly from labour and socialist groups. The
demand for the provision of work by public authorities was,
however, less exclusive and covered a wide spectrum of

[1] C. 7421/1894, *RC on Labour, Minority Report*, pp. 131–4.

[2] The exception was Will Thorne, Secretary of the National Union of Gasworkers and General Labourers of Great Britain and Ireland, who had organized
the successful strike for an eight-hours day at Beckton Gasworks in 1889. Even
Thorne admitted that the eight-hours day would merely afford temporary relief
to the unemployed in certain types of industry (H. of C. 365/1895, QQ. 10832–3;
W. Thorne, *My Life's Battles*, pp. 64–72).

[3] e.g. *22nd Quarterly Report of the General Federation of Trade Unions*, Dec.
1904, Report on Unemployment, pp. 5–6.

[4] J. M. Robertson, *The Eight Hours Question* (1899 ed., first publ. 1893),
pp. v–vii. Robertson's book was mainly devoted to proving that the limitation
of hours was useless without control of population.

political principles and practical reforms. It was clearly linked with the socialist doctrine of the 'right to work';[1] but to the non-socialist it might mean little more than *ad hoc* relief works or a revival of the Elizabethan statutes relating to the employment of the able-bodied poor.[2] During the unemployment crisis of 1886 the *Pall Mall Gazette* urged the Local Government Board to start a public-works scheme modelled on the Lancashire experiment of 1863.[3] The leader of the Conservative party, Lord Salisbury, declared his support for a national housing programme;[4] and the correspondence of his followers after the Trafalgar Square incident made it clear that many rank-and-file Conservatives had no doctrinaire reservations about the dangers of public works.[5] In 1887 the philosopher and economist, Henry Sidgwick, came to the conclusion that a recognition of the 'right to work' should be incorporated in the Poor Law;[6] and in March 1888 the veteran administrator, Edwin Chadwick, assured Lord Salisbury that public works for the unemployed in an emergency were in no way incompatible with the intentions of the Poor Law commissioners of 1834.[7]

Socialist pressure for the provision of work by public authorities therefore took place in a political atmosphere that was by no means uniformly hostile to the principle of 'public

[1] H. Russell Smart, *The Right to Work*, 1895. Smart, the parliamentary Labour candidate for Huddersfield, claimed that 'economic freedom may now be defined as the Right to demand common labour from the local authority for a maximum working day of eight hours and a minimum wage of twenty-four shillings a week' (ibid., p. 4). On the history of the 'right to work' doctrine see J. H. Jones, 'The Unemployed Workmen Bill of the Labour Party', *Transactions of the Liverpool Economic and Statistical Society*, 1908–9, pp. 11–18.

[2] J. Theodore Dodd, *To Boards of Guardians in Rural Districts. The Winter's Distress—How to Provide for the Unemployed*, 1894.

[3] *Pall Mall Gazette*, 8 Feb. 1886, p. 1.

[4] Ibid., 6 Feb. 1886, p. 3.

[5] Salisbury MSS., Class M, Box 22, *passim*. Conservative proposals for state intervention to increase employment included the widening of London streets, the provision of electric lighting, the construction of imperial railways, the restriction of immigration, and protective tariffs.

[6] A. S. and E. M. S., *Henry Sidgwick: A Memoir*, p. 481, Henry Sidgwick to John Addington Symonds, 1 Dec. 1887. Earlier in the same year, however, Sidgwick had written that even though 'the Right to Labour had been conceded in Germany, I do not know any means by which it could be attained in a community like our own, without a grave danger of disastrous consequences' (*The Principles of Political Economy*, 2nd ed., 1887, p. 533).

[7] Salisbury MSS., Class E, Edwin Chadwick to Lord Salisbury, 7 Mar. 1888.

works'. But, nevertheless, many public authorities were reluctant to accept any practical obligation to employ the unemployed.[1] Early in February 1886 the Metropolitan Board of Works claimed that it had no power to create relief employment.[2] The President of the Local Government Board, Joseph Chamberlain, denied that distress from unemployment was severe enough to warrant state intervention; and even after the demonstration of 8 February, he maintained that 'the question of public works is not one within the province of the L.G.B.'[3]

A series of parliamentary debates in February and March 1886 revealed, however, that pressure for public employment was by no means confined to the socialists, and that a number of Liberal, Conservative, and Nationalist back-benchers were in favour of public works. Proposals put forward by private members for increasing employment by state intervention included prison, road, harbour, and house building;[4] the provision of free seed-corn for distressed agricultural areas;[5] an increase in armaments production;[6] and a 'day of humiliation and prayer as a National Appeal to Almighty God'.[7] The Prime Minister, Mr. Gladstone, replied with the orthodox economic argument that public works would 'paralyze . . . private enterprise, and very possibly lead to the dismissal of a number of persons now employed';[8] and in a debate on harbour construction on 12 March 1886 Chamberlain implied that few of the 'unemployed' would accept public employment, since they were mainly of the class that shunned work at any price.[9] He admitted, however, that 'he, for one, had no idea of pretending that the House and Government were not responsible for some measures to deal with . . . distress';[10] and three days later he issued a circular to local authorities, urging them to schedule necessary public works for periods of depression and to co-operate with Poor Law

[1] H. H. Champion, *The Facts about the Unemployed. An Appeal and a Warning*, pp. 10–16.
[2] *Pall Mall Gazette*, 6 Feb. 1886, p. 7.   [3] Ibid., 10 Feb. 1886, p. 7.
[4] *Hansard*, 3rd series, vol. 302, col. 1035; vol. 303, cols. 102–3.
[5] Ibid., vol. 303, col. 123.   [6] Ibid., vol. 302, col. 1912.
[7] Ibid., vol. 303, col. 1182.
[8] Ibid., vol. 302, col. 713.
[9] Ibid., vol. 303, col. 663.   [10] Ibid., vol. 303, col. 661.

guardians in providing temporary non-pauperizing employment for the deserving unemployed.[1]

The issue of the Chamberlain circular attracted much attention from subsequent students of the unemployment problem as the first major breach in the exclusively 'Poor Law' treatment of the unemployed.[2] But this view exaggerates both the historical and practical significance of the circular. Some local authorities already deliberately postponed certain kinds of public work until periods when trade was slack;[3] and the circular merely gave formal encouragement to such a policy. Moreover, Chamberlain's chief motive in authorizing public employment was not to supersede but to strengthen the Poor Law's capacity for dealing with the problem, as was revealed by his contemporary correspondence with the future Mrs. Webb. 'It will remove one great danger,' he wrote on 5 March 1886, 'viz. that public sentiment should go wholly over to the unemployed, and render impossible that state sternness to which you and I equally attach importance. By offering reasonable work at low wages we may secure the power of being very strict with the loafer and confirmed pauper.'[4]

The Chamberlain circular was issued five times between 1886 and 1893.[5] From a practical point of view, however, it was for several reasons an almost complete failure. In the first place many local authorities already employed a permanent staff of workmen adequate for all construction purposes.[6] Many authorities complained that the kind of work envisaged in the circular—painting, roadmending and other forms of unskilled labour which would not compete with private enterprise—was unsuitable for the winter months when unemployment was most severe.[7] Work undertaken in an emergency with an untrained labour force proved in almost every

---

[1] 16th Annual Report of L.G.B. (1886–7), pp. 5–7.

[2] S. and B. Webb, English Poor Law History, II. ii. 645–7.

[3] H. of C. 111/1895, SC on Distress from Want of Employment, Q. 181.

[4] P. Fraser, Joseph Chamberlain: Radicalism and Empire, 1886–1914, p. 125; J. Chamberlain to B. Webb, 5 Mar. 1886. For Beatrice Webb's attitude to unemployment at this time see Pall Mall Gazette, 18 Feb. 1886, p. 11, 'A Lady's View of the Unemployed at the East'.   [5] In 1886, 1887, 1891, 1892, and 1893.

[6] H. of C. 365/1895, SC on Distress from Want of Employment, Minutes of Evidence, Q. 10,408.

[7] H. of C. 253/1895, SC on Distress from Want of Employment, Second Report, Appendix 5, pp. 204–5.

case to be expensive, inefficient and 'demoralising' to the workmen employed.[1] An exceptional case was Battersea where, after 'elaborate and careful provision' by the vestry surveyor, certain kinds of work were reserved for periods of distress, and the unemployed were successfully employed in snow clearance, roadbuilding, and the construction of the Thames foreshore.[2] The experience of Battersea showed that relief works needed careful preparation and a regular role in vestry policy; but the permanent guarantee of public employment every winter was precisely what most authorities wished to avoid.

Secondly, the work largely failed to attract the type of unemployed workmen for whom it was introduced, namely 'artisans and others' who 'make personal great sacrifices in order to avoid the stigma of pauperism' and to maintain their independence.[3] Much of the evidence about applicants for work authorized by the circular suggested that the majority were chronically irregular workmen, who resorted to relief works merely as another form of casual employment. A special report of the Kensington vestry for 1892–3 showed that of 1,056 applicants for relief work, 913 were 'general labourers';[4] and in every locality the applicants for public employment were primarily composed of casual and unskilled workmen.[5]

Thirdly, the central government made little attempt to enforce the principles contained in the circular or to recommend them to other employing bodies. C. T. Ritchie, who succeeded Chamberlain as President of the Local Government Board in the new Conservative administration of March 1886, was convinced that 'nothing can be done by Government' to provide work for the unemployed. He thought that 'the only real remedy was greater thrift, less drunkenness, more industry and fewer early marriages';[6] and he insisted

---

[1] C. 7182/1893, *Report on Agencies and Methods for Dealing with the Unemployed*, pp. 216–17; *Charity Organisation Review*, 9 (1893), 395–6.
[2] H. of C. 365/1895, *SC on Distress from Want of Employment*, QQ. 5020–5154.
[3] *16th Annual Report of the L.G.B.* (1886–7), pp. 5–6.
[4] *Charity Organisation Review*, 10 (Feb. 1894), 84.
[5] C. 7182/1893, *Report on Agencies and Methods for Dealing with the Unemployed*, pp. 224–8, 235–7, 408.
[6] Salisbury MSS., Class E, C. T. Ritchie to Lord Salisbury, 1 Nov. 1887.

that his department had no power to compel reluctant local
authorities to undertake public works.[1] In November 1887
the Local Government Board was urged by the Fabian
Society to extend the regulation of employment to government departments and to the London dock companies, and
to rectify the 'attitude of inaction' adopted by the Board of
Works.[2] The Permanent Secretary, Sir Hugh Owen, promised to forward the Fabian proposals to the Post Office, War
Office, and Admiralty; but he protested that any interference
in the management of private companies would be 'quite
unprecedented', and he disclaimed any jurisdiction over the
Metropolitan Board of Works.'[3]

The most serious impediment to the policy authorized by
the Chamberlain circular arose, however, from the difficulty
of raising the necessary funds. No special provision was made
for financing works to relieve the unemployed. But the normal
method by which local authorities raised extraordinary loans
—through the Public Works Loans Commissioners after
approval by the L.G.B.—involved so much delay that the
need for relief work had often elapsed by the time the money
was granted.[4] Moreover, the Local Government Board was
reluctant to authorize loans for minor works or works that
had not been carefully planned; but emergency relief works
for the unemployed were necessarily of this character. Local
authorities were therefore compelled to rely mainly on the
rates. But often the areas with most unemployment and the
greatest need for public improvements were those that could
least afford the cost of public works; and the rateable value of
districts in the metropolis tended to vary inversely with the
extent of local distress.[5]

This was a problem that was to hamper the development of
all aspects of social administration at a local-government level
before 1914.[6] As part of the local-government reforms of

[1] Salisbury MSS., Class E, C. T. Ritchie to Lord Salisbury, 29 Oct. 1887.
[2] Fabian Society MSS., Box 19, Sydney Olivier to C. T. Ritchie, 12 Nov. 1887.
[3] Ibid., Sir Hugh Owen to Sydney Olivier, 2 Dec. 1887.
[4] H. of C. 111/1895, *SC on Distress from Want of Employment, Minutes of Evidence*, QQ. 1012, 1019.
[5] H.L.G. 29/42, vol. 42, ff. 3–5, Memorandum by B. F. C. Costello (Chairman
of the Local Government and Taxation Committee of the L.C.C.), 2 Feb. 1893,
revised 10 Feb. 1893.
[6] See Appendix A, pp. 369–70.

1888 local authorities were given annual subsidies from the national exchequer in the form of assigned revenues; but these were allocated on a purely historical basis rather than on local poverty or social need. The situation in London had been partially relieved by the creation of the Metropolitan Common Poor Fund in 1867, which made certain items of Poor Law expenditure a charge on the metropolis as a whole.[1] In 1888 Ritchie introduced an annual contribution to Poor Law expenditure out of the county rate;[2] and in 1894 an Equalisation of Rates Act empowered the London County Council to levy a uniform 6d. rate which would be distributed for sanitary purposes 'according to the needs and poverty of the various localities'.[3] Nevertheless, ratepayers in metropolitan districts with a low rateable value were still paying nearly twice as much in the pound as ratepayers in districts with a high rateable value in 1895;[4] and as John Williams Benn pointed out to the Select Committee on Distress from Want of Employment, the policy of the Chamberlain circular had in many areas been rendered virtually meaningless by the geographical imbalance of poverty and wealth. 'So far as this circular is concerned it ended with itself in many parishes of London ... the position in London is this; that in the districts where the unemployed congregate, where such works are most necessary, the rates are highest and the machinery least effective for carrying out such works.'[5]

Organized pressure for an extension of public responsibility for the relief of the unemployed was revived during the depression of 1892–5. On this occasion the extent and duration of unemployment were less severe than in the mid 1880s; but the forces of protest were in a better position to influence central and local authorities and political parties. In London this pressure took three main forms. Firstly, a revival of the kind of unemployed demonstration which had so alarmed the

---

[1] Cd. 4499/1909, *RC on the Poor Laws, Majority Report*, p. 128, para. 171.

[2] Ibid., p. 129, para. 175.

[3] H.L.G. 29/42, vol. 42, f. 7, Sir Hugh Owen to Shaw Lefevre, n.d.; f. 46, Mr. Dalton to Sir Hugh Owen, 16 June 1894.

[4] H. of C. 111/1895, *SC on Distress from Want of Employment, Minutes of Evidence*, QQ. 3089–90. Before the Equalisation of Rates Act, metropolitan rates varied from 4s. 1d. in the £ in St. James's, Westminster, to 7s. 11d. in the £ in Bow. After the Act they ranged from 4s. 5½d. in St. James's to 7s. 5d. in Bow.

[5] Ibid., Q. 3082.

metropolis in 1886–7; but now the restriction on the right
of assembly in the West End confined such meetings mainly
to the area east of St. Paul's. Secondly, pressure was brought
to bear on local authorities to implement the Chamberlain
circular, even where no funds were available for this purpose;
the focal points of this activity were Poplar and Bermondsey
within the metropolis and West Ham outside its borders.
Thirdly, there was the campaign conducted in Parliament to
promote the intervention of the central government, a campaign in which the central figure was James Keir Hardie, the
Independent Labour member for South West Ham.

In September 1892 the London Trades Council, having
tried unsuccessfully to conduct a survey on the extent of
unemployment, petitioned the Local Government Board for
an official inquiry into the 'prevailing destitution' among the
unemployed.[1] The L.G.B. had no precise information about
the dimensions of the unemployment problem;[2] but a public
inquiry was nevertheless refused and the Government took
the view that 'such distress as exists can best be dealt with by
local authorities'.[3] In the autumn of 1892 an Unemployed
Organisation Committee was therefore established in London, consisting of representatives of the S.D.F., the London
Trades Council, and 'many other working-class or political
organisations'.[4] Its aims were to investigate the unemployed
problem, to persuade local authorities to engage in public
works, and to restore the right of public meeting in Trafalgar
Square. Unemployed protest meetings recommenced on
Tower Hill in November 1892, and a delegation headed by
Edward Aveling waited on Shaw Lefevre at the Office of
Works and persuaded him to reserve part of the work on the
demolition of Millbank prison for the unemployed.[5] A similar

[1] C. 7182/1893, *Report on Agencies and Methods for Dealing with the Unemployed*, p. 182.   [2] *Hansard*, 4th series, vol. 9, col. 1449.
[3] Add. MS. 44516, ff. 281–2, H. H. Asquith to W. E. Gladstone, 30 Nov.
1892. This letter mentioned that the Prince of Wales had expressed a desire to be
appointed to a commission of inquiry on the unemployed.
[4] C. 7182/1893, p. 182; Julius Jacobs (ed.), *The London Trades Council, 1860–
1950*, pp. 79–80.
[5] *Standard*, 2 Dec. 1892. 122 unemployed workmen were employed on the Millbank demolition site between Dec. 1892 and July 1893, their wages amounting
to £1,644 (H. of C. 419/1894, *Report of the Surveyor on the Demolition of Millbank
Prison by the Unemployed to H.M. Office of Works and Public Buildings*, 28 July 1894).

deputation to the Postmaster General urged him to 'produce
. . . plans for extra buildings which would provide work for
some of the unemployed';[1] and on 14 November, the President of the L.G.B., Henry Fowler, reissued the Chamberlain
circular and asked to be kept informed of the progress of local
works.[2]

The minister most immediately concerned with the unemployed, however, was, as in 1886-7, the Home Secretary;
and, in so far as it was a political problem, unemployment was
still seen primarily as a threat to public order rather than as
a source of social distress. The demonstrations on Tower Hill
initially came under the jurisdiction of the Corporation of
London; but in December 1892 they were brought to the
attention of the Home Office, because the unemployed
planned to hold a torchlight procession in the Metropolitan
Police area.[3] The Home Secretary, Herbert Asquith, was
advised to forbid the procession;[4] and the midnight march
which set out for the West End on 1 December was broken
up by the united action of Metropolitan and City police.[5]

Asquith was disinclined to impose a permanent limitation
on the right of assembly, and on taking office he had partially
relaxed the restrictions that had been in force since 1887.[6] In
1893, however, threats of violence at unemployed meetings
led to a widespread and occasionally hysterical demand for
their suppression.[7] In August 1893 the Cabinet rejected a
demand from the 'Central Committee of the London Unemployed Fund' that £1,000,000 should be made available
for public works;[8] and in September the daily demonstrations
were reconvened on Tower Hill. A series of deputations was
dispatched to the L.G.B.;[9] but Fowler refused to seek addi-

[1] *Standard*, 2 Dec. 1892.

[2] *22nd Annual Report of the L.G.B.* (1892-3), pp. 36-8.

[3] H.O. 45/9861/B13077A/1, Sir Evelyn Bradford, Chief Commissioner of the
Metropolitan Police, to Sir Godfrey Lushington, Permanent Under-Secretary at
the Home Office, 29 Nov. 1892.

[4] Ibid., Sir Godfrey Lushington to H. H. Asquith; Lushington privately admitted to Asquith that the Home Office had 'no legal power' to prevent the demonstration.   [5] *Standard*, 2 Dec. 1892.

[6] R. Jenkins, *Asquith*, pp. 64-5. Asquith had acted as counsel for Robert Cunninghame-Graham after the Bloody Sunday riot of 13 Nov. 1887.

[7] e.g. H.O. 45/9861/B13077/23; C. Dudley Ward to H. H. Asquith, 30 Dec.
1893.   [8] CAB 41/22/46, W. E. Gladstone to the Queen, 18 Aug. 1893.

[9] H.O. 45/9861/B13077A/9, Report to C.I.D. Commissioners, 2 Sept. 1893.

tional statutory powers for relieving the unemployed.[1] This rebuff was followed by the failure of Keir Hardie to persuade the House of Commons to hold a special debate on unemployed distress;[2] whilst in St. Pancras the S.D.F. tried unsuccessfully to bring a summons against local guardians for failing to implement the Elizabethan Poor Law statutes whereby the able-bodied destitute were entitled to remunerative work.[3]

The leaders of the unemployed demonstrations were therefore able to point convincingly to the futility of legal and constitutional methods of influencing public authorities.[4] In December 1893 Jack Williams, the Secretary of the Unemployed Organisation Committee, claimed to have founded an unemployed 'secret society';[5] and as winter dragged on and the authorities took no action he began to urge his audiences that the only way to get public money for the unemployed was by a 'reign of terror' and 'open revolt'.[6] All meetings of the unemployed were patrolled by police, who took shorthand notes in anticipation of arrests. Early in February 1894 several persons were injured when the police forcibly disbanded a march to the West End;[7] and a few days later Williams threatened to call up a contingent of international revolutionaries on behalf of the unemployed to 'send the police to heaven by Chemical parcel-post'.[8]

Asquith, as minister responsible for public order, was assailed in Parliament and the press both for countenancing 'police brutality' and for his 'foolhardy indulgence of these rowdies and larrikins masquerading as the unemployed'.[9]

---

[1] *The Times*, 13 and 16 Sept. 1893.

[2] *Hansard*, 4th series, vol. 18, cols. 1706–7.

[3] H.O. 45/9861/B13077D/12, H. Knibbs, Hon. Sec. of the Kentish Town branch of the S.D.F., to H. H. Asquith, 7 Nov. 1893.

[4] *Standard*, 14 Dec. 1893.

[5] H.O. 45/9861/B13077/18, Papers relating to Mr. Knatchbull-Hugesson's question to the Home Secretary in the House of Commons, 18 Dec. 1893.

[6] H.O. 45/9861/B13077C/5, Report from New Scotland Yard on unemployed meeting on 12th inst., 15 Feb. 1894.

[7] *The Times*, 5 Feb. 1894.

[8] H.O. 45/9861/B13077C/2, Police statements made to the Treasury Solicitor, 8 Feb. 1894.

[9] *The Times*, 1 Dec. 1893, 6 Feb. 1894, 16 Feb. 1894. Although *The Times* blamed Asquith for his leniency, the unemployed blamed him for public indifference to their cause (Margot Asquith, *Autobiography* (1962 ed.), p. 197).

In February 1894 Williams's threat of explosives, which coincided with a series of continental assassinations, drove him to take legal advice;[1] and Home Office consultations were held with the Chief Commissioner of Metropolitan Police, Sir Evelyn Bradford, and the law officers of the Crown. A collection of police statements and shorthand accounts of demonstrations was found to contain evidence for several convictions for breach of the peace.[2] But Sir Evelyn Bradford insisted that 'from the police point of view the whole movement is absolutely insignificant; they talk as they have always talked any time these past twenty years, stark rebellion and riot, but they don't mean to do anything.'[3] And Asquith himself took the view that an indictment of the agitators would merely give them 'artificial importance', by furnishing undesirable publicity for their cause.[4]

In so far as the maintenance of public order was concerned the Home Secretary's diplomatic inertia proved justified. Under the surveillance of two police forces the threats of violence remained no more than threats. Impassioned appeals to the East End's starving unemployed to overthrow society fell for the most part on sceptical or indifferent ears.[5] Even in the winter of 1894–5 when the Thames was frozen over and the Port of London closed for three months, the unemployed on Tower Hill failed to evolve into a revolutionary force. The campaign of the Unemployed Organisation Committee once again demonstrated the central government's indifference to 'unconstitutional' pressure and the severe limitations of political action based on the support of the unemployed.[6]

[1] H.O. 45/9861/B13077C/1, H. H. Asquith to Sir Godfrey Lushington, 6 Feb. 1894 (2nd letter). Asquith was anxious to avoid turning the problem into a 'question of policy'.

[2] H.O. 45/9861/B13077C/2, Police statements, 8 and 9 Feb. 1894.

[3] H.O. 45/9861/B13077C/1, Sir Godfrey Lushington to H. H. Asquith, 6 Feb. 1894.

[4] Ibid., H. H. Asquith to Sir Godfrey Lushington, 6 Feb. 1894 (1st letter).

[5] H.O. 45/9861/B13077/9, Report of a meeting (1 Sept. 1893) by Assistant Commissioner Anderson, C.I.D. to Sir Godfrey Lushington.

[6] The subsequent history of the Unemployed Organisation Committee is obscure. It passed into the control of the Anarchists and was reconstituted in 1895 as the Central Unemployed Association, under the patronage of the Reverend Dr. William Thackeray of Greenwich. One of its members, Charles Cooper, an unemployed 'pharmaceutical journalist' gave evidence to the Select Committee on Distress from Want of Employment of 1896 (H. of C. 321/1896, QQ. 440–629). A London

At a local-government level, however, unemployed pressure was rather more effective in the early 1890s than it had been in 1886. Thirty-three metropolitan local authorities undertook relief works in the winter of 1892–3 and sixty-three elsewhere.[1] In thirty-five cases relief works were accompanied by some kind of labour registry, ranging from full-scale 'labour bureaux' to lists of vacancies kept at the town hall.[2] In Wandsworth, Clapham, Southwark, Bermondsey, and Rotherhithe, organized demonstrations were directly responsible for persuading guardians to open stoneyards in the winter of 1894–5;[3] and early in 1895 Keir Hardie claimed that 'unemployed committees' throughout the country were agitating for public funds to provide work for the unemployed.[4]

An area in which the problems of the labour market and the limitations on the power of local authorities were acutely demonstrated was the borough of West Ham. West Ham was in many ways abnormal, since it suffered from all the metropolitan difficulties of casual employment, labour congestion, overcrowding, and rural immigration, without being inside the metropolitan boundary and without therefore sharing even the small amount of financial relief afforded by the Metropolitan Common Poor Fund and Equalisation of Rates Act.[5] Demographically it was placed at a confluence between the 'ebbing tide from the overcrowded districts of East London, and the flowing tide from the agricultural districts' of the eastern and home counties.[6] Occupationally it shared in both the decay of agriculture and the disorganization of the

Central Workers' Committee on Unemployment was established by the S.D.F. under Harry Quelch during the depression of 1904–5 (*C.U.B. Minutes*, i, 1 Dec. 1905, unpaginated. Add. MS. 46323, Burns Diary, 13 Dec. 1905).

[1] C. 7182/1893, *Report on Agencies and Methods for Dealing with the Unemployed*, p. 212.

[2] Ibid., pp. 118–208, 213–35.

[3] H. of C. 111/1895, Q. 2811; H. of C. 365/1895, QQ. 5198–5200.

[4] H. of C. 111/1895, QQ. 700–42, 762–3. The committees were composed of 'delegates from trade unions, socialist organisations and branches of the I.L.P. . . . clergymen and outside sympathisers . . .' (Q. 701).

[5] C. Howarth and M. Wilson, *West Ham*, pp. 399–409.

[6] H. of C. 111/1895, Q. 2486. West Ham consisted of seven parishes with a mixed urban and rural population of 365,130 at the census of 1891. At Michaelmas 1894 its rateable value was £1,627,485; and rates were paid at about 8s. in the £ (ibid., QQ. 1244, 1368).

London docks.[1] It was not covered by the Mansion House relief scheme of 1892–3 which gave employment to distressed waterside labourers in the East End.[2] It was, however, the only borough in the country with an Independent Labour member of Parliament—James Keir Hardie, who was returned by the constituency of South West Ham in the General Election of 1892.

Hardie's first major speech in February 1893 was an amendment to the address asking for public provision for the unemployed.[3] In July 1893 he asked the President of the L.G.B. to revive statutes passed in the reigns of George III and William IV, which empowered guardians to acquire land for employing the unemployed.[4] Fowler, having consulted the Attorney- and Solicitor-General, admitted that these statutes had never been repealed, but was doubtful whether after half a century of obsolescence the L.G.B. would be justified in resurrecting them without fresh legislation.[5] Hardie therefore planned to introduce a private bill restating these old powers, and meanwhile proceeded to act as though they were still in force.[6] He collaborated with Archibald Grove, the Liberal member for North West Ham, in drawing the attention of local guardians to their unused powers;[7] and he supported Will Thorne, the local dockers' leader and branch secretary of the S.D.F., in urging both guardians and vestry to take action. Hardie fully endorsed the unemployment policy of the S.D.F. 'It was said that if work was found for the unemployed it will mean the breaking-up of old industrial systems' he told an open-air meeting in August 1893. 'He believed this was so, and was one reason why he advocated this solution so strongly.' He insisted that guardians already had power to acquire land, and suggested that a 'home colonisation' scheme should be developed on the lines envisaged by the Reverend Herbert Mills.[8] In conjunction with Thorne he

---

[1] Ibid., Q. 2007.   [2] Below, p. 113.

[3] *Hansard*, 4th series, vol. 8, cols. 724–32. His maiden speech, on 18 Aug. 1892, had been an unsuccessful plea for an autumn session of Parliament to discuss social reform (ibid., vol. 7, cols. 450–2).   [4] Ibid., vol. 14, cols. 807–8, 1143–4.

[5] Ibid., vol. 17, cols. 940–1; CAB 37/34/55, Employment of the Poor by Boards of Guardians, by H. H. Fowler, 29 Nov. 1893.

[6] *West Ham Herald*, 5 Aug. 1893.   [7] Ibid., 22 and 29 July 1893.

[8] Ibid., 19 Aug. 1893. Hardie revealed in this speech that he had read Mills's book on home colonization, *Poverty and the State* (see below, pp. 119–20).

persuaded local unions to form an unemployed committee to fight local elections. At a meeting of trade-union delegates at Mansfield Hall on 12 September Hardie estimated that 1,300,000 adult workers, or 10 per cent of the adult labour force were unemployed; and as immediate remedies he proposed the eight-hours day and the provision of public employment at 6d. an hour or the 'dockers' tanner'.[1] On 16 December he issued a manifesto to his constituents, condemning the Government's inactivity and asking for a mandate to devise a new unemployment policy and 'in a constitutional way to force it to the front'. He announced a series of public meetings, at which he would sound out the views of his constituents; and he promised that if even a substantial minority were against him he would resign and re-contest his seat.[2]

Meanwhile the Mayor of West Ham announced his intention of opening a relief fund to provide work for the unemployed. Arnold Hills, the proprietor of the Thames Ironworks, promised to contribute £1,000 if the council could raise the same amount.[3] A similar offer was made to the neighbouring borough of Poplar, where the unemployed were demanding snow-clearance work from the guardians and district board.[4] Applicants for relief in both boroughs were set to work on Wanstead flats, preparing football and cricket pitches and an artificial lake; they were paid at 6d. an hour for six hours a day and for three days a week.[5]

The accounts of the West Ham relief scheme, published in May 1894, showed that 2,152 men had been given employment at a cost of just over £1 per head.[6] Nevertheless, the work performed was very uneconomical, as Arnold Hills told the Select Committee on Distress from Want of Employment in 1895.

Four months experience . . . proved to me conclusively that unless some very strong line was taken, so as to separate relief or assisted

---

[1] *West Ham Herald*, 16 Sept. 1893.   [2] Ibid., 16 Dec. 1893.
[3] Ibid., 18 Nov. 1893; George Haw, *The Life Story of Will Crooks M.P. From Workhouse to Westminster*, pp. 224–5.
[4] Lansbury MSS., vol. 1, f. 197, Press-cutting entitled 'Scenes at Poplar', 11 Jan. 1894. Will Crooks, the local dockers' leader, advised the unemployed to apply for a writ compelling the District Board to remove the snow.
[5] H. of C. 365/1895, *SC on Distress from Want of Employment, Minutes of Evidence*, Q. 10804.   [6] *West Ham Herald*, 19 May 1894.

labour from the labour in the ordinary industrial channels, not only was there great danger to the industry of the district, but that there would be very great cost in the works carried out. In West Ham the work carried out cost about 50% more than it would have cost if it had been done under ordinary conditions. In Poplar it cost about 100% more. . . .

Hills therefore came to the conclusion that it was necessary to penalize inefficiency by paying piece-rates rather than time-rates to the unemployed;[1] and in November 1894 he offered a further £1,000 to the West Ham Council on condition that employment was given at a basic wage of only 4*d*. an hour, or two-thirds the local trade-union rate for unskilled labour.[2] The council at first accepted this offer by a majority of twenty-five votes to three;[3] but this acceptance alienated local unions and caused a major crisis of relief-giving principle in West Ham. Will Thorne gave notice of a motion to reverse the decision at the next council meeting and organized a series of protest meetings among local workmen. 'I went to street corners and exposed the whole business,' he told the Select Committee on Distress from Want of Employment, '. . . I did not believe in it as I thought it was economically unsound.'[2] A fortnight later Thorne, together with representatives of the dockers', gasworkers', and compositors' unions, led a deputation to the town council. They 'brought a red flag into the Council Chamber . . . and a large crowd came down, who stood outside and . . . brought a great deal of terrorism to play on the Town Council, and they rescinded that resolution by exactly the same majority as that which had accepted the offer a fortnight before.'[3]

The West Ham relief works programme henceforth split into two parts, that under the council giving time-wages of 6*d*. an hour, and that under Hills's 'Unemployed Relief Committee' paying piece-rates at a minimum of 4*d*. an hour. The Committee claimed that relief works at piece-rates were a 'form of legitimate trade socialism which deserves to be encouraged and extended';[4] and in 1895 Hills proposed that

[1] H. of C. 111/1895, *SC on Distress from Want of Employment, Minutes of Evidence*, Q. 2534.   [2] H. of C. 365/1895, Q. 10804.
[3] H. of C. 111/1895, Q. 2535.
[4] H. of C. 365/1895, *Appendix 9*, p. 495.

similar schemes should be introduced by legislation in all
distressed areas, to raise the 'morale of the working class' and
to restore unemployed workmen to the 'ordinary channels of
trade'.[1] Special relief works of this kind were, however,
objectionable to the 'orthodox' Poor Law school, on the
ground that they tended to displace employed workmen and
to sap working-class independence;[2] and, whatever the merits
of Hills's scheme, it was clearly incompatible with the trade-
union principles that had been successfully imposed by
labour and socialist pressure on the council of West Ham.

The most significant feature of the West Ham councillors'
policy, however, was not their implicit recognition of the
'right to work', nor their adoption of standard rates, but their
failure to relieve more than a fraction of the local unemployed.
They were trapped by the dilemma that troubled many
authorities during the frozen winter of 1894–5;[3] that the
work suggested by the L.G.B. circular was impossible to
perform, and yet the issue of the circular had created expectations in the minds of unemployed workmen that could not be
ignored.[4] Hence the growing conviction of many councillors
and local government officials that the problem should be
taken over by the central government and that it was 'the
supreme duty of the state to find work for the unemployed'.[5]

Nevertheless, the central government was sceptical about
the extent of unemployed distress. When Lord Rosebery
received a deputation at Stratford Town Hall in December
1894, he queried their estimate that 5,000 workmen were
unemployed in West Ham.[6] The local trade unions, together

[1] H. of C. 111/1895, QQ. 2567–74. Hills circulated a draft bill to the 1895
Select Committee, proposing that certain areas should be scheduled as Employment
Relief Districts; relief works modelled on his experiment in West Ham would be
organized by joint committees, representing charities and local authorities, and
financed by private subscriptions and by local and national taxation.

[2] Ibid., QQ. 395–6, Evidence of Sir Hugh Owen, Permanent Secretary of the
L.G.B.

[3] National Review, 25 (Mar. 1895), 14. The Thames was frozen across, and some
of the unemployed were used to sweep the ice and to prepare the river for skaters.

[4] H. of C. 111/1895, Q. 1366, Evidence of F. Hilleary, clerk to both the guardians
and the council of West Ham, who stated that the guardians could deal with any
distress that arose, but that the unemployed wanted employment not poor relief.

[5] For a criticism of this view see C. S. Loch, 'Relief Not Charitable Employment
the Better Method of Dealing with the Unemployed', Charity Organisation Review,
9 (Feb. 1895), 68–9.   [6] H. of C. 111/1895, Q. 1979.

with the I.L.P. and the S.D.F., therefore conducted an unemployed census under the supervision of Percy Alden, the Warden of the Mansfield House settlement in Canning Town. They found that over 10,000 persons were out of work, of whom nine-tenths were unskilled or casual workmen, and half were in distress.[1] To relieve this situation the West Ham guardians, having failed to get permission to provide work at wages for the unemployed, gave liberal outdoor relief; and the number of workmen employed in the union stoneyard enormously increased during the first fortnight of February 1895.[2] Many of the unemployed, however, would no longer accept even outdoor relief from the guardians; and in West Ham they established their own 'Executive Committee', which urged the borough council to press the central government for the nationalization of land.[3] Early in February 1895 the council petitioned Parliament that 'the duty of finding employment for able-bodied persons out of work should be undertaken by the state, which alone possesses the means for properly dealing with a question of such vast magnitude.'[4] The Government was also being urged by the Independent Labour party 'to co-operate with local bodies to provide work for the unemployed';[5] and on 13 February 1895 a 'strong and powerful' Select Committee of the House of Commons was set up, to inquire into the powers of local authorities and the extent of unemployed distress.[6] This was the first national inquiry into unemployment and the first

[1] Ibid., QQ. 1982, 2113-15.

[2] Ibid., Q. 1261. 8,618 men were relieved in the West Ham stoneyard between 31 Jan. and 19 Feb. 1895, compared with 2,589 during the two previous months. This rise was reflected throughout London, metropolitan pauperism in Feb. 1895 reaching its highest level since 1871 (*26th Annual Report of the L.G.B.* (1895-6), p. 162).

[3] H. of C. 365/1895, QQ. 11005-62, Evidence of James Morton, an unemployed seaman, chairman of the Unemployed Executive Committee of West Ham.

[4] H. of C. 111/1895, Q. 1296.

[5] H.O. 45/9861/B13077D/28, A. Hopson, Secretary of the Kentish I.L.P., to H. H. Asquith, 7 Feb. 1895.

[6] *Hansard*, 4th series, vol. 30, cols. 637-43. Keir Hardie had asked for a Select Committee in Nov.-Dec. 1893 to consider the provision of work by central government departments; but Gladstone replied that unemployment was a question for local authorities, and that it was contrary to correct financial procedure to vote money to government departments for creating artificial work (*Hansard*, 4th series, vol. 18, cols. 1915-18; vol. 19, col. 1769).

official recognition that it might be a fit subject for remedial legislation.

THE SELECT COMMITTEES OF 1895 AND 1896

The Select Committees on Distress from Want of Employment were set up to deal with a situation that had been largely created by the policy of the Local Government Board. Although the depression had imposed a certain amount of strain on the Poor Law, particularly in unions under the Outdoor Labour Test Order,[1] there was little evidence to suggest that the unemployed had exhausted the resources of the Poor Rate, even in the poorest and most depressed areas.[2] But since 1870 the L.G.B. had encouraged the unemployed to look to charity, and since 1886 to the sanitary authorities, rather than to the guardians for the relief of distress.[3]

The depression of 1892–5 had shown, however, that relief works were usually inefficient and expensive, and often no less demoralizing than work in the stoneyards; and that many vestries, councils, and boards of works had neither the staff nor the funds nor the practical facilities for providing work which complied with the conditions laid down by the Local Government Board. Moreover, local authorities were least able to fulfil this responsibility in areas where unemployment was most severe and where demotic pressure on behalf of the unemployed was increasingly powerful.[4] But since the central government had conceded that the unemployed should be given non-pauperizing employment by the local authority, it was virtually impossible for local authorities in such areas to insist that the board of guardians was the proper instrument of unemployment relief.

Nevertheless, the Select Committee appointed in February 1895 was singularly ill fitted to cut the administrative knot fastened by the Chamberlain circular. The Committee's terms of reference specifically invited it to suggest administrative

---

[1] 25th *Annual Report of the L.G.B.* (1895–6), pp. 162–7.

[2] H. of C. 111/1895, QQ. 1368; 1569, Evidence of F. Hilleary, who thought that the school-rate of 2s. 3d. was more onerous than the Poor Rate of 1s. 1d. in West Ham.

[3] Ibid., Q. 1570.

[4] Ibid., QQ. 3082–9.

and legislative reforms;[1] but although a majority of the twenty-five members were in favour of some kind of adjustment of the *status quo*, they shared few common principles or policies. Although the Committee contained twelve Liberals, twelve Unionists, and one Labour member, differences of opinion did not necessarily coincide with differences of party. Chief among the Unionists were George Bartley, Chairman of the Islington COS, founder of the National Penny Bank, and champion of the principles of 1834; William Lawrence, a Liverpool business-man and spokesman of the 'rate-paying' interest; and William Bousfield, an amateur psychologist, who believed that social problems were 'branches of experimental science'.[2] The Liberals included John Burns, the ex-social democrat, John Williams Benn, the municipal reformer, and William Mather, the Salford iron-master and protagonist of the eight-hours day. Each of these men saw unemployment as part of a wider spectrum of social problems, and each of them tried to fit unemployment into their own particular scheme for social reform. There is little evidence, however, of any behind-the-scenes collaboration between either those who favoured or those who opposed an extension of state responsibility for the unemployed. The chairman, Sir Henry Campbell-Bannerman, had a positive distaste for 'Poor Law subjects' and failed to define the problems facing the Committee, or to give positive direction to its investigations.[3]

Hence, although the Committee conducted what was the most extensive inquiry into unemployment prior to the Royal Commission on the Poor Laws of 1905–9, very little of the

[1] *Hansard*, 4th series, vol. 30, col. 367. Nevertheless, Sir William Harcourt, the Leader of the Commons, stated that 'he did not desire that the Committee should be involved in political issues'.

[2] George Bartley (1842–1910), M.P. for Islington North, knighted 1902; an advocate of Old Age Pensions; author of *The Seven Ages of a Village Pauper*. William Lawrence (1844–1935), M.P. for Abercromby, barrister, Director of Imperial British East Africa Company. William Bousfield (1854–1943), M.P. for Hackney North, barrister, author of *The Mind and its Mechanism*. For his views on unemployment see W. Bousfield 'The Unemployed', *Contemporary Review*, 70 (Dec. 1896), 835–52.

[3] J. A. Spender, op. cit. i. 166; Campbell-Bannerman to James Campbell, 12 Feb. 1895: 'They are going to put me on as chairman of this unemployed committee—a horrible thing. I protested and said I knew nothing about poor laws subjects—I had never picked oakum in my life. The grim reply was, "My dear fellow, you'll wish you were picking oakum before you are done with this job".'

evidence collected about the causes of unemployment or about practical reforms was reflected in its final recommendations. The Committee at first concentrated on finding a scheme for the immediate relief of distress, which 'Parliament might be reasonably expected to accept . . . or even take . . . into serious consideration, without further inquiry into the facts upon which it is founded or the principle it embodies'.[1] But in view of the controversial nature of the problem, a policy that would command such unquestioning parliamentary support was scarcely conceivable. It was not therefore surprising that the Committee's first report came to the rather lame conclusion that 'no plan has been suggested which fulfils these conditions'.[2]

When the Committee proceeded to a more exhaustive inquiry into unemployment, however, it proved even more difficult to reach a practical consensus. Much of the cross-examination of witnesses was concerned less with practical remedies than with the discussion of first principles of economic policy and social administration. Thomas Mackay, the Poor Law historian, claimed that unemployment could only be cured by the complete restoration of a free market in which labour found its natural price;[3] whereas Percy Alden saw 'no final remedy for the unemployed question other than the entire re-organisation of industry upon an ethical instead of a competitive basis'.[4] But statements of this kind were affirmations of belief rather than recommendations of policy. Of the practical reforms suggested to the Committee, only three were seriously considered, presumably because each of them already had persuasive advocates within the House of Commons.

The first and most important set of proposals were those which concerned the relaxation of pauper disfranchisement and the revival of the ancient powers of the guardians. The boards of guardians had been newly elected under the extended Poor Law franchise of 1894, which the Local Government Board inspectorate had expected to cause a 'general

---

[1] H. of C. 111/1895, *First Report*, p. iv, para. 9 (b).   [2] Ibid., para. 10.
[3] H. of C. 365/1895, Q. 5357. Thomas Mackay, author of *The English Poor*: Hon. Sec. of the COS in St. George's-in-the-East; a leading protagonist of the movement for the abolition of outdoor relief in the East End.
[4] H. of C. 111/1895, Q. 2301.

bouleversement of accepted methods and doctrines' towards
outdoor relief, the aged poor, and the unemployed.[1] Nevertheless, the guardians consulted by the Select Committee
were remarkably cautious in their views. Many guardians in
distressed areas of England and Wales believed that the
unemployment crisis had been caused solely by the severity
of the weather and that their existing powers were quite
adequate for dealing with all kinds of distress.[2] The guardians
cross-examined by the Committee, however, held conflicting
views on the question of disfranchisement. Some thought that
the loss of the vote was a matter of indifference to most of the
unemployed;[3] whereas Mr. Hilleary, the clerk to the guardians of West Ham, thought that it was the 'crux' of the unemployed problem;[4] and George Lansbury, the representative of the Poplar guardians, thought that the relaxation of
disfranchisement was the only legislative reform on behalf of
the unemployed that was likely to be accepted by the House
of Commons.[5] The Committee also considered Keir Hardie's
proposal that the guardians should be encouraged to use their
powers of acquiring land and providing work under the
statutes of George III and William IV, which Henry Fowler
had admitted had never been repealed.[6] Both proposals were
strongly criticized by Sir Hugh Owen, the Permanent Secretary of the Local Government Board;[7] but nevertheless in
its final report the Committee recommended that the franchise should be retained by 'deserving men' who sought
relief in 'exceptional distress'; and that the Board should
frame rules to enable guardians to exercise their obsolete
statutory powers to acquire land and thus to provide 'work
at wages' for the unemployed.[8]

The second set of reforms considered by the Committee
referred to the readjustment of financial and administrative
responsibility for the unemployed. Keir Hardie proposed a

---

[1] 25th Annual Report of the L.G.B. (1895–6), p. 166.
[2] H. of C. 253/1895, SC on Distress from Want of Employment, Appendices 5 and
6.   [3] H. of C. 365/1895, QQ. 3759, 5265.
[4] H. of C. 111/1895, QQ. 1286–7.   [5] H. of C. 365/1895, Q. 10408.
[6] Ibid., QQ. 7537–46, Evidence of John Dodd, counsel to the Birmingham
Allotments Association.
[7] H. of C. 111/1895, QQ. 390–7, 457–88. H. of C. 321/1896, QQ. 1229–31.
[8] H. of C. 365/1895, Third Report, pp. iv–v.

tripartite system, whereby the national exchequer would double the value of contributions raised locally from charity and from the rates. He urged that the Government should instantly make available £100,000 for this purpose—a sum which was clearly inadequate, if his estimate that the unemployed and their dependants numbered six millions was correct.[1] But Hardie was less concerned to devise a practical scheme than to lay down a principle; he admitted that his main purpose was 'to establish a precedent through this committee'—the precedent that the State should take action on behalf of the unemployed.[2]

A less ambitious scheme was put forward by John Williams Benn, who thought that distress from unemployment could be relieved by extending the unit of local administration. He suggested that expenditure on relief works in London should be paid out of the Metropolitan Common Poor Fund, and that the work should be organized by specialist officials for the whole city. Both urban and rural areas would be brought into the scheme; the wasteland around London would be fertilized by street refuse, and the city's unemployed would be engaged on the reclamation of the Thames estuary and Hackney and Plumstead marshes. Such a plan would enable London to deal with its own unemployed and make assistance from the central government unnecessary; it was 'based on the principle that the wealth of London should enable the poorer districts to cope . . . with such pressure as we are at the present time considering'.[3] Benn's proposal had the support of guardians in many poor unions;[4] and, in spite of opposition from the L.G.B.,[5] it was partially accepted by the Select Committee, which recommended that half the cost of relief works incurred by sanitary authorities in London should be borne by the metropolis as a whole.[6]

Thirdly, the Committee considered several schemes for settling the unemployed on the land. The most comprehensive plan was outlined by William Mather, who proposed that the State should finance agricultural training homes for the unemployed and that county councils should provide

---

[1] H. of C. 111/1895, QQ. 906–12, 1186.   [2] Ibid., QQ. 974–5.
[3] Ibid., QQ. 3106, 3109–12.   [4] Ibid., Q. 401.
[5] Ibid., Q. 405.   [6] H. of C. 365/1895, *Third Report*, p. v.

land for smallholdings and allotments. Like many contemporary land revivalists, Mather saw his scheme as a panacea for a multitude of social evils—for urban congestion and 'corruption', rural depopulation, physical degeneracy, rising Poor Law expenditure, and foreign competition.[1] Similar schemes were advanced by witnesses from the Salvation Army and Church Army, who hoped that the Government would subsidize their own farm colonies for the unemployed.[2] These schemes were sympathetically considered by the Committee, probably because the Government and the Liberal party were already to a certain extent committed to policies of land revival.[3] But in July 1895 its work was cut short by the dissolution of Parliament and a change of government; and the Committee concluded its final report with the hope that unemployment reforms would be further considered by a similar inquiry in the following session.[4]

A second Committee on Distress from Want of Employment was set up by the Conservatives at the beginning of 1896. Its terms of reference were the same as those of the previous Committee, except that it was also asked to consider ways of discriminating between 'regular' and 'exceptional' unemployed applicants for poor relief. More than half the members of the old Committee were reappointed, but Conservatives now outnumbered Liberals and those who in 1895 had put forward the most radical proposals were no longer members of Parliament.[5] The chairmanship was again offered to Campbell-Bannerman, indicating that unemployment was a 'non-party' issue;[6] but he declined, and the new Committee was presided over by Thomas Russell, the Parliamentary Secretary to the Local Government Board.[7] The 1896 Com-

---

[1] H. of C. 365/1895, Appendix 23, pp. 535–42.

[2] Ibid., QQ. 9941, 10190, Evidence of Bramwell Booth and Revd. William Hunt. See below, Chapter III.

[3] *Report of a Conference on the Condition of the Rural Population*, held by the National Liberal Federation at the Memorial Hall, Farringdon Street, 10 Dec. 1891.

[4] H. of C. 365/1895, *Third Report*, p. v.

[5] Keir Hardie and John Williams Benn had lost their seats. William Mather did not stand for Parliament in 1895.

[6] Add. MS. 41233, ff. 282–3, Henry Chaplin to Campbell-Bannerman, 17 Aug. 1895.

[7] Thomas Russell (1841–1920), Unionist M.P. for South Tyrone, promoter of Irish land reform, and founder of the 'New Land Movement' in Ulster.

mittee was noticeably more pragmatic than its predecessor; it avoided discussions of principle and concentrated on practical reforms. It was also more cautious in its final proposals, although this was probably a reflection, not of the change of government, but of the declining urgency of the problem in the blossoming trade boom of 1896.

Several schemes were considered for distinguishing the habitual pauper from the 'capable, willing, sober and industrious unemployed';[1] and the Committee came to the significant conclusion that this could not be done on a personal basis, since the investigation of character was necessarily subjective and difficult to prescribe by statute.[2] Several East End guardians proposed that the question should be settled geographically—that the Local Government Board should be empowered to schedule areas of high unemployment as 'distressed districts', in which relief might be given automatically without disfranchisement.[3] But the Committee accepted Sir Hugh Owen's objection that the execution of such a power would be politically invidious and administratively impracticable. They therefore recommended that poor relief should be given without disfranchisement to unemployed workmen who had not been relieved in two preceding years; and that guardians should as far as possible arrange for the employment of respectable workmen on relief works rather than in the workhouse.[4]

The members of the 1896 Committee were in fact largely content with the compromise policy of the Chamberlain circular, although they suggested certain important modifications. They thought that efficient workmen employed on relief works should be paid at the standard hourly rate, and that 'less eligibility' should be maintained by working shorter hours per day, which would also give the unemployed some

---

[1] H. of C. 321/1896, *Report*, p. iv.

[2] Ibid., pp. xiv–xv.

[3] On 2 Nov. 1894 a conference of guardians from Bethnal Green, Fulham, Hackney, Holborn, Kensington, Lambeth, Mile End, Poplar, Strand, Stepney, Wandsworth, and Clapham had called for a formation of a 'London Unemployed District' in which relief could be given to the unemployed 'without debt to citizenship'. This resolution was submitted to the L.G.B., which denied that it had power to authorize the creation of such a district, nor could it hold out any hope of introducing legislation to this effect (H. of C. 111/1895, QQ. 400–1).

[4] H. of C. 321/1896, *Report*, pp. ix, xv.

free time in which to look for normal employment.[1] They proposed that local authorities who started relief works should be assisted by special loans at cheap rates from the Public Works Loans Commissioners; and that relief works in London should be subsidized from the Metropolitan Common Poor Fund.[2]

These were the limits of the Committee's proposals for positive reform. Grants from the national exchequer were dismissed as demoralizing and unnecessary. Farm colonies were rejected as inappropriate to the needs of the urban unemployed. Cautious approval was given to 'wise co-operation' between guardians and charity, to working-class mutual thrift, and to local and voluntary experiments in the provision of employment. Finally, the Committee, which had been brought into existence because the unemployed shunned the Poor Law, concluded on a note of complacent satisfaction with the Poor Law's powers of dealing with the unemployed; and it thrust the onus of relief back upon the 'popularly elected guardians' who were responsible for applying those powers 'within the limits that the law and the regulations have prescribed'.[3]

The report of the Select Committee of 1896 marked the end of the first phase of the movement to promote public action on behalf of the unemployed. The recommendations of both Committees were virtually ignored by the Government, by Parliament, and by the press.[4] Early in 1896 the Conservative cabinet discussed the possibility of legislation dealing with the 'unemployed poor', but this never materialized.[5] Nothing was done to retain the vote for the pauperized unemployed;[6] and the recommendation that guardians should

[1] Ibid., p. xi.   [2] Ibid., pp. xi–xiii.   [3] Ibid., p. xvi.

[4] J. A. Spender, the editor of the *Westminster Gazette*, was apparently unaware that the Select Committee of 1896 had been appointed or that either Committee reported (J. A. Spender, op. cit. i. 166). Beveridge in 1908 remarked that the 'net result' of the depression of 1893–4 was 'the appointment of a House of Commons committee which did not report, and the acquisition of much negative experience as to the value of stoneyards and temporary relief works' (W. H. Beveridge, 'Public Labour Exchanges in Germany', *Economic Journal*, 18, no. 69 (Mar. 1908), 1). The *Nation*, 21 Mar. 1908, referred to the Select Committee of 1895 'which never reported'.

[5] CAB 41/23/43, Lord Salisbury to the Queen, 18 Jan. 1896.

[6] A private member's bill to this effect, introduced by Samuel Hoare, did not reach a second reading.

revive their powers of acquiring land foundered on the caution
of the Local Government Board. The Board refused to issue
rules for the acquisition of land until the guardians had submitted specific schemes for approval; while the guardians
were understandably reluctant to frame such schemes until
they knew the rules to which they were supposed to conform.[1]
The proposal that relief works should be financed by cheap
loans might have enabled the haphazard policy authorized
by the Chamberlain circular to be developed into a systematic
long-term programme for stabilizing employment by means
of public works. But this possibility was scotched in 1897
by the refusal of the Treasury to sanction cheap loans repay-
able over long periods for works for the unemployed.[2] Without the assistance of such loans, relief works continued to be
financed out of the local rates, which tended to be most
inadequate in areas where the effects of unemployment were
most severe. In 1896 a Light Railways Commission was
established which gave grants and loans to local authorities in
distressed agricultural areas, for the purpose of improving
rural transport, attracting industry to the countryside, and
arresting the migration of labour to the towns. This measure
was historically significant as an assertion of the principle that
the central government should assist depressed areas; but
politically it was designed as a form of relief to the agricultural
community and had no direct bearing on the problem of the
unemployed.[3]

Nevertheless, for the time being the policy of official
inaction paid off. The revival of trade and the temporary
eclipse of the nascent parliamentary labour movement in
1895 relieved pressure on the central government. Low
interest rates and the collapse of certain forms of foreign
investment assisted the boom in public and private building

[1] H. of C. 321/1896, pp. viii–ix.
[2] T. 168/37, Miscellaneous memoranda, vol. vii, Edward Hamilton to Sir
Michael Hicks-Beach on 'Relief of Distress Loans', 14 Jan. 1897.
[3] Gerald Balfour MSS., P.R.O. 30/60/45, Memorandum on 'Light Railways and
Tramways and the Light Railways Act 1896' (n.d. ? 1903). Although introduced
under a Conservative government, this Act was the outcome of a conference summoned by James Bryce, then President of the Board of Trade, in 1894, which was
attended by representatives of local authorities, railway companies, and Chambers
of Agriculture and Commerce. The Act made available £1,000,000, of which the
Treasury had allocated £186,000 in 'free grants' and £43,000 in loans by 1903.

which reacted on the whole economy.[1] In London, the dispersal of some of the casual poor from the central area to new working-class suburbs had, by the turn of the century, helped to dispel any immediate fear of unemployed mobs.[2] In 1899 the outbreak of the Boer war forestalled the depression that the Treasury was predicting for the following year;[3] and the level of unemployment recorded by the skilled trade unions fell to its lowest point since 1874.[4] Discussion of the subject vanished completely from the deliberations of the Trades Union Congress between 1896 and 1903—a crude measure of the extent to which unemployment among skilled workmen was a cyclical rather than a chronic problem. In London and other large towns each winter brought a recurrence of unemployment among the casual labouring class; but the fact that such people could not find work even at the height of prosperity tended to confirm the view that they were morally and physically degenerate.[5] The government of Lord Salisbury, which had no special commitment to social reform and was preoccupied with the situation in Ireland and South Africa, was content to ignore a subject that, temporarily at least, had ceased to be a political problem.

After ten years political pressure on behalf of the unemployed had therefore achieved virtually nothing. Engels's prediction that the unemployed would take the law into their own hands had, except in a few isolated local incidents, conspicuously failed to come to fruition. One by one the groups that had championed the unemployed had suffered political eclipse or had been diverted to other social problems. This was true of the social democrats, the trade unionists, the Eight-Hours League, the London radicals, and the independent labour movement. One by one the remedies proposed— the limitation of hours, the 'right to work', the relief of local

[1] Sidney Homer, *History of Interest Rates*, p. 209; J. Blackman and E. M. Sigsworth, 'The Home Boom of the 1890s', *Yorkshire Bulletin of Economic and Social Research*, 17, no. 1 (May 1965), 81–2, 93–6.

[2] E. G. Howarth and M. Wilson, *West Ham*, pp. 20–2; C. F. G. Masterman, 'Realities at Home', in *The Heart of the Empire* (1901), pp. 4–5.

[3] T. 168/40, Financial Papers 1897–8, Memorandum by T. Llewellyn Davies on 'Cycles of Good and Bad Times', Apr. 1898.

[4] Appendix B, Table 2, p. 374.

[5] *Report of a COS Special Committee on the Relief of Distress Due to Want of Employment*, 1904, pp. 10–11.

expenditure, the acquisition of land by the guardians—had either proved irrelevant to the problem or had been shelved by the central government. Other more drastic solutions—such as nationalization, home colonization, and the imposition of protective tariffs—had not been given serious political consideration. Despite the ambitious claims of their supporters, their efficacy as remedies for unemployment was by no means self-evident, although they had the merit of seeing it as a problem of general economic policy rather than as an isolated problem of social distress. In spite of consistent pressure from socialist, labour, and radical groups, neither the Liberal nor the Conservative party had any specific plan for preventing or relieving unemployment; and Mr. Gladstone had stated categorically in 1893 that such action was outside the legitimate sphere not merely of legislation but of parliamentary discussion.[1]

In the central government there had been not merely indifference but positive hostility to the demand for non-pauperizing public assistance for the unemployed. The Poor Law inspectors admitted that there was a class of unemployed persons without visible means of support who nevertheless refused to apply for poor relief.[2] But Local Government Board officials were luke-warm in their execution of the Chamberlain circular, fearing that increased expenditure out of the rates would discourage thrift and deplete local 'wages-funds'[3] and that local relief works ran 'a very great risk of depriving of employment some of those who are already at work'.[4] The Home Office confined itself to the surveillance of unemployed demonstrations. The only other central office concerned with the unemployed was the Labour Bureau of the Board of Trade, which was incorporated into the Commercial, Labour and Statistical department in 1893.[5] But at this time the Board of Trade's interest in unemployment was purely statistical. Llewellyn Smith, the head of the

---

[1] B. Webb, *Our Partnership*, p. 104. *Hansard*, 4th series, vol. 16, col. 1735.
[2] *16th Annual Report of the L.G.B.* (1886–7), pp. 173–4.
[3] *19th Annual Report of the L.G.B.* (1889–90), pp. 119–21, report of Mr. Baldwyn Fleming.
[4] H. of C. 111/1895, Q. 396, Evidence of Sir Hugh Owen.
[5] J. A. M. Caldwell, 'The Genesis of the Ministry of Labour', *Public Administration*, 37 (winter 1959), 367.

Labour department, took an optimistic view of the extent of unemployed distress;[1] and early in 1895 he considered and rejected a proposal from W. T. Stead and H. C. Burdett that the Government should sanction an 'unemployed census' as a preliminary to 'a gigantic scheme of relief'.[2]

Only in the sphere of local government and of voluntary effort had certain innovations in the prevention and relief of unemployment occurred. Many local authorities had ignored the Chamberlain and Fowler circulars. Others had acquiesced unwillingly in a duty which they were ill suited to perform. But a small minority had responded to the unemployed movement by initiating their own unemployment policies. They had opened relief funds and labour registries and had introduced the eight-hours day and contracts of employment into their regular public works. At the same time certain voluntary organizations had been experimenting with relief funds, relief works, and institutional schemes for providing employment. The Select Committee of 1896 gave the seal of its approval to these 'experimental efforts'; and it was to local and voluntary schemes for the unemployed that it advised the Local Government Board to look for guidance in the formation of future policy.[3]

[1] CAB 37/38/2, 'A Recent Estimate of the Number of Unemployed', 8 Jan. 1895.
[2] CAB 37/38/10, 'The Unemployed', 23 Jan. 1895, pp. 7–9. See also above, p. 8.
[3] H. of C. 321/1896, *Report*, pp. xv–xvi.

III

THE REGIMENTATION OF THE UNEMPLOYED

In the mid 1880s the inadequacy of traditional forms of unemployment relief gave rise to many local and charitable experiments on behalf of the unemployed. These experiments varied widely in their aims, principles, and practical scope. Some merely gave doles in cash and kind, whilst others provided work, board and lodging, industrial training, and religious and moral instruction for the unemployed. Some were purely philanthropic, whereas others had semi-commercial aims. Some tried to segregate the unemployed from the labour market, others to prepare them for the resumption of normal industrial work. Some dealt with the unemployed workmen in isolation, whilst others attempted to reform the whole of 'family life'. Some were designed to prove the sufficiency of voluntary action, others to provoke intervention by the State. Relief schemes were often administrative hybrids, involving uneasy co-operation and compromise between individual philanthropists, charitable organizations, local vestries and councils, and the guardians of the poor. Among social reformers there was much disagreement about the merits of 'casework' and 'artificial employment', of 'institutional' and 'domiciliary' relief. But from the mid 1880s onwards there was a widespread reaction against unconditional almsgiving in favour of methods of treatment that involved the 'reform' or 'regimentation' of workmen who were unable to support themselves whilst unemployed.

UNEMPLOYMENT AND CHARITY

To many nineteenth-century social reformers the relationship of charity to the relief of unemployment was paradoxical —as indeed it was to all kinds of social distress. Within the strictest limits of political economy and Poor Law theory, charity was neither useful nor necessary. It was guilty of the economic crimes of depressing wages, encouraging reproduction, and discouraging thrift; and Malthusians in the 1830s[1] and social Darwinists in the 1880s feared that the maintenance of the unemployed and the artificial provision of employment merely reinforced and multiplied the social evils that they were designed to redress. 'We hear the death groans of the 100,' wrote Beatrice Webb in her Spencerian days, 'we do not hear the life groans of the 500 until it is too late!'[2]

Yet throughout the nineteenth century, economists and Poor Law administrators had recognized that, without the supplementation of charity, the rigours of the Poor Law would have been intolerable.[3] Malthus himself had enjoined the duty of charity, subject to 'the criterion of utility';[4] and his disciple Thomas Chalmers had proposed that the Poor Law be confined to the 'worthless self-centred and immoral poor', leaving the rest to cautious private almsgiving of which the purpose was 'not relief, but a process of improvement which would render it unnecessary'.[5] Dr. Arnold, who had condemned the New Poor Law for 'driving the poor into economy by terror', proposed that it should be supplemented by an 'organised system of church charity'.[6] Moreover, charity had a theological as well as a social significance, and there was clearly a conflict between the utilitarian principle of maximizing benefits to receiver and donor and the Christian duty of giving without counting the cost. The 'organised charity' movement, of which Chalmers was one of the founders, had tried from the 1820s onwards to achieve a reconciliation between the two.[7]

There were two main kinds of formal charity—endowed

---

[1] N. Masterman, *Chalmers on Charity*, p. 175.

[2] P. Fraser, *Joseph Chamberlain, Radicalism and Empire 1868–1914*, p. 124, B. Webb to J. Chamberlain, 4 Mar. 1886. For a discussion of the Darwinist dilemma see J. W. Slaughter's review of C. S. Loch's 'Charity and Social Life', *Eugenics Review*, 2, no. 3 (Nov. 1910), 249–50.

[3] Henry Sidgwick, *Principles of Political Economy*, p. 536.

[4] T. Malthus, *Essay on Population* (1826 ed.), ii. 351–73.

[5] N. Masterman, op. cit., pp. 109, 227.

[6] Quoted by Charles Bosanquet, *London: Some account of its Growth, Charitable Agencies and Wants*, pp. 196, 199.

[7] C. S. Loch's Diary, 3 June 1887, ff. 127, 134; K. Heasman, *Evangelicals in Action*, p. 289; H. Bosanquet, *Rich and Poor*, pp. 43–50.

charities and those that relied on public subscription. Few of the endowed charities that came under the scrutiny of the Charity Commissioners after 1853 specifically mentioned the unemployed; but there were thousands of 'dole charities' for the destitute, and bequests for the training and apprenticeship of poor children.[1] One of the aims of the Charity Commissioners was to transform obsolete charities into schemes for 'the encouragement among the poor of habits of providence, thrift and self-help';[2] and in 1885 a Mansion House Committee on Exceptional Distress proposed that the proceeds of charitable endowments should be transferred to a provident fund for the unemployed.[3] In the 1890s, however, the Charity Commissioners were accused of having converted bequests, which had been designed to assist the unemployed, into middle-class educational endowments;[4] and Jesse Collings complained to a Select Committee of 1894 that the commissioners had not merely misappropriated charitable funds, but that they had sabotaged the smallholdings and allotments movement by refusing to allow bequests of land to be held in trust.[5]

The unemployed were more directly affected, however, by charities that had no regular source of income but relied on public subscriptions and the services of charitable volunteers. Charitable societies for the relief of unemployment had a history going back at least to the end of the eighteenth century.[6] They were nearly always local and often ephemeral, depending on the enthusiasm of a single individual or responding to what appeared to be a merely temporary social

---

[1] H. of C. 221/1894, SC on the Charity Commission, Minutes of Evidence, Q. 460.

[2] C. 6960/1893–4, Fortieth Report of the Charity Commissioners (1892), pp. 19–20.

[3] Pall Mall Gazette, 5 Feb. 1886.

[4] H. of C. 365/1895, SC on Distress from Want of Employment, Minutes of Evidence, QQ. 7490–7.

[5] H. of C. 221/1894, SC on the Charity Commission, QQ. 3706–8, 3738–3741. Charitable bequests of land were prohibited by 9 Geo. II, cap. 36, 'An Act to restrain the disposition of lands whereby the same become inalienable'. This prohibition was lifted by the Mortmain and Charitable Uses Act, 1891, which permitted the transfer of land by will to charity; but since all such land was to be converted to personalty within a year of the testator's death, it did nothing to promote the private provision of allotments (54 and 55. Vict. cap. 73, sections 5–6).

[6] Annual Return of Subscription Charities . . . in and about the Metropolis. Examples of such charities were the Ladies' Charity of Wardour Street, Soho; the Thames Rivermen Society; and the Refuge for the Destitute in Hackney Road.

need.[1] They were notorious for encouraging professional mendicants;[2] but since the 1830s the 'organised charity' movement had tried to counteract indiscriminate almsgiving by imposing a clear line of demarcation between charity and the Poor Law, by preventing the duplication of charitable effort, and by laying down norms of 'scientific' casework for the dispensation of relief.[3]

The success of organized charity in persuading local guardians to adopt a stricter policy towards 'casual' and 'outdoor' pauperism was partly responsible in the mid 1880s for the failure of conventional methods of relieving 'unemployed' distress.[4] At the same time, however, the Charity Organisation Society, and kindred associations formed a powerful and articulate pressure group against local or central government intervention to fill the administrative hiatus to which this policy had given rise.

The resistance of the COS to public relief of unemployment has been ascribed partly to a dread of undermining working-class independence,[5] and partly to a residual attachment to Ricardian economics. 'Only with the greatest reluctance and then in qualified terms', remarked a recent historian of philanthropy, 'would the COS ... admit unemployment to be a fact.'[6] This interpretation of the COS attitude is partially, but not wholly, correct. C. S. Loch, it is true, compared the unemployed to a giant sea anemone, insatiably devouring all forms of philanthropy, but unable to survive in the 'thin smooth unsalted water' of social independence;[7] and he believed that in the last resort labour congestion could only be relieved by persuading surplus workmen to 'follow the market'.[8] But nevertheless, by the end of the 1880s many

---

[1] e.g. Robert Owen's Association for the Relief of the Manufacturing and Labouring Poor (David Owen, *English Philanthropy 1660–1960*, p. 97).

[2] Henry Mayhew, *London Labour and London Poor*, i. 416.

[3] A. F. Young and E. T. Ashton, *British Social Work in the Nineteenth Century*, pp. 75–114.

[4] On the COS theory of Poor Law administration see C. S. Loch, *Charity Organisation* (reprint of a paper read to the Congrès Internationale d'Assistance in Paris, July/Aug. 1889, pp. 13–31). Above, p. 54.

[5] C. L. Mowat, *The Charity Organisation Society 1869–1913*, p. 132.

[6] David Owen, op. cit., p. 242.

[7] C. S. Loch's Diary, 13 Sept. 1888, p. 137. Loch was Secretary of the London COS 1875–1914.

[8] Ibid., 15 July 1887. Loch suggested that this should be done by a strict applica-

members of the COS had rejected the doctrines of the 'wage-fund' and 'perfect competition'.[1] The Society as a whole was less concerned with either promoting working-class self-sufficiency, or with denying the economic fact of unemployment, than with keeping the treatment of the problem under its own control; and two distinct lines of development can be discerned in the Society's treatment of unemployment, both of which belied its reputation for neglect of or indifference to the distress of the unemployed.

The first approach, which dated from the foundation of the Society in 1869, was simply one of classification, and of 'face to face work among the poor'.[2] The unemployed were divided into 'thrifty and careful men'; 'men of differing grades of respectability, with a decent home'; and 'the idle, loafing class or those brought low by drink and vice'.[3] The latter were referred to the Poor Law, and suitable cases in the first two classes were helped by the payment of benefit club arrears, by the redemption of tools from pawn, by 'private influence' and by 'careful advertising in suitable papers'.[4] Families who wished to emigrate and girls looking for domestic posts were referred to two societies affiliated to the COS, the East End Emigration Fund and the Metropolitan Association for Befriending Young Servants.[5] Within this threefold framework, local COS committees in the 1880s declared that 'distress from want of employment should be dealt with case by case. They can give no other answer to the "unemployed" problem'.[6]

However, charitable casework by its very nature was unable to cope with more than a small proportion of unemployed distress. The number of cases demanding relief in a period of

tion of the 'workhouse test'. See also the evidence of Thomas Mackay to the Select Committee on Distress from Want of Employment, H. of C. 365/1895, Q. 5353.

[1] *Charity Organisation Review*, 4 (1888), 146–8.
[2] C. S. Loch's Diary, ? Dec. 1879, p. 71.
[3] *Report of the COS Special Committee on the Best Means of Dealing with Exceptional Distress*, 1886, p. xxi. The same classification was used by the Mansion House committees of 1887–8 and 1893–4; and by the COS Special Committee on the Relief of Distress Due to Want of Employment, 1904.
[4] C. S. Loch, *How to Help Cases of Distress* (1895 ed.), p. lxxix.
[5] On the work of the MABYS, see *20th Annual Report of the L.G.B.* (1890–1), p. lxxxvi.
[6] C. S. Loch, *How to Help Cases of Distress* (1895 ed.), p. lxxx.

high unemployment far exceeded the financial and administrative resources of organized charity, particularly since the COS sought to limit the caseload of its almoners to three or four families.[1] Hence between 1886 and 1896 an average of fewer than 800 cases a year were 'assisted to find employment' by the London COS;[2] and funds were raised by temporary and haphazard methods such as the holding of 'garden parties' for the unemployed.[3] Secondly, the systematic scrutiny of an individual's character and private affairs was almost as objectionable to independent workmen as the receipt of poor relief;[4] and, thirdly, a classification of the unemployed that was based on moral probity proved almost impossible to apply, particularly since the 'undeserving' were often those most in need of charitable relief.[5] 'In these days of trade depression, agricultural distress, over-population, immigration and like evils,' declared a speaker at a meeting of the COS Council in 1895,

we have on all sides a large number of people to deal with who are not quite typical COS cases, and yet who would be deteriorated and pauperised by being forced on the parish. . . . They are *not* thrifty; the wife is *not* a good manager; the man does *not* think of the future; they live extravagantly to our ideas when they have the money, but when bad times come they rub along uncomplainingly and suffer anything sooner than go to the parish.[6]

Cases of this kind defied moral classification; and by 1890 the COS had been forced to renounce the criterion of 'desert' as a basic principle of casework, and had begun to supplement casework with other less personal methods of relieving distress.[7]

This tacit recognition of the limitations of casework led to

[1] C. S. Loch, *The Elberfeld System* (COS Occasional Papers, No. 20, 3rd series).
[2] Appendix B, Table 3, p. 375.
[3] *Charity Organisation Review*, 12 (1896), 273.
[4] H. of C. 111/1895, *SC on Distress from Want of Employment, Minutes of Evidence*, Q. 1046.
[5] C. L. Mowat, op. cit., p. 37.
[6] Sister Constance of St. Frideswide's Mission, Poplar, 'How to Organise Relief in a Parish in Time of Unusual Distress', *Charity Organisation Review*, 9 (1895), 520.
[7] C. S. Loch, *How to Help Cases of Distress* (1894 ed.), p. xlii: '. . . "deserving", the favourite word of thoughtless almsgivers, implies a wrong test. Strictly used, it is merciless; loosely used, it is meaningless. Almoners should assist in order to cure and not in order to reward.'

the second stage in the COS approach to unemployment.[1] In 1886 a COS Special Committee on Exceptional Distress admitted that there were 'permanent causes of distress which it is impossible for philanthropy to cope with or even in any sufficient degree to palliate by schemes of direct relief'.[2] This Committee therefore recommended as long-term solutions the development of new industries, the extension of education, and the encouragement of working-class thrift.[3] In the short term it proposed that local committees, representing guardians, charities, and local authorities, should investigate cases, organize relief funds, and arrange temporary public employment for the normally independent unemployed.[4]

This report laid down three principles—preliminary casework, the limited provision of work for the unemployed, and co-operation between different kinds of local organization— which were to dominate the treatment of unemployment for the next twenty years.[5] But although the COS continued unwaveringly to uphold the need for casework, its attitude to relief works during this period varied according to whether or not they could be kept under charitable control. Charles Loch, the Secretary of the London COS, played a prominent role in arranging relief employment for the Mansion House committees of 1887 and 1893–4; yet he condemned the Toynbee Hall Conference of 1892, which proposed to introduce an identical relief scheme in conjunction with local vestries and the London County Council.[6] The non-pauperizing work authorized by the Chamberlain circular was denounced as 'dole' employment which pandered to 'middle-class sentimentality';[7] but in 1904 a COS committee on Distress Due to Want of Employment recommended that in an

---

[1] Schemes for the artificial provision of work had been considered by COS members since the 1860s. Alsager Hill, *Our Unemployed: An Attempt to Point out some of the Best Means of Providing Occupation for Distressed Labourers etc.* (1868); C. L. Mowat, op. cit., p. 52; A. F. Young and E. T. Ashton, op. cit., p. 96, suggest that it was possibly the discussion of proposals for relieving the unemployed that led to the establishment of the COS in 1869.

[2] *Report of a COS Special Committee on the Best Means of Dealing with Exceptional Distress*, 1886, p. i.

[3] Ibid., p. x.   [4] Ibid., pp. xx–xxi.

[5] Substantially the same principles were embodied in the Unemployed Workmen Act of 1905. Below, Chapter 5.

[6] Below, pp. 112–13.

[7] *Charity Organisation Review*, 9 (1893), 395–6.

emergency local 'joint committees' of guardians, councillors, and charitable agencies should be able to provide employment on 'public or other works'.[1] In the Royal Commission on the Poor Laws of 1905–9, some COS representatives were prepared to go as far as the socialists in recommending anti-depressive public works and industrial retraining for the unemployed;[2] but they were concerned that at least a share in the direction of such schemes should be kept in the hands of charity and that 'the treatment of distress should be entirely dissociated from the municipalities and from municipal employment'.[3]

It was the desire to retain administrative control of unemployment policies that throughout this period primarily determined the attitude of organized charity to the treatment of the unemployed. The apparent inconsistencies in the COS view of relief works were rationalized with the argument that schemes that succeeded under wise voluntary management were likely to 'fail entirely' when placed in the hands of public officials and boards.[4] As Alfred Marshall remarked to the Aberdare commission, the members of the COS were an 'oligarchy', who had 'taken upon themselves' some of the most important functions of the State.[5] But at the same time the Society felt itself threatened by obsolescence;[6] and its performance of these functions was secondary to, and modified by, its desire to ward off competition for administrative control. Thus in the 1880s and 1890s representatives of organized charity bestowed qualified approval on relief works, labour colonies, labour bureaux, and even the promotion of new industries on behalf of the unemployed[7]—provided that

---

[1] Report of a Special Committee of the Council of the COS on the Relief of Distress Due to Want of Employment, Nov. 1904, Appendix 2, p. 50.

[2] Cd. 4499/1909, RC on the Poor Laws, Memorandum by Mr. T. Nunn in Regard to Unemployment, pp. 712–18.

[3] Ibid., Memorandum by C. S. Loch and Mrs. Bosanquet on The Treatment of Unemployed Persons, p. 677.

[4] C. S. Loch's Diary, 1 Oct. 1888, p. 150.

[5] C. 7684—I and II/1893, RC on the Aged, Poor, Minutes of Evidence, Q. 10210.

[6] C. S. Loch's Diary, 1 Nov. 1888, p. 165. Loch mentioned the need to recruit more young men into the COS, and his 'fear lest COS be a matter for our generation only'.

[7] Charity Organisation Review, 4 (1888), 43–5; Report of the COS Special Committee on the Best Means of Dealing with Exceptional Distress, 1886, p. x. See also below, p. 118.

such schemes were at least partially under the direction of the COS. The Society consistently opposed direct action by local authorities, like the relief schemes of West Ham and Poplar, apparently fearing that if such action spread throughout the country, 'there would be no need for the Charity Organisation Society to exist'.[1] Their approval of co-operation with public authorities was confined to local 'joint committees' at a vestry or district council level, where the influence of COS members, with their well-formulated theories and practical experience of charitable casework, was almost bound to prevail.[2] Even so, the COS thought that applicants for employment should be referred to charitable associations and not relieved directly by 'joint committees'. They were jealous of attempts to transfer the level of co-operation to the county or to the central government, since under a purely local system the COS with its federal structure was inevitably the main channel of communication. When a central body was essential they preferred a 'strong and experienced executive' of voluntary workers rather than a representative institution such as the L.C.C.[3]

The questions of whether the unemployed should be assisted by casework or relief work and whether direction should be exercised by charitable or public authorities was fought out in a series of Mansion House committees, convened by the Lord Mayor of London to administer unemployed relief funds between 1886 and 1894.[4] The antagonists on these committees included London M.P.s, local councillors, philanthropists, clergymen, socialists, and settlement-dwellers, as well as the COS;[5] and the balance of power within the

---

1 W. C. Steadman to the *Daily Chronicle*; quoted in the *Charity Organisation Review*, 10 (1894), 92.

2 See e.g. *Charity Organisation Review*, N.S. 16 (1904), 345, where COS members were urged to join local distress committees in order to influence the administration of relief.

3 *Report of a COS Special Committee on the Relief of Distress Due to Want of Employment* (1904), pp. 51–2, 11–12.

4 On the Mansion House as a traditional source of charitable relief see Walter Besant, *London in the Nineteenth Century* (1909), pp. 86, 144–5; W. H. Beveridge, 'Emergency Funds for the Relief of the Unemployed: A Note on their Historical Development', *Clare Market Review*, 1, no. 3 (May 1906), 73–8; H. Bosanquet, *Past Experience in Relief Works* (COS pamphlet, 1903).

5 *Report of the Mansion House Conference on the Condition of the Unemployed*, 1892–3, p. 4.

committees constantly shifted to and fro between supporters of casework and supporters of relief work, and between those who favoured a public and those who favoured a voluntary system of control.

The Mansion House relief fund and newspaper charities of 1886 caused a widespread revulsion against 'unconditional' and indiscriminate almsgiving to the unemployed. Public opinion was shocked by the open trafficking in relief tickets, the exploitation of clerical patronage, and misapplication of charitable funds. 'Every penny has eternal issues upon the characters of the recipients,' wrote Samuel Barnett, 'yet pounds are given and scattered without prayer or thought whether those issues end in heaven or hell.'[1] Robert Giffen proposed that henceforward 'mischievous charity ought in some way or other to be made a penal offence';[2] and twenty years later it was believed that the relief funds of 1886 had debauched a whole generation of residents in London's East End.[3] But more significant than any supposed effect on the morals of the poor was their demonstration of the futility of random almsgiving and the complete absence of any evidence of permanent improvement in the 'condition of the people' as a result of the distribution of nearly £100,000.

This free distribution had come about, however, mainly as a result of panic induced by the riots in the West End. A Mansion House inquiry into distress, which reported in December 1885, had specifically condemned indiscriminate charity and had recommended dock reorganization, state-aided emigration, charitable advice on self-help and home management, 'education, recreation, culture and temperance', as remedies for unemployment in the East End.[4] Moreover, a small proportion of the men relieved by the Mansion House Fund were not merely given bread-tickets but were employed by the Metropolitan Public Gardens Association on the conversion of derelict churchyards into pleasure-

[1] *Charity Organisation Review,* 2 (1886), 99–100.
[2] Giffen MSS., vol. iii, item 24, f. 140.
[3] Beveridge, *Unemployment* (1930 ed.), p. 158.
[4] *Report of the Mansion House Committee, appointed March 1885 to Inquire into the Causes of Permanent Distress in London and the Best Means of Remedying the Same,* pp. 11–15.

grounds in Stepney and Bow.[1] In the winter of 1887–8 a Mansion House conference decided to extend the policy of giving employment rather than relief. COS representatives urged that the experiment should be conducted privately; but other members of the conference thought that a public scheme was necessary in order 'to show working-men that we really intended to tackle the question of the unemployed'.[2] Over £5,000 was raised,[3] and 456 unemployed labourers were given temporary employment in Camberwell Park. The subsequent case histories of all the workmen employed were carefully investigated by a 'reference committee' under C. S. Loch; but it was found that only 17 per cent of the cases had been permanently 'improved'. The committee therefore concluded that amateur relief works afforded no solution to the problem and actually aggravated distress from unemployment by encouraging rural immigration and discouraging thrift. They recommended that future policy should be confined to charitable casework and 'local co-operation' in the management of relief.[4]

When distress from unemployment revived in the winter of 1892–3, Samuel Barnett convened a further conference of clergymen, socialists, labour leaders, and social workers, which aimed initially to 'do something permanent for the displaced dock labourers' and 'to stave off a "Mansion House Fund"'.[5] At the end of December 1892 this conference called for the provision of work by local authorities, supplemented by a 'small, voluntary committee' to investigate cases, compile statistics, and appeal for charitable funds.[6] The COS delegates to the conference dissented strongly from this report on the grounds that 'the acceptance of an obligation on the

---

[1] Helen Bosanquet, *Past Experience in Relief Works* (COS pamphlet, 1903), pp. 1–2; W. Collison, *The Apostle of Free Labour*, pp. 19–21.

[2] C. S. Loch's Diary, 1 Oct. 1888, p. 152.

[3] Helen Bosanquet, op. cit., p. 2.

[4] *Mansion House Council, Report of Reference Committee to the Lord Mayor,* 1887–8, pp. 1–4.

[5] Buxton MSS., unsorted, Sidney Webb to Sydney Buxton, 19 Dec. 1892.

[6] Samuel Barnett and others to the Editor of *The Times*, 28 Dec. 1892, reprinted in the *Report of the Mansion House Conference on the Condition of the Unemployed,* 1892–3, pp. 5–11. Signatories included John Williams Benn, Sydney Buxton, Hugh Price Hughes, Canon Scott Holland, W. C. Steadman, Sidney and Beatrice Webb.

part of the community to provide work to those who are out of employment is both morally and socially injurious'; and that 'Local Authorities have nothing to do with the relief of the unemployed'.[1] Nevertheless, it was typical of the strategy of 'organised charity' that when the plan was taken over by a Mansion House conference in January 1893, six of the fifteen members of the executive committee were members of the Council of the COS.[2]

The COS representatives effectively prevented this committee from using funds raised by the Lord Mayor to subsidize the work of local authorities, even though several London vestries were unable to afford further expenditure on necessary relief works out of the rates.[3] The work provided by the committee was very limited in aim and scope. It was confined to waterside labourers living in Stepney, Poplar, St. George's, and Mile End, and its main function was to test 'their industry and strength of purpose' in order to identify those who could be given permanent help. Forty acres of wasteland at Abbey Mills in West Ham were borrowed from the L.C.C., and 253 out of 716 applicants were offered work converting the land to allotments at 6d. an hour for 45 hours a week.[4]

The Mansion House conference was reconvened in October 1893 and the Abbey Mills relief works were re-opened in January 1894. On this occasion the policy pursued by the executive committee aroused so much bitter controversy that it is difficult to arrive at an objective assessment of the working of the scheme. In December 1893 the COS condemned the revival of the conference as 'pointless';[5] but nevertheless, Arnold Hills, the chief subscriber to the scheme, complained that its administration 'practically was taken over by the COS, and the result was that a negative report was issued, generally recommending nothing and taking up the position that nothing could be done'.[6] Hugh Price Hughes,

[1] Reprinted in the *Report of the Mansion House Conference on the Condition of the Unemployed*, 1892–3, pp. 12–14.
[2] Ibid., p. 4. i.e. C. H. Turner, C. S. Loch, G. Gretton, J. Allen, M. A. S. Walrond, and Sir Charles Fremantle.
[3] *Charity Organisation Review*, 10 (1894), 76–94.
[4] *Report of the Mansion House Conference on the Condition of the Unemployed*, 1892–3, pp. 15–16.
[5] *Charity Organisation Review*, 9 (1893), 485.
[6] H. of C. 111/1895, Q. 2534.

the Congregational leader, criticized the committee for adopting the 'police and detective' casework system of the COS, and Samuel Barnett condemned its report for using language about 'the poor' that was ten years out of date.[1] These judgements were not entirely justified since, although the members of the committee recorded that many cases were too 'demoralised' to benefit from constructive assistance, they conceded that the majority of applicants 'were *willing to work* if they could obtain it, and that relief employment was a superior alternative to the workhouse test'. They claimed, moreover, that in spite of complaints about the system of inquiry, relief had been dispensed with a maximum of security to charitable donors and a minimum of irritation to the unemployed.[2]

The main practical objection to the Mansion House relief schemes, however, was not their system of classification but their extremely limited application to the problems of the unemployed. In 1893–4 the Abbey Mills relief works were thrown open to the whole of Tower Hamlets and to all unemployed workmen irrespective of trade; but even so only 141 workmen were offered employment out of 414 applicants and only 49 cases were permanently 'improved'.[3] In 1894–5 the scheme was extended to the whole of London, but only eighty-five workmen were given employment at a time when it was estimated that throughout the United Kingdom over one and three quarter million workmen were unemployed.[4]

Similar relief schemes, financed by voluntary subscription and managed by charitable or municipal committees, were organized in many provincial cities during the depressions of 1886–8 and 1892–5—particularly in commercial centres with large philanthropic communities, such as Birmingham, Nottingham, and Glasgow.[5] These schemes afforded a con-

---

[1] *Charity Organisation Review*, 10 (1894), 92–3.

[2] *Report of the Executive of the Mansion House Conference on the Unemployed*, 1894, pp. 7–10.

[3] C. S. Loch, 'Manufacturing a New Pauperism', *Nineteenth Century*, 37 (Apr. 1895), 697–708.

[4] Helen Bosanquet, *Past Experience in Relief Works* (COS pamphlet, Feb. 1903), p. 3; H. of C. 111/1895, QQ. 906–10.

[5] Richard Simon, 'Relief Works at Nottingham', *Charity Organisation Review*, 1 (1885), 98–101; Beveridge, *Unemployment* (1930 ed.), p. 66; F. Tillyard, 'Three

venient opportunity for administrative experiment and the airing of disagreements between different reforming groups; but constructive action was usually frustrated by the wide cross-section of interests represented in the management of such schemes, since the need for a consensus inevitably favoured those who thought that nothing could be done. Their dependence on charitable donations gave undue weight to those who preferred voluntary action to state intervention; and their emphasis on preliminary 'casework' meant that they involved an expenditure of administrative effort out of all proportion to the numbers relieved. They were therefore more useful as evidence of the limitations of such a policy than as immediate remedies for the distress of the unemployed. Even as experiments they were hampered by the difficulty of finding work at short notice which was appropriate to the capacity of a random collection of men. It was feared that they ran the risk of increasing labour congestion by attracting rural immigration and by encouraging irresponsible employers to dismiss their workmen during a temporary depression.[1] But measures to promote labour mobility were frustrated by the stubborn reluctance of the vast majority of redundant workmen to change their occupation or, in the case of London, to move out of the East End.[2] 'Difficult as it is to make up one's mind to leave these poor fellows to their present condition or to the Poor Law,' concluded the Secretary of the Mansion House committee of 1892–3, 'it must be remembered that it would be cruel kindness to take any steps which, while acting as a palliative to the misery of the present generation, would tend to perpetuate the existence of a class whose labour is no longer required.'[3]

'HOME COLONISATION' AND THE UNEMPLOYED

During the nineteenth century there had been many experiments in co-operative farming and 'home colonisation', some-

Birmingham Relief Funds—1885, 1886 and 1905', *Economic Journal*, 15 (Dec. 1905), 505–20.

[1] *Mansion House Council Report of Reference Committee*, 1887–8, p. 4.
[2] Ibid., pp. 2–3.
[3] H. V. Toynbee, 'A Winter's Experiment', *Macmillan's Magazine*, 69 (1893–4), 54–8.

times merely to provide employment and sometimes with an ulterior religious or moral aim.[1] It is arguable that such experiments had a history in democratic politics dating back to the diggers and levellers of the commonwealth period; but in the nineteenth century they were mostly emanations of middle-class philanthropy rather than of working-class self-help. Since the 1830s they had been economically discredited;[2] and in the 'high farming' era of the 1850s and 1860s they had had no obvious economic function. They were usually eccentric and ephemeral, hampered by inefficient management, poor opportunities for marketing, and chronic shortage of funds.[3] In the 1880s, however, the contraction of land under cultivation, the shortage of labour in some rural districts, and the congestion of urban labour markets re-awakened an interest among social reformers in schemes for 'home colonisation' and for settling unemployed workmen on vacant agricultural land.[4]

The influences on this movement were complex and must be seen as part of a much wider movement for the reconstruction of urban and rural life. Firstly, the demonstrable futility of casework and relief works forced administrators to look for more permanent methods of maintaining and employing the urban unemployed. Secondly, the contemporary radical movement for the reform of land tenure and the creation of a free peasantry suggested that the solution to urban problems lay in the regeneration of village life. Thirdly, there was the example of continental labour colonies, public and voluntary, penal and commercial, which over the previous half-century had attempted to deal with unemployment and vagrancy by settling surplus labourers to work on the land. And, fourthly, the growing belief that the urban unemployed were physically and morally degenerate led to a widespread desire for some

[1] James Mavor, 'Setting the Poor on Work', *Nineteenth Century*, 34 (Oct. 1893), 523–32.

[2] N. Masterman, op. cit., pp. 173–83.

[3] W. H. G. Armytage, *Heavens Below. Utopian Experiments in England 1560–1960*, pp. 77–358.

[4] On the agricultural depression see Lord Ernle, *English Farming Past and Present* (6th ed. introduced by G. E. Fussell and O. R. McGregor), pp. 377–92; C. S. Orwin and E. H. Whetman, *History of British Agriculture 1846–1914*, pp. 240–87. On migration from the countryside, John Saville, *Rural Depopulation in England and Wales, 1851–1951*, pp. 8–20.

form of institutional treatment, whereby they could be either reformed or repressed.

The historian of 'home colonisation' has warned against the fallacy of associating it too closely with the political movement for 'three acres and a cow';[1] and it is true that supporters of Chamberlain's Radical Programme were more concerned with promoting the economic freedom of the agricultural labourer than with employing or reforming the urban unemployed.[2] Nevertheless, there was a certain amount of convergence in the aims of the two movements. The supporters of smallholdings and allotments claimed that they would incidentally reduce unemployment, by arresting migration from the countryside[3] and by reducing England's dependence on food imported from abroad.[4] Similar claims were made for other items in the 'back to the land' programme, such as rural rehousing, co-operative marketing, and agricultural credit banks.[5] Unemployment reformers, on the other hand, hoped that smallholdings and allotments would not merely relieve pressure on the urban labour market, but provide a permanent livelihood for the urban unemployed.[6]

The methods and principles of continental labour colonies were well known to English social reformers through the exhaustive inquiries conducted by the COS, the Local Government Board, and the Board of Trade.[7] These colonies had been established partly to suppress mendicancy, and partly to maintain and employ the unemployed. The Swiss

[1] W. H. Armytage, op. cit., p. 325.

[2] 'The Radical Programme', with a preface by Joseph Chamberlain (reprinted, July 1885, from the *Fortnightly Review*), especially pp. 92–125.

[3] H. of C. 223/1890, *SC on Smallholdings, Report* (chairman, Joseph Chamberlain), p. iii; H. of C. 365/1895, *SC on Distress from Want of Employment*, QQ. 6948–70; 7588–9.

[4] *Newcastle Daily Chronicle*, 27 Mar. 1897, Report of a Conference of Guardians held at Gateshead on 'How to Find Work for the Unemployed'.

[5] *To Colonise England: A Plea for a Policy*, Reprints of *Daily News* articles by C. F. G. Masterman and Others, pp. 33–4.

[6] J. W. Southern, *The Unemployed: Causes and Remedies of Poverty*, n.d., p. 13; D. Campbell, *The Unemployed Problem—the Socialist Solution* (S.D.F. pamphlet, Dec. 1892), p. 15.

[7] H. Willinck, 'The Dutch Labour Colonies', *Charity Organisation Review*, 4 (1888), 241–58; and 'Agricultural Beneficent Colonies of Belgium etc.', ibid. 7 (1891), 6–20; C. 5341/1888, Reports by J. S. Davy on *The Elberfeld Poor Law System* and *German Workmen's Colonies*; C. 7182/1893, *Report on Agencies and Methods for dealing with the Unemployed*, pp. 268–339.

colonies, which were the most repressive, had more or less succeeded in their limited objective of eliminating vagrancy by the end of the nineteenth century.[1] The German colonies had been founded by evangelical pastors as refuges for 'any one who had suffered inward or outward shipwreck'; and since 1883 they had received state assistance through a Labour Colony Central Board, which had the avowed purpose of outflanking socialist pressure for a public recognition of the 'right to work'.[2] In Holland labour colonies offered a permanent settlement on the land to unemployed workmen and their families;[3] and in Belgium a comprehensive colony system was controlled by the Department of Justice, with special centres for convicted beggars and for unemployed workmen 'whose poverty had arisen from circumstances beyond their own control'.[4]

The continental system met with a mixed reception from English social reformers who were searching for new methods of relieving the unemployed. Critics of the system complained that the colonies were primarily 'deterrent' and not reformatory or economically profitable, and that they prolonged rather that reduced the period during which men were without work.[5] The Charity Organisation Society, which had given tentative approval to 'home colonisation' in 1887,[6] withdrew its support after adverse reports on the continental system had been received from Sir Henry Willinck and C. S. Loch.[7] In 1892 Professor James Mavor advised the Board of Trade that colonies might be useful for detaining beggars and as a temporary 'sanatorium for discouraged single workmen'; but he thought that they offered no permanent solution to the problems of the urban unemployed.[8]

Nevertheless, from the 1880s onwards a growing body of expert social and economic opinion proposed that labour

[1] H. Preston Thomas, *Work and Play of a Government Inspector*, pp. 337–43.
[2] C. 7182/1893, *Report on Agencies and Methods for Dealing with the Unemployed*, pp. 269–70.
[3] Ibid., pp. 308–19.   [4] Ibid., p. 321.
[5] H. Moore, 'The Unemployed and the Land', *Contemporary Review*, 63 (Mar. 1893), 423–38.
[6] *Charity Organisation Review*, 4 (1888), 43–5.
[7] *The State and the Unemployed: With Notes regarding the action of vestries in Different Parts of London* (COS pamphlet), Nov. 1893.
[8] C. 7182/1893, p. 339.

colonies should be established either inside or outside the
Poor Law to train and reform the dependent unemployed. In
1884 Alfred Marshall suggested that such colonies should be
used to remove surplus workmen from the London labour
market, and that 'being without the means of livelihood must
be treated, not as a crime, but as a cause for uncompromising
inspection and inquiry'.[1] In 1886 Robert Giffen urged that
the 'unemployed, thriftless and semi-criminal class' should be
segregated from the rest of society, and deprived of civil
liberties and the right to bear children;[2] and in 1892 Charles
Booth put forward a plan for buttressing the 'free market', by
providing state labour colonies for socially dependent families
and thereby liberating efficient and independent workmen
from the harmful competition of the irregularly employed.[3]

There were in fact two distinct strands in the labour colony
movement. At one extreme reformers dreamed of colonies in
the Utopian tradition, which would be economically self-
supporting and establish a new moral and political ideal. At
the other extreme colonies were seen as the last word in
material 'less eligibility', as refuges for the misfits of capitalism and as disciplinary institutions for the recalcitrant unemployed.

The first practical attempt during this period to relate land
revival to the relief of unemployment was made by the
London Congregational Union, which arranged for the employment of 150 London workmen by a Lincolnshire farmer
in February 1886.[4] In the same year more ambitious proposals for co-operative 'home colonisation' were put forward
by the Reverend Herbert Mills's *Poverty and the State*. A disillusioned Unitarian minister and member of the Liverpool
Central Relief Society, Mills was described by an unsympa-
thetic contemporary as 'of the slightly hectic type; not of the

[1] Alfred Marshall, 'The Housing of the London Poor', *Contemporary Review*,
45 (Feb. 1884), 224–31.
[2] Giffen MSS., vol. iii, item 24, ff. 141–3.
[3] Charles Booth, *Life and Labour of the People in London*, i (1892), 167–8.
Booth's aim was by 'thorough interference on the part of the state with the lives
of a small fraction of the population . . . to make it possible ultimately to dispense
with any socialistic interference with the lives of all the rest'.
[4] *Report of the Mansion House Conference on the Condition of the Unemployed*,
1887–8, Revd. Andrew Mearns to the Sub-Committee on Agricultural Colonies,
29 Dec. 1887.

sagacious but of the sensitive kind'.[1] He was nevertheless a key figure in liberating the discussion of poverty from the straitjacket of orthodox political economy, and in linking charity with collectivist experiments and the relief of unemployment with 'back to the land'. Without conceding the 'right to work' he suggested that the State had a duty towards the unemployed, and diagnosed the cause of the unemployment problem as shortage of purchasing power. 'The marvel of it all is that men are starving and wanting employment because there is abundance and because commodities are cheap. . . . Food, and clothing, are rotting because they are not consumed and yet men and women are on the verge of starvation.'[2] He prescribed a remedy which he hoped would not offend against private interests—the formation of state-controlled, self-supporting co-operative estates, which would employ the unemployed in agriculture and domestic industries and send into the market nothing that would compete with existing producers. These colonies would be financed by a capital sum equal to two years' national Poor Law expenditure; they would gradually supersede the existing workhouse system, and would aim at the eventual recreation of social as well as industrial life.[3]

In the spring of 1887 Mills discussed his plans with C. S. Loch, the secretary of the London COS. Loch was mildly interested in the idea that the unemployed might be set to work for 'self-support' and not for the open market; but he disliked the economic primitivism of Mills's scheme, and doubted whether agricultural colonies could be made economically independent. He feared, moreover, that a community that was sheltered from the effects of competition would encourage an unwarranted increase of population; and he concluded that 'the whole thing is airy, unsubstantial; the refuge of the destitute of religion. Schemes instead of theologies. The life in either case set aside for the fanciful and easy. . . .'[4]

Nevertheless, in 1887 Mills founded the Home Colonisation Society, to publicize co-operative ideas and to conduct

---

[1] C. S. Loch's Diary, 13 Apr. 1887, f. 120.
[2] H. V. Mills, *Poverty and the State*, pp. 103-4.   [3] Ibid., pp. 184-213.
[4] C. S. Loch's Diary, 14 Apr. 1887, pp. 120-2.

co-operative experiments—which, if successful, were to be taken over by the State. He claimed that his proposals were attracting 'favourable notice' from clergymen, philanthropists, economists, public officials, and organized labour; but his initial appeal for £25,000 met with little response, and in December 1887 he submitted his scheme to the Mansion House conference, asking for £50,000 to found a colony for 1,000 persons. He proposed to purchase an arable estate in southern England, which would become as far as possible economically self-contained.

The wages of the ordinary workers should not consist of a money payment but of a rent-free house, three good meals daily supplied in the dining hall of the Society, a suit of clothing annually, education for the children, an allotment of land to each family consisting at first of one-third of an acre, with lessons in the art of bee-keeping, mushroom culture, basket-making, mat-making and various handicrafts.

The rules of the colony would not be 'unduly oppressive or inquisitorial . . . but should include prompt dismissal for disobedience, idleness, drunkenness and immorality'; and special provision was to be made for unemployed women, whom Mills believed were an exceptionally vulnerable and neglected class.[1]

The Mansion House authorities had neither the power nor the funds to undertake this kind of experiment in social reconstruction, and Mills's scheme was rejected by the conference. By 1892, however, he had raised £5,000 and an estate was purchased at Starnthwaite in Westmorland for the realization of his plan.[2]

Highly conflicting reports survive of the subsequent history of the Starnthwaite colony. In September 1893 Llewellyn Smith found twenty-two colonists in residence, but thought it was too early to attempt to pass judgement on the results of the scheme. He remarked, however, that 'the progress of the colony has been seriously impeded during its first year by internal dissensions turning chiefly on the mode of

---

[1] *Report of the Mansion House Conference on the Condition of the Unemployed,* 1887–8, Revd. Herbert Mills to the Sub-Committee on Agricultural Colonies, 26 Dec. 1887.

[2] 'Utopia Limited', *Charity Organisation Review,* N.S. 2 (Sept. 1897), 123.

government of the village.'[1] A year later Mills presented an optimistic report to a Co-operative conference at Holborn Town Hall, claiming that 'at the present the colony is nearly self-supporting and self-contained.'[2] In September 1897, however, the *Charity Organisation Review* published a damning account of the Starnthwaite scheme, ascribing its failure partly to the false expectations of the colonists and partly to their lack of discipline and skill.[3] In 1893 Llewellyn Smith had found that 'some of the first colonists appear not to have belonged to the ordinary unemployed class, but to have been attracted to the colony by the expectation of taking part in a communal experiment';[4] and Mills's own original intention had almost certainly been to found, not merely an asylum for the unemployed, but a model community that would eventually become self-governing. In the meantime, however, 'the pigs fell ill from want of care; the sluices at the mill were left open, leading to a serious overflow of water; the carpenters were in the habit of spending their working hours in bed, and altogether the whole management of the farm was far from admirable.'[5]

Mills therefore introduced professional foremen, at which the colonists rose in revolt; they proposed that henceforward the colony should be governed by an elected committee and that colonists who chose to leave should receive compensation for loss of earnings whilst in residence. Mills rejected this plan, reduced the wages of the colonists to 2s. 6d. a week, and had the ring-leaders forcibly ejected. The colonists then held a protest meeting in the neighbouring town of Kendal, complaining that they had been 'misled by socialist promises' into working at a blackleg price.

Mills eventually restored order with the help of the local magistrates; but his Annual Report for 1893 concluded that 'self-government' was not essential to the principle of co-operation. The goal of self-maintenance was equally elusive,

---

[1] C. 7182/1893, *Report on Agencies and Methods for Dealing with the Unemployed*, pp. 179–80.
[2] J. A. Hobson (ed.), *Co-operative Labour Upon the Land*, pp. 64–9.
[3] 'Utopia Limited', *Charity Organisation Review*, N.S. 2 (Sept. 1897), 121–34.
[4] C. 7182/1893, p. 180.
[5] 'Utopia Limited', loc. cit., p. 126. The following description of the Starnthwaite colony is based on this article.

since quite apart from the initial capital expenditure of
£5,000 his accounts showed a deficit of 7s. per colonist per
week. He therefore opened a guest house for lakeland tourists
to absorb the produce of the colony; and the colonists were
increasingly employed in profit-making industrial pursuits.
By 1897 none of the colonists had been permanently settled
on the land, and the COS deplored their inability either to
make themselves independent within the colony system or to
re-enter the open market.

In 1900 the Starnthwaite colony was taken over by Dr.
John Paton's English Land Colonisation Society, which was
affiliated to the interdenominational Christian Social Service
Union.[1] Paton was an advocate of 'training' rather than
'labour' colonies, and of smallholdings, allotments, and Co-
operative Banks.[2] He realized more clearly than most reformers that the problems of rural revival were cultural as well
as material—that labourers left the countryside not merely
because of low wages but because of the 'bovine dullness' of
village life.[3] But his ideas on practical farming were even more
naïve than those of Mills, since he believed that 'any man
who can handle a hoe, a fork, a spade or a rake' could find
employment on the land, and that such employment could be
financed at very little capital cost.[4] Under Paton's direction
the aim of establishing a permanent co-operative estate at
Starnthwaite faded; and the colony became a refuge for unemployables and for persons who were physically or mentally
unsuitable for ordinary industrial life.[5]

In 1891 a colony of a rather different kind was founded by
the Self-Help Emigration Society at Langley in Essex. The
purpose of this scheme was to provide agricultural training
for the 'hopeless unemployed' and to test their suitability for

---

[1] The National Union for Christian Social Service was founded in 1895, under
the presidency of Lord Meath, for 'the promotion, encouragement and carrying
on of Christian social service . . . the upraising of the fallen and submerged . . .
and the training, maintenance, care or assistance of persons in need'. It supported a
training colony for epileptics at Lingfield in Surrey (Encyclopedia of Social Reform,
i (1909), xv–xvi).

[2] James Marchant, J. B. Paton. Educational and Social Pioneer, pp. 162–76.

[3] Ibid., p. 178.

[4] J. A. Hobson (ed.), Co-operative Labour on the Land, pp. 82–3.

[5] Percy Alden, 'Labour Colonies', Encyclopedia of Social Reform, ii (1909),
678.

permanent settlement overseas. The farm was managed by
a bailiff, who aimed to reproduce as far as possible the conditions of 'Christian family life'. As a training establishment it
was largely ineffective, only forty-three men being successfully emigrated between 1891 and 1896.[1] Nevertheless,
Walter Hazell, the founder and chief promoter of the scheme
claimed that it was justified by its 'social and moral results';[2]
and Wilson Gates, the Secretary of the Society, proposed to
the Russell Committee on Distress from Want of Employment that the farm should be used as a model for a massive
public emigration scheme, financed and managed by a central
department of state.[3]

SOCIAL SALVATION AND THE UNEMPLOYED

The most controversial scheme for employing and 'restoring'
the unemployed workman was that of the Salvation Army,
based on General William Booth's best-seller, *In Darkest
England and the Way Out*. In the autumn of 1890 Booth, with
the help of W. T. Stead, outlined the most detailed and comprehensive scheme for social redemption since the days of
Robert Owen.[4] It was born of the conviction, after twenty
years of evangelistic experience, that spiritual and material
destitution were inextricably interdependent. His scheme
applied not only to the unemployed but the whole of the
'submerged tenth' of English society—the 3,000,000 inhabitants of the jungle of poverty broken up into three drink-
sodden circles, the criminal, the vicious, and the starving but
honest.[5] Booth did not suggest that all these were unemployed, but he did suggest that want of employment was the
crux of the material aspects of the 'social problem'. All plans
for social improvement were liable to founder on 'the bottom-
less bog of the stratum of the workless';[6] and in the frontis-

[1] *Charity Organisation Review*, 12 (1896), 270.
[2] J. A. Hobson (ed.), *Co-operative Labour on the Land*, p. 62.
[3] H. of C. 321/1896, Q. 994.
[4] For a discussion of the genesis of Booth's scheme and the controversy it aroused
see Herman Ausubel, 'General Booth's Scheme of Social Salvation', *American
Historical Review*, 56, no. 3 (Apr. 1951), 519–25.
[5] William Booth, *In Darkest England and the Way Out*, p. 24.
[6] Ibid., p. 34.

piece to Booth's book the unemployed floundered graphically in a sea of betting, beggary, vice, and crime. Booth was one of the first writers to suggest that unemployment might be a cause as well as a result of sickness and physical degeneracy; and he drew attention to a factor subsequently emphasized by Hobson and the Webbs, that unemployment was not a prerequisite of industrial efficiency, but a form of chronic personal and social waste.[1]

Intellectually, Booth's approach to the unemployed and to all kinds of social failure was ambivalent. He was committed to the doctrine of the total depravity of man; and for Booth, as for many nineteenth-century evangelicals, this was more than a mere theological abstraction; it was a condition that made itself manifest in the social behaviour of individuals.[2] At the same time he was smitten with the helplessness of the work-man, practically, spiritually, and morally, in the face of a hostile industrial environment. For many years he had believed that this condition could be altered only by the pro-cess of conversion; but in the 1880s he came to the conclusion that conversion itself was being frustrated by the effects of material and social distress.[3]

Once Booth had become convinced that social reform was necessary to spiritual revival there was nothing ambiguous in his plan of action. He dismissed existing panaceas as hope-lessly inadequate—charity and combination as too narrow in scope, emigration as impracticable without prior training, and thrift as a mere mockery to those who were living on the margins of subsistence.[4] Instead he proposed reforms that would change both the character and the environment of the distressed individual. For the rescue of the unemployed he outlined a threefold scheme of 'self-helping and self-sustain-ing communities, each being a kind of co-operative society or patriarchal family, governed and disciplined on the principles which have proved so effective in the Salvation Army'.[5] The unemployed would first be received into a 'City colony', where they would receive food and shelter in return for work,

[1] Ibid., p. 32.
[2] Salisbury MSS., Class E, William Booth to Lord Salisbury, Apr. 1884.
[3] St. John Ervine, *God's Soldier: General William Booth*, p. 674.
[4] William Booth, op. cit., pp. 67–8.   [5] Ibid., pp. 91, 94–111.

and be prepared either for restoration to their former employment or for transference to the second stage, the rural labour colony.[1] There they would receive agricultural and industrial training; and finally would be emigrated to a Salvation Army colony overseas, to begin a new life of economic independence.[2] At every stage employment and maintenance would be conditional upon submission to rigid discipline, total abstinence, and exposure to daily prayer; and the scheme would be both a means of employing the unemployed, and a 'Great Machine' for crushing all the vices of a corrupt society.[3]

The 'social wing' of the Salvation Army began its work in London in 1888, under the supervision of Commissioner Frank Smith, who later abandoned the Army to join the I.L.P.[4] The first Salvation Army 'elevators' or city workshops were opened in Whitechapel and Battersea in the spring of 1890. Unemployed workmen who applied for relief at the Army's night shelters and cheap food depots, were invited to enter the 'elevators', where they were given board and lodging in return for eight hours' work a day. The employment consisted mainly of carpentry, upholstery, cobbling, painting, and 'scavenging' in the London streets. Efficiency, 'deportment and cleanliness' were rewarded by graduation of the food-ration; and after a time deserving workmen received an allowance of 5s. a week with which to equip themselves with tools and clothing for re-entry to the normal labour market.[5] Attached to the elevators was a 'labour bureau', designed to put men in touch with potential employers, which had ten metropolitan branches in 1893.[6]

In 1891 the Army acquired a freehold estate of nearly 3,000 acres at Hadleigh in the Essex marshes, to implement the second cycle of the Darkest England scheme. This was the farm colony, to which men were sent from the elevators and labour bureau who could not be reabsorbed by the commercial labour market. Men who entered the colony were required to sign a pledge, agreeing to observe the discipline

---

[1] William Booth, op. cit., pp. 124–42.
[2] Ibid., p. 143.   [3] Ibid., p. 93.
[4] E. I. Champness, *Frank Smith, M.P., Pioneer and Modern Mystic*, pp. 17–25.
[5] C. 7182/1893, pp. 164–7.   [6] Ibid., p. 161.

and to perform the tasks of work imposed. Employment was provided in land-reclamation, brick-making, and market-gardening; and maintenance without wages was given for the first month, after which the men were paid at the discretion of the superintendent.[1] Participation in the Army's religious life was 'voluntary', but officials of the colony clearly tended to equate industrial improvement with spiritual reform.[2]

The third stage of the Darkest England scheme involved more complex problems, political, logistical, and financial. Booth's plan was to acquire a tract of land from the Canadian, American, or Australian government and to settle colonists on co-operative smallholdings, modelled on the 'Owenite' experiment at Ralahine sixty years before.[3] He hoped that the colonists would become a self-supporting salvationist sect, sealed off like the Doukhobors from the corruption of frontier civilization. He claimed that the Salvation Army with its international organization was singularly well fitted to manage an enterprise of this kind; but in 1891 the Select Committee on Colonisation recommended group colonization only for the inhabitants of 'congested districts' in Western Ireland and the Highlands and Islands.[4] The Salvation Army therefore concentrated initially on the emigration of individual workmen and families, who were provided with suitable clothes and assisted passages. They were met at the port of disembarkation by Salvation Army officers, and placed in agricultural and domestic employment. By 1906 the Army claimed to be assisting the emigration of 10,000 persons a year, of whom 45 per cent were solitary workmen, 20 per cent were married couples, and 24 per cent were children under 18 years of age.[5] This kind of settlement was regarded as a substitute for General Booth's vision of overseas colonization; and in 1908 he concluded that 'of all the remedies propounded for the solution of the recurring problem of unemployment, I am satisfied that for the immediate

---

[1] Ibid., pp. 167–72.
[2] H. of C. 365/1895, Q. 9864; *COS Special Committee on the Relief of Distress Due to Want of Employment (1904). Minutes of Evidence*, QQ. 584–5.
[3] William Booth, op. cit., pp. 142–51.
[4] H. of C. 152/1890–1, *SC on Colonisation, Report*, pp. x, xvi.
[5] Commissioner A. Nicol, 'The New Emigration', *Sketches of Salvation Army Social Work* (1906), pp. 43–52.

and permanent relief of thousands of the selected surplus, *Emigration still holds the field'*.[1]

How successful were the Salvation Army schemes in either assisting or illuminating the problem of the unemployed? A recent historian of the Army's social work has suggested that the Darkest England scheme was from the start rent by dissension between officers who feared that spiritual ends would be eclipsed by material concerns and officers who went further along the road of social reform than Booth himself was prepared to go. It is suggested that the resignation of Frank Smith as head of the 'social wing' and his replacement by Commissioner Elijah Cadman marked a victory for the former group over the latter; and that by 1893 the public had lost interest in the scheme and Booth himself was weary of the whole business of social salvation.[2]

But this account is misleading. It was true that social reform did not have the effect which had been hoped for in swelling support for the Army and in bringing souls to Christ. It was also true that articulate public interest subsided, and the press moved on to more pristine sensations. In the religious history of the Salvation Army, social reform was never again so important as in the early 1890s, when it was hoped that a new window had been opened into lost souls. But, nevertheless, the writings of Booth and his son Bramwell do not suggest that their disillusionment was more than temporary, although possibly they became more reconciled to social improvement not as a means to conversion but as an end in itself.[3] The statistics of Salvation Army work, imperfect as they are, suggest that it continued to grow in volume between 1892 and 1908;[4] and during that period the 'social wing' acted as both a 'laboratory' for social administration and as a pressure group for intervention by the State. The Army's colony and workshop were frequently inspected by politicians

---

[1] William Booth, *The Surplus* (1908), p. 5.

[2] K. Inglis, *Churches and the Working Classes in Victorian England*, pp. 209–12; E. I. Champness, op. cit., pp. 14–16, suggests, however, that Smith resigned not over the question of how far the Army should be committed to social reform, but because Booth refused to let him direct the 'Social Wing' as an autonomous organization.

[3] *Sketches of Salvation Army Social Work*, 1906, foreword by General Booth, pp. vii–xv.   [4] Appendix B, Table 4, p. 376.

and social reformers as evidence of what could and what could not be done to rehabilitate the unemployed;[1] and the first direct appeal for state assistance was made by Elijah Cadman, supposedly the leader of the anti-reformers, in 1892.[2] In 1895 Bramwell Booth urged the Select Committee on Distress from Want of Employment to recommend government subsidies for the Army's work in emigration; and he claimed that one of the purposes of the Social Wing was to lay down prescriptive models for action by the State: '. . . what we . . . aim at is to carry out experiments so that we may show the country that these principles are applicable to all conditions of the unemployed and unfortunate classes. . . .'[3]

The two spheres of policy in which the Booths were most anxious to obtain state assistance were the establishment of penal labour colonies for vagrants and unemployables and the foundation of an overseas colony for the surplus unemployed.[4] The officers in charge of the Darkest England scheme found that their work was hampered by a class of vagrants and social parasites who discredited the 'unemployed' as a class and demoralized the 'genuine worker'.[5] This problem was accentuated by the outbreak of vagrancy after the South African war, when thousands of discharged soldiers were reported to be wandering the countryside in search of work.[6] In February 1904 Colonel David Lamb, the superintendent of Hadleigh, impressed upon General Booth that the existing vagrancy laws and casual ward system were obsolete and ineffective;[7] and in the following month Sir

[1] e.g. *Report on Labour Colonies*, by Professor James Mavor and others, to the Glasgow Labour centres committee, 1892.

[2] H.O. 45/9861/13077, Commissioner E. Cadman to H. H. Asquith, 1 Sept. 1892 and 22 Sept. 1892. Cadman sought an interview with the Home Secretary in order to submit proposals 'not only on behalf of the submerged, but also on behalf of the respectable unemployed working man who has his little home and which he is anxious to retain'.

[3] H. of C. 365/1895, *SC on Distress from Want of Employment, Minutes of Evidence*, Q. 9941.

[4] William Booth, *Emigration–Colonisation* (1905), pp. 11–12.

[5] William Booth, *The Vagrant and the Unemployable. A Proposal for the Extension of the Land and Industrial Colony system, whereby vagrants may be detained under suitable conditions and compelled to work* (1904), pp. 5–8, 21–3, 28–31.

[6] Cd. 2891/1906, *Departmental Committee on Vagrancy, Minutes of Evidence*, QQ. 909, 4649–51, 7555.

[7] William Booth, *The Vagrant and the Unemployable*, pp. 9–20.

John Gorst introduced into Parliament a Bill prepared by the Salvation Army, giving power to magistrates to confine vagrants in state-subsidized labour colonies for periods not exceeding twelve months.[1] General Booth thought that the new system might be supervised by a new central authority, modelled on the Lunacy Commissioners. 'If an effective system of State control were established under reasonable and benevolent conditions', he wrote, 'I am led to believe that a considerable proportion of these wandering vagrants would be reformed.'[2] This Bill was rejected by the House of Commons, but the plan was approved by the Local Government Board's Departmental Committee on Vagrancy, which reported in 1906.[3] Colonel Lamb, giving evidence before the Committee, stated that the Army would be prepared to use Hadleigh for such a purpose; and he suggested that colonies should be run by dedicated voluntary officers, who were in a better position to exert a 'reformatory' influence than salaried employees of the State.[4] This proposal was endorsed by the Committee, with the proviso that local authorities should share in the management of colonies which should be subject to Home Office inspection and control.[5]

In 1905 the Hadleigh colony and the Salvation Army settlements in America were inspected on behalf of the Colonial Secretary by the agricultural reformer, Henry Rider Haggard. At this date three colonies had been established in America, one for farm labourers, one for inebriates, and one for the unemployed. Haggard reported that he was favourably impressed with these colonies, as a barrier against urbanization and consequent racial decay; and he urged the imperial government to introduce legislation extending the system of labour colonies in England, and to assist Booth's schemes of overseas colonization with a 6 per cent exchequer loan.[6]

Nevertheless, there were many practical objections to the

[1] *A Bill to amend the Vagrancy Act 1824, and to facilitate the establishment of Labour colonies,* 10 Mar. 1904.

[2] William Booth, *The Vagrant and the Unemployable,* p. 6.

[3] Cd. 2852/1906, *Departmental Committee on Vagrancy, Report,* para. 435.

[4] Cd. 2891/1906; *Departmental Committee on Vagrancy, Minutes of Evidence,* Q. 7099.

[5] Cd. 2852/1906, *Report,* para. 146.

[6] Cd. 2562/1905, *Report on the Salvation Army Colonies in the U.S.A. and Hadleigh, England, with a scheme of National Land Settlement,* paras. 10–17, 34–8.

adoption by the Government of General Booth's proposals—
the first of these objections being the chronic financial dis-
order of the Darkest England scheme. Harold Moore, the
designer of Hadleigh, resigned from the scheme in protest
against the futile squandering of funds;[1] and the finances of
the scheme were the subject of a special committee of inquiry
under Sir Henry James, the ex-Attorney-General, in 1892–
3.[2] This committee acquitted Booth of the charge of using
funds subscribed for the relief of distress for evangelistic
purposes; but, nevertheless, it was impossible to trace precisely the expenditure of the £130,000 subscribed to the
'Darkest England Fund'.[3] Bramwell Booth in 1895 estimated
that £94,000 had been invested in Hadleigh;[4] but Harold
Moore claimed that less than £7,000 had actually been spent
on employing the unemployed.[5] Even Rider Haggard admitted that there had been a 'slight failure of finance' in the
overseas colonies;[6] and in 1906 a Colonial Office committee
under Lord Tennyson rejected his proposal that the Army
should be entrusted with the management of imperial funds.[7]

A second objection stemmed from the failure of the Army
to keep accurate statistical records of the workmen treated
under the Darkest England scheme. During the course of
1892 the Army claimed that 10,473 applicants for employment had been registered by its labour bureau, of whom
6,654 were recorded as having been placed in work. But, as
Llewellyn Smith pointed out, these figures were highly misleading, since only 421 of the successful applicants went to
permanent posts. Of the rest the vast majority were transferred to 'elevators' or placed in temporary situations—the
largest single group being employed as 'sandwich-men', one
of the most notoriously precarious of casual jobs.[8] In 1895

---

[1] H. of C. 365/1895, *SC on Distress from Want of Employment, Minutes of Evidence*, QQ. 9605–6.   [2] H. Begbie, *Life of William Booth*, ii. 172–4.
[3] Booth had originally appealed for £100,000 and had raised £130,000 by
Sept. 1892 (St. John Ervine, *God's Soldier: General William Booth*, ii. 725).
[4] H. of C. 365/1895, Q. 9880.   [5] H. of C. 365/1895, Q. 9563.
[6] Cd. 2562/1905, *Report on the Salvation Army Colonies in the U.S.A. and at
Hadleigh, England, etc.*, p. 14, para. 10.
[7] Cd. 2978/1906, *Report of the Departmental Committee appointed to consider
Mr. Rider Haggard's Report on Agricultural Settlements in British Colonies*, para. 38.
[8] C. 7182/1893, *Report on Agencies and Methods for Dealing with the Unemployed*,
pp. 163–4.

Bramwell Booth told the Select Committee on Distress from Want of Employment that 86 per cent of criminals, 75 per cent of prostitutes, and 80 per cent of the unemployed who passed through the Army's hands had proved to be 'reclaimable'.[1] Of the inmates of Hadleigh he stated that 60 per cent were unskilled, 30 per cent skilled, and 10 per cent 'professional';[2] and he claimed that the success of rehabilitation varied directly with the length of a workman's stay.[3] But these figures were almost entirely conjectural since, as Colonel Lamb admitted to a COS committee of inquiry, the Army kept no records of the subsequent case histories of workmen relieved.[4] Unreliable statistics of this kind made it difficult for contemporary reformers to accept the Army's proposals, even when they were favourably disposed towards its work for the unemployed.[5]

Thirdly, the colony and workshops evoked many protests from trade unionists, farmers, and social workers who accused the Army of underselling their products, sweating their labour, and undermining the standard rate.[6] General Booth opened his accounts to public inspection to prove that this was not the case;[7] but even so it was argued that by intensifying local competition the Hadleigh colony was displacing local workmen and forcing prices down. Bramwell Booth replied in 1895 that residents in the surrounding countryside actually benefited from having a prosperous colony settled in their midst. 'It must be borne in mind . . .', he told the Campbell-Bannerman committee, 'that all these people . . . who benefit become much larger consumers than they were before we took hold of them; they take the produce of other men's labour in exchange for their own.'[8] The Booths claimed, moreover, that the men employed in their workshops and colonies were incapable of earning an economic wage, and

[1] H. of C. 365/1895, SC on Distress from Want of Employment, Minutes of Evidence, QQ. 9841, 9843, 9862.

[2] Ibid., Q. 9858.   [3] Ibid., Q. 9839.

[4] COS Special Committee on the Relief of Distress Due to Want of Employment, 1904, Minutes of Evidence, QQ. 698–70.

[5] COS Special Committee on the Relief of Distress Due to Want of Employment, 1904, Report, pp. 41–2.

[6] St. John Ervine, op. cit. ii. 694, 728–9; S. Higenbottam, Our Society's History, pp. 176–9.

[7] H. of C. 365/1895, QQ. 9911–13.   [8] Ibid., Q. 9920.

that any payments they received were merely gratuitous rewards for good behaviour.[1] But trade unionists replied that the Salvation Army simply did not understand the normal use of the terms 'sweating' and 'underselling'; and that the published price-lists and annual reports of the Social Wing proved conclusively that the Army was guilty of both these offences. They claimed, moreover,

that . . . by no possible stretch of imagination, even the fanciful evangelical imagination of the Salvation Army, can men be 'elevated' by just placing them in what is virtually a compound, situated in one of the most squalid and drink-ridden districts of a big city, and then systematically depriving them of any purchasing power that should legitimately accrue to them as the result of their labour.[2]

The battle between the salvationists and the organized labour movement was finally resolved in 1909 when, after a long series of negotiations between Booth and the Parliamentary Committee of the T.U.C., the Army agreed to withdraw from competitive production.[3] Goods henceforward were to be produced only for the Army's internal consumption, thereby seriously restricting the further development of salvationist colonies and workshops as methods of employing the unemployed.

Finally, it is difficult to see how any government of the period could have justified the delegation of coercive or administrative powers over unemployed workmen to any unestablished religious sect, let alone a body whose methods and motives aroused so much controversy as those of the Salvation Army. The Inebriates Act of 1898 authorized the Home Secretary to delegate powers of detention over confirmed alcoholics; and the Salvation Army was licensed under this Act.[4] But the treatment of alcoholics was scarcely an adequate precedent for the treatment of able-bodied workmen, even if 'wilful' unemployment had been made a criminal offence.

[1] H. Rider Haggard, *Regeneration: Being an Account of the Social Work of the Salvation Army in Great Britain*, 1910, p. 66.

[2] Beveridge MSS., Coll. B, vol. iv, item 15, Manifesto by the United Workers' Anti-Sweating Committee on 'Salvation Army Sweating', Oct. 1908.

[3] *Report of the Ipswich T.U.C.*, 1909, pp. 75–85; *Report of the Sheffield T.U.C.*, 1910, pp. 58–60.

[4] *Report of a Special Committee of the COS on the Relief of Distress Due to Want of Employment*, 1904, Q. 565.

Moreover, the Army did not envisage that state assistance would impose any constraints upon the religious aspects of its social activities. Thomas Mackay, cross-examining on behalf of the COS Committee on the Relief of Distress in 1904, pointed out to Colonel Lamb that a state-supported scheme of relief would not allow much scope for 'religious influence'. 'Why not?' replied Colonel Lamb, assuring the Committee that the Army had 'people in thousands and thousands' able and anxious to exert such influence.[1] But the religious aspect of the Army's work recommended itself to only a small minority of persons interested in social reform. William Beveridge, who approved of penal labour colonies, objected strongly to the economic blackmail of the 'penitent form', and thought that the Army was only really interested in the conversion of economically productive workmen.[2] Puritan agnostics like the Webbs, who admired the Army's social work, admired also the selflessness and self-discipline that salvationist convictions seemed to inspire.[3] But they were unwilling to accept that the spreading of such convictions was a necessary feature of the regeneration of the unemployed.[4]

In view of these manifold objections, what is perhaps most surprising about proposals for making the Army an organ of public administration is not that they were ultimately rejected, but that they should ever have been considered at all. Yet Lord Rosebery in 1905 suggested that General Booth should be given a government contract to deal with the 'residuum';[5] and the *Westminster Review* proposed that the Army should be made part of a newly constituted department of 'National Health'.[6] In 1908 Beatrice Webb seriously contemplated the possibility of incorporating the Salvation Army into a new statutory 'Drainage System' for reclaiming surplus labour;[3] and although the revivalist pressures of Hadleigh

---

[1] *Report of a Special Committee of the COS on the Relief of Distress Due to Want of Employment*, QQ. 619–20.

[2] Beveridge MSS., Coll. B, vol. iii, item 38, Beveridge's MS. annotations to 'The Vagrant and the Unemployable etc.'

[3] Passfield MSS., ii. 4, d, item 2, B. Webb to Mary Playne, 2 Feb. 1908.

[4] B. Webb, *Our Partnership*, p. 401. See also H. Preston Thomas, op. cit., p. 349.

[5] *Liberal Magazine*, Dec. 1905, p. 671.

[6] Frederick Thoresby, 'How to Deal with the Unemployed', *Westminster Review*, 165 (Jan. 1906), 39.

deterred her from proposing that the Army should be 'state-or-rate-aided',[1] the Minority Report of the Poor Law Commission strongly recommended that public authorities should work in conjunction with voluntary religious organizations in training and reforming the recalcitrant unemployed.[2]

POOR LAW LABOUR COLONIES

The kind of labour colony that most clearly invited state intervention was the colony for the employment of able-bodied paupers under the direction of guardians of the poor. Certain groups of late-nineteenth-century social reformers were very conscious of historical precedents for interpreting literally the concept of the workhouse and 'setting the poor on work'. It was difficult to read the mind of the Tudor legislators, but the wording of the Poor Law statute of 1601 suggested that it had been intended to provide the able-bodied unemployed with work at wages in their own homes; and this historical point was conceded by the Secretary of the Local Government Board, Sir Hugh Owen, in 1895.[3] At the end of the eighteenth century Poor Law farms had been created in certain parishes, or able-bodied paupers had been hired out to private contractors;[4] and legislation in 1819 and 1834 had empowered guardians to acquire land on which to provide work at wages. Keir Hardie's parliamentary cam-

---

[1] B. Webb, *Our Partnership*, p. 401.

[2] Cd. 4499/1909, *RC on the Poor Laws, Minority Report*, p. 1214. This recommendation was not confined to the Salvation Army; but the only other religious body that organized relief for the unemployed on a comparable scale was the Church Army, which had been founded by Wilson Carlile for evangelism and general 'rescue work' within the Church of England. By 1907 it had an income of £247,000 p.a., and by 1909 claimed that its 'labour homes' were giving half a million days of relief work a year (*Encyclopaedia of Social Reform*, i (1909), p. xvi). Like the Salvation Army the Church Army continually pressed for state intervention. In 1895 the secretary of its 'Social department', the Revd. William Hunt, called for the creation of a Ministry of Labour to publish unemployment statistics and to persuade government departments to start counter-depressive public works (H. of C. 365/1895, *SC on Distress from Want of Employment, Minutes of Evidence*, QQ. 10184-8). In 1905 he urged the L.G.B. to establish graded labour colonies, worked by philanthropic bodies, but financed and controlled by the State (W. Hunt, *Labour Colonies, What are They? What Can They Do?*, pp. 3-4, 8-11).

[3] H. of C. 365/1895, *Appendix 30*, p. 561.

[4] B. Kirkman Gray, *A History of English Philanthropy*, pp. 215-21; M. K. Ashby, *Joseph Ashby of Tysoe*, p. 279.

paign of 1893–5 to revive these statutes was a signal failure.[1]
But in some metropolitan unions a less spectacular movement
to utilize these obsolete powers had already been set on foot
by local guardians several years before.

The first serious advocate of Poor Law labour colonies was
Canon Barnett, a member of the Whitechapel board of
guardians and Warden of Toynbee Hall. In November 1886
Barnett assured the Government that the existing powers of
the Poor Law, together with private charity, were quite
adequate to relieve distress in the East End.[2] In the following
winter, however, Barnett submitted proposals to the Mansion
House conference for rejuvenating the Poor Law by permit-
ting guardians to provide remunerative work.[3] He suggested
that guardians should be allowed to purchase country estates
and to select 'able-bodied men of good character and apparently solid determination' for agricultural training 'without forcing them to conform to rules which are suggestive of
the workhouse'. The men would be prepared for permanent
settlement overseas or in the English countryside, and their
wages would be paid and their families relieved out of
charitable funds.[4]

This proposal was strongly criticized by C. S. Loch, who
accused Barnett of trying 'to create a new status of pauper—
the unemployed—making want of employment a qualification for relief, as well as destitution'. It was, he objected,
'absurd to think that charitable bodies could keep the families
of resident unemployed persons for two or three months at

---

[1] Above, p. 85. On the guardians' powers to provide work see J. Theodore
Dodd, *To Boards of Guardians in Rural Districts. The Winter's Distress—how to
provide for the unemployed.*

[2] Salisbury MSS., Class E, C. T. Ritchie to Lord Salisbury, 4 Nov. 1886.
Ritchie reported that Barnett's only proposal for a change in the Poor Law was that
'the Guardians should have given to them the power of giving outdoor relief for
the family of a man who himself became an inmate of the workhouse'. He thought
that otherwise 'with such private charity, as has . . . never failed him and others in
the East End the distress will be coped with'.

[3] *Report of the Mansion House Conference on the Condition of the Unemployed,*
1887–8.

[4] See also S. Barnett, 'A Scheme for the Unemployed', *Nineteenth Century,* 24
(Nov. 1888), 753–63, in which Barnett stated that guardians would be the most
efficient authority to manage labour colonies; but he suggested that, in view of the
'poor repute' into which they had fallen, they should co-operate with voluntary
organizations (pp. 761–2).

a time'. They would therefore have to be supported by the
guardians, which could only mean an extension of outdoor
relief—an outcome that COS representatives could only
deplore.[1]

Barnett's scheme was therefore rejected by the Mansion
House conference—an event that marked the beginning of
his estrangement from more orthodox members of the COS.[2]
In 1888, however, the Whitechapel guardians set up a committee to consider 'whether Poor Law administration in this
country is capable of development in the direction of training
the unemployed and destitute poor in agricultural pursuits'.
This committee reported that in London and other large
urban centres there was a surplus of unskilled labour 'becoming enervated and demoralised by enforced idleness and the
conditions of city life... whilst in the countryside agricultural
land is rapidly going out of cultivation ... the production of
wealth is being diminished' and 'millions of money is annually
paid for . . . produce from abroad.' It recommended that
guardians should attempt to solve this problem by 'adapting
their administration to modern needs and making it as
remedial and as helpful as possible'; but concluded that the
creation of an agricultural settlement would only be justified
where there was an actual shortage of workhouse accommodation, which was not the case in Whitechapel.[3] The question
was therefore submitted to a conference of all metropolitan
guardians in July 1888. This conference issued a circular to
guardians in the Essex unions, asking whether vacant land
was available, whether there were openings for unemployed
workmen and whether a Poor Law farm was a commercial
proposition. The replies were almost universally unfavourable, since although vacant land was available in abundance,
local guardians asserted that it could not yield a profit while
food prices remained so low. They thought, moreover, that
the crux of the labour problem was not to 'colonize' the urban
unemployed but to prevent rural workmen from migrating
to the towns.[4]

[1] C. S. Loch's Diary, 1 Oct. 1888, pp. 150–2.
[2] C. L. Mowat, op. cit., p. 127.
[3] H. of C. 365/1895, *Appendix 29*, pp. 548–9, 'Report of Committee of the
Whitechapel Guardians on Agricultural Training Homes for the Unemployed',
13 Mar. 1888.   [4] Ibid., pp. 549–52.

In 1892 the Whitechapel guardians appointed a second committee, to consider the establishment of 'agricultural training homes for the unemployed'. The committee was urged to take action by Barnett, Walter Hazell, and Captain le Mesurier Gretton, the secretary of the East London Emigration Fund; but discouraging evidence was again received from farmers, clergymen, and rural guardians. A cautious report was drafted by the clerk of the guardians, stating that agricultural settlements might be a useful means of employing the 'self-respecting . . . aristocracy of the poor', but deprecating any attempt to save them 'from the Poor Law by the machinery of the Poor Law'. It was therefore recommended that settlements should be established by voluntary rather than statutory authorities.[1] Samuel Barnett protested, however, that for the guardians to support voluntary agricultural settlements would be in no way inconsistent with the strict Poor Law policy pursued in Whitechapel during the previous twenty-three years;[2] and the guardians therefore compromised by sending respectable unemployed workmen to the Self-Help Emigration Society's training farm at Langley in Essex.[3]

Whitechapel was, however, an unsuitable union for experiments of this kind since the influence of 'organized charity' was too deeply entrenched on the board of guardians to allow major deviations from a strict interpretation of the Poor Law; and the initiative in pressing for more ambitious Poor Law reforms passed to other metropolitan unions. In May 1893 the London and District Poor Law Officers' Association called for the establishment of penal labour colonies in which 'habitual' paupers might be confined by order of a magistrate for periods not exceeding three months.[4] In November 1894 a conference of metropolitan guardians proposed to the Local Government Board that a special authority should be created to supervise the employment of unemployed workmen on the land. The Board replied that such an authority could not be established without special legislation, which it was not prepared to introduce. But it would consider applications from individual boards of guar-

---

[1] H. of C. 365/1895, *Appendix 29*, p. 553. [2] Ibid., pp. 554–5.
[3] Ibid., pp. 555–6. [4] Ibid., p. 556.

dians for permission to acquire land to test 'the necessities of persons who apply to them on grounds of their being unemployed'.[1]

The union that responded most enthusiastically to this invitation was the union of Poplar under the influence of George Lansbury, who was elected as a socialist member of the board of guardians in 1894. Lansbury's attitude to labour colonies was initially rather ambiguous, and he was clearly uncertain about whether the existing Poor Law system should be abolished or merely reformed. His manifesto for the guardians' election called for the 'formation of Labour Colonies for the treatment of the habitual casual and repression of the loafer';[2] but as S.D.F. candidate in the Walworth parliamentary by-election early in 1895 he denied that he was in favour of 'pauper colonies or in any way perpetuating the workhouse system'.[3] Before the Campbell-Bannerman committee he advocated 'self-supporting colonies', but thought that they were unlikely to be successful without the abolition of free competition;[4] and in 1897 he told the Central Poor Law Conference: 'I do not wish to be misunderstood. I do not wish for penal settlements, for you will never drive out wickedness by wickedness, you cannot do good work with the devil's tools.'[5]

Nevertheless, Lansbury became convinced that, with only minor adjustments in the existing law, local authorities could employ unemployed workmen on 'self-supporting co-operative farms';[6] and after the failure of his bid for Parliament, the Poplar guardians became the instrument of his programme for settling the unemployed on the land. In 1895 the guardians applied for permission to purchase a 280-acre farm in Essex,[7] on which unemployed residents of Poplar, both male

[1] *Report on The Poplar Labour Colony* (by G. H. Lough, Clerk to the Guardians), Oct. 1904, p. 5.

[2] Lansbury MSS., vol. 1, f. 186, 'Advertisement for Bow and Bromley S.D.F.'s programme in the election of Guardians for the parish of Bow', issued by G. Lansbury and W. Purdy, Nov.–Dec. 1893.

[3] Ibid., ff. 204–8, G. Lansbury to the electors of the Walworth division of Newington.

[4] H. of C. 365/1895, QQ. 10408, 10508–10.

[5] G. Lansbury, *The Principles of the English Poor Law* (printed as a pamphlet by the Twentieth Century Press), 1897.

[6] Lansbury MSS., vol. 1, ff. 204–7.

[7] *Report on Poplar Labour Colony*, pp. 5–6.

and female, could be given work and maintenance for a minimum of one year.[1] W. E. Knollys, an Assistant Secretary at the L.G.B., replied that the powers conferred by the statutes of George III and William IV ,were confined to land in or adjacent to the union and to 50 acres per parish.[2] After lengthy correspondence, the L.G.B. decided that it would in any case sanction the purchase of land for additional workhouse accommodation only; and 'this not being in accord with the original intention of the Guardians, it was decided not to proceed further in the matter'.[3]

In 1903 the question was raised again in Poplar, largely provoked by a shortage of accommodation in the union workhouse. Walter Long, the President of the L.G.B., was thought to be sympathetic to the home colonization movement;[4] and in July 1903 the guardians sought permission from the L.G.B. to establish a 'country workhouse' for the Poplar unemployed.[5] While the correspondence was in progress, Lansbury was introduced to Joseph Fels, the American soap millionaire and patron of the Philadelphia Vacant Land Cultivation Society.[6] Fels was looking for an English outlet for his theories of home colonization;[7] and in November 1903 he offered to purchase an estate and to lease it to the Poplar guardians at a peppercorn rent for three years, after which they would have the option of buying it at the original purchase price.[8] An estate was selected at Laindon in Essex, which was formally taken over by the Poplar guardians in 1904;[9] and at the end of 1904 Fels purchased a second estate

[1] Lansbury MSS., vol. 1, ff. 252–3, Farm Employment, regulations suggested by G. Lansbury, c. June 1895.

[2] Ibid., ff. 250–2, W. E. Knollys to G. Lansbury, 6 June 1895.

[3] Report on Poplar Labour Colony, p. 6.

[4] Lansbury MSS., vol. 7, f. 164, G. Lansbury to M. Fels, 30 June 1914. But see below, p. 154.   [5] Report on Poplar Labour Colony, p. 9.

[6] M. Fels, Joseph Fels: His Life-Work, pp. 60–2. Joseph Fels (1853–1914), owner of the Fels-Naphtha Soap Company; and financier of Lloyd George's budget campaign of 1909.

[7] Ibid., pp. 123–4. Vacant Land Cultivation Societies were established in 1904 in London, Middlesbrough, Edinburgh, Belfast, and Dublin.

[8] Lansbury MSS., vol. 7, ff. 160–8, G. Lansbury to Mary Fels, 30 June 1914; Report on Poplar Labour Colony, p. 10.

[9] Ibid., p. 12. The L.G.B. was apparently persuaded to authorize the venture by the argument that it 'provided an extremely favourable opportunity of dealing with the problem of finding suitable work for able-bodied men without committing the ratepayers of the union to any serious expenditure'.

at Hollesley Bay near the mouth of the Thames estuary, which was transferred under similar conditions to the trustees of the London Unemployed Fund early in 1905.[1]

With the acquisition of Hollesley Bay the history of labour colonies became involved with the development of an unemployment policy on the part of the central government. By 1904 the promoters of both Poor Law and private labour colonies were convinced that they had found a solution to the problems of the labour market, if they could only get financial assistance from the State; and this belief was echoed by reformers in all political parties. The growth of support for the labour colony movement was a reflection of two trends which have already been noticed in late Victorian and Edwardian social ideas. On the one hand there was a hardening of feeling against industrial inefficiency and social inadequacy —a belief that the upward mean of progress should be liberated from those who could not or would not conform. And on the other hand there was a reaction against urbanization and against the social confusion and personal corruption that an urban environment was supposed inexorably to entail. Both these attitudes were reinforced in 1904 by the revelations of the Interdepartmental Committee on Physical Deterioration, which recommended labour colonies for reclaiming the 'waste elements of society', and rural revival and agricultural resettlement for those who had been 'crushed and broken by the wheels of city life'.[2]

On a more abstract level, both types of colony were a reaction against old-fashioned liberalism; the one being primarily designed to impose constraints on the deviant individual, the other to conduct experiments in collectivist production under popular control. It has been shown that colonies of a reformatory kind came close to getting both financial support and powers of compulsory detention from the State. This was recommended by the Departmental Committee on

---

[1] Lansbury MSS., vol. 2, ff. 22–4, Draft of a letter by G. Lansbury, p.p. Joseph Fels, to the Committee of the London Unemployed Fund, 24 Nov. 1904. Fels made the offer with the stipulation that the estate should be transferred to any statutory unemployment authority constituted in the next three years (*Report of the Central Executive Committee of the London Unemployed Fund*, 1904–5).

[2] Cd. 2175/1904, *Report of the Interdepartmental Committee on Physical Deterioration*, paras. 91, 191–8.

Vagrancy in 1906, and endorsed by the Poor Law inspectorate and many local boards of guardians.[1] John Burns, who became President of the Local Government Board in 1905, regarded the whole 'home colonisation' movement with ill-disguised contempt.[2] But other prominent Liberal politicians were not averse to labour colonies, and were interested in Booth's plans for assisted emigration.[3] Both the Majority and the Minority Reports of the Royal Commission on the Poor Laws suggested that labour colonies should be used as a means of repressing the 'loafer' and retraining those whom the organization of the labour market had displaced.[4] If the Salvation Army had been more skilful at pressure-group diplomacy, if it had managed to allay fears of religious bigotry and of 'unfair competition', then labour colonies run on salvationist lines and under the direction of Salvation Army officers might have been adopted as part of the Liberal social programme. But political compromise was completely alien to Salvation Army ideals; and the Salvationists therefore failed to take advantage of public feeling in their favour. They failed in their attempts to cultivate the co-operation of politicians and social reformers, who might have secured for the Army permanent public support; and when the popular journalist, Arnold White, investigated the Salvation Army's social work in 1910 he found that the public controversy that it had aroused a few years previously had largely subsided into public indifference.[5]

Co-operative colonies of the kind envisaged by the Poplar guardians were, however, free from many of the objections that applied to penal colonies run by an authoritarian body like the Salvation Army. They were neutral in religion, self-

---

[1] *36th Annual Report of the L.G.B.* (1906–7), pp. 285, 311; *Report of the Vagrancy Committee appointed at the Midsummer Sessions, 1903, in Parts of Lindsey, Lincolnshire,* 23 Oct. 1903; *Report of the Proceedings of a Conference of Representatives of Lancashire, Cheshire, Westmorland and Cumberland Unions on Vagrancy,* 1 Sept. 1905.

[2] Add. MS. 46324, Burns Diary, 5 June 1906.

[3] H. Begbie, op. cit. ii. 362–6; St. John Ervine, op. cit., pp. 691–3. In 1907 General Booth was interviewed by several members of the Liberal cabinet on the prospects of a colony in Rhodesia; but the scheme 'though never abandoned, was sickeningly deferred'.

[4] Cd. 4499/1909, *Majority Report,* p. 633; *Minority Report,* pp. 1206–8.

[5] Arnold White, *The Great Idea* (1909–10), p. ix.

supporting rather than competitive, and designed to afford opportunities for self-improvement rather than to coerce and reform the wayward unemployed. 'Home aliment' was provided for the wives and children of colonists, and it was hoped that eventually not merely individuals but whole families would be settled on the land. Co-operative colonies had the sympathy and often active support of organized labour;[1] and not only to socialists and labour representatives, but to many Liberal and Conservative reformers, co-operative colonies which offered both training and permanent agricultural settlement seemed to point the way towards a solution of the problem of surplus labour and a revival of rural life.[2] The movement was not without its critics. The COS, fearing a resurgence of the rate-aided employment schemes of the 1820s,[3] objected that colonies were 'but relief works methodised; and so methodised indeed that they tend to conceal the evils which they produce'.[4] And in 1904, Harold Mann, a disciple of Seebohm Rowntree, questioned the whole rationale of the 'back to the land' movement by revealing that in a specimen agricultural community 50 per cent of the working-class were in 'primary' or 'secondary' poverty, even though labourers' wages were supplemented by village allotments and the practice of rural crafts.[5] But nevertheless, in the early 1900s the creation of farm colonies with a view to restoring workmen permanently to the land was the most widely canvassed solution to the problem of surplus labour;[6] and under

[1] In the autumn of 1904 a committee representing Labour M.P.s, the parliamentary committee of the T.U.C. and the General Federation of Trade Unions issued a report which proposed among other items that distress committees should acquire land 'to which the unemployed may be drafted with a view to the workers and their families eventually becoming self-supporting through a system of Co-operative Farming' (*Report of the Hanley T.U.C.*, 1905, p. 62).

[2] Christian Social Service Union, *Notice of a conference on Labour and Training Colonies* at the Mansion House, 5 June 1905. The delegates included fifteen Conservative, twenty-four Liberal, two Liberal Unionist, and two Labour M.P.s, of whom fifteen represented London or Middlesex constituencies.

[3] C. S. Loch, *Employment Relief. A letter to the Hon. Sec. of the Winchester COS*, 15 Nov. 1905. (COS Occasional Papers, No. 23, Fourth Series), p. 2.

[4] *The Times*, 17 May 1905, C. S. Loch to the Editor on 'The Unemployed Workmen Act'.

[5] P. H. Mann, 'Life in an Agricultural Village in England', *Sociological Papers*, i (1904), 163–93.

[6] *Liberal Magazine*, Dec. 1905, p. 667, Memorandum by Sydney Buxton to Arthur and Gerald Balfour, 7 Nov. 1905.

the provisions of the Unemployed Workmen Act of 1905, it seemed likely that co-operative labour colonies would become a permanent feature of English public administration. It is necessary therefore to turn to the political and administrative history of this Act to discover why they became not a central theme but a permanent cul-de-sac in the development of English social policy.

IV

UNEMPLOYMENT AND LOCAL ADMINISTRATION 1903–1908

THE Unemployed Workmen Act was born in the shadow of a major crisis in English social administration. National expenditure on poor relief, which had been reduced and stabilized in the 1870s and 1880s, rose in the 1890s and swung dramatically upwards between 1901 and 1906.[1] Between 1870–1 and 1905–6 the average annual cost of maintaining an individual pauper increased by over 100 per cent,[2] even though the cost of living was stable or falling for most of this period.[3] The rise in the number of indoor paupers since the 1880s suggested that Poor Law institutions were ceasing to fulfil their deterrent function;[4] and the Poor Law guardians, who since 1894 had been subject to a quasi-democratic franchise, were increasingly on the defensive against reformers who wished to deprive them of their administrative functions and clients who demanded a more liberal scale of relief.[5]

Much of the alarm evoked by the apparent increase of pauperism was in fact misplaced, since the figures referred mainly to the relief of old age and sickness, rather than to the 'able-bodied' destitution which the New Poor Law had been mainly designed to repress. The ratio of paupers to total population continued to fall after the franchise reform of 1894 and rose only slightly during the depression of 1903–5.[4] The increased average expenditure on individual

[1] Annual Poor Law expenditure in England and Wales was £7,886,724 in 1871; £8,102,136 in 1881; £8,643,318 in 1891; and £11,548,885 in 1901. By 1906 it had risen to £14,035,888 (*29th Annual Report of the L.G.B.* (1899–1900), pp. 432–5; *41st Annual Report of the L.G.B.* (1911–12), pp. 180–4).

[2] From £7. 12s. 0¾d. to £15. 12s. 6¼d.

[3] B. R. Mitchell and Phyllis Deane, *Abstract of British Historical Statistics*, pp. 344–5.

[4] Cd. 5077/1911, *RC on the Poor Laws. Statistical Memoranda and Tables*, pp. 24–5, 29.

[5] *The Times*, 2 May 1906, report of a lecture by Sidney Webb at the London School of Economics.

paupers was largely accounted for by the shift from outdoor to indoor pauperism, since it cost nearly four times as much to maintain a destitute person inside as outside the work-house.[1] Poor Law expenditure per head of population in 1904–5 was still 10 per cent lower than it had been in 1833–4, although real income per head had doubled during the previous fifty years.[2] The country as a whole was well able to bear such a burden, although it fell heavily on certain localities. Nevertheless, the apparent slackening of discipline within the Poor Law, and the failure of education, thrift, and sanitary reforms to eliminate pauperism tended to re-inforce the view of conservative administrators that the situation could only be remedied by a strict revival of the deterrent principles of 1834.[3]

The Local Government Board's Poor Law inspectors emphasized that a high level of pauperism was nearly always a reflection of administrative laxity rather than local poverty;[4] and the rise in pauperism and Poor Law expenditure was widely ascribed to the extension of the franchise and abolition of property qualifications for guardians in 1894.[5] The election of guardians for the first time acquired potential political significance; and members of the COS deplored the fact that guardians would henceforth be chosen by, and even chosen from, the friends and relatives of those whom they were empowered to relieve.[6] But while conservative critics accused the guardians of succumbing to popular pressure, radicals accused them of deliberately encouraging the poor to accept relief in order to get them struck off the electoral register.[7] The influence of the reform of 1894 was almost certainly exaggerated, since the rise in Poor Law

[1] Cd. 4499/1909, *Royal Commission on the Poor Laws, Majority Report*, Part II, p. 31, para. 65. Other causes of the increase were the extension of infirmaries, and 'cottage homes' for pauper children; increased building costs; and an ageing population.

[2] Ibid., p. 31, para. 67; B. R. Mitchell and Phyllis Deane, op. cit., pp. 367–8.

[3] *The Times*, 19 Jan. 1905, G. C. Bartley to the Editor on 'London and the Unemployed Problem' (reprinted as COS Occasional Paper, No. 5, Fourth Series).

[4] *26th Annual Report of the L.G.B.* (1896–7), p. 85.

[5] Asquith MSS., vol. 78, f. 89, Typescript notes by Harold Baker on the Majority and Minority Reports of the Royal Commission on the Poor Laws (early 1909).

[6] C. S. Loch, *Employment Relief. A Letter to the Hon. Sec. of the Winchester COS*, 15 Nov. 1895 (COS Occasional Paper, No. 23, Fourth Series), p. 3.

[7] H. of C. 321/1896, Q. 253.

expenditure preceded the widening of the franchise; and only in areas with a well-organized labour movement did it make any significant difference to the personnel and policies of boards of guardians.[1] Indeed, in certain areas the conservative element among the guardians was strengthened, since in the 1900s the COS pursued a deliberate policy of putting up candidates to defend strict Poor Law principles —a policy that, in an atmosphere of widespread electoral lethargy, met with conspicuous success.[2]

In fact, although political change cannot be discounted, the crisis in the Poor Law was brought about by factors that were more deep-seated than the extension and relaxation of political control. The administration of the Poor Law was based on the principle that the condition of a pauper should always be 'less eligible' than that of the lowest-paid independent labourer. But social investigation in the 1890s revealed that the lowest-paid independent labourers lived in conditions that were harmful to physical and industrial efficiency;[3] and this was confirmed by official inquiries in 1904 and 1908.[4] This new slant on the social effects of poverty had important repercussions on the administration of poor relief. Officials of the Local Government Board, whilst seeking to discourage outdoor relief, nevertheless insisted that such relief should be adequate for 'decent living', if it was given at all.[5] And progressive Poor Law guardians were increasingly reluctant to reproduce inside the workhouse conditions that when brought about

---

[1] *36th Annual Report of the L.G.B.* (1906–7), pp. 338–41.

[2] In certain northern unions the number of COS supporters who were members of boards of guardians increased between 1898 and 1908 by over 350 per cent (E. W. Wakefield, 'The Growth of Charity Organisation in the North of England', *Charity Organisation Review*, 24 (July 1908), compare tables on pp. 48–9). On public indifference to guardians' elections see W. Bailward, 'Local Government and Popular Elections', ibid., 18 (Oct. 1905), 183–94.

[3] S. Rowntree, *Poverty: A Study of Town Life*, pp. 198–221.

[4] Cd. 2175/1904, *Report of the Inter-Departmental Committee on Physical Deterioration*, para. 142; H. of C. 246/1908, *SC on Homework, Report*, pp. iii–iv.

[5] *27th Annual Report of the L.G.B.* (1897–8), p. 143; *29th Annual Report of the L.G.B.* (1899–1900), pp. 105–6. A circular issued to guardians on 4 Aug. 1900, authorizing guardians to give out-relief to the 'aged deserving poor', remarked that 'too frequently such relief is not adequate in amount. [The Board] are desirous of pressing upon the guardians that such relief should when granted be always adequate' (Gerald Balfour MSS., PRO 30/60/48).

by private enterprise were widely condemned.[1] 'It is frankly
impossible,' wrote William Beveridge in 1906, 'for any
public committee openly to give those dependent on it
conditions of life approaching in badness and harmfulness
the conditions which . . . public thoughtlessness passes by
as "inevitable" for large sections of a free and independent
proletariat.'[2] Hence, 'as a means of meeting unusual or
exceptional distress, it must be admitted that the Poor Law
as administered in certain unions has failed entirely' reported
a COS committee in 1904. 'The workhouse under its
altered conditions is to many people hardly less "eligible"
than outdoor relief. Consequently the Poor Law is converted practically into an enormously rich, rate-endowed
charity open to all comers.'[3]

How did these changes affect the relief of unemployment?
The confidence of the Russell committee of 1896 that in the
last resort the Poor Law was well able to deal with unemployment was misplaced, since it ignored the fact that,
at least since 1870, the Poor Law had never been a major
source of relief to the unemployed. Genuinely unemployed
workmen shunned outdoor relief because the labour test
in the stoneyard was liable to impair their industrial skill
and prevented them from looking for work elsewhere. Able-
bodied workers accounted for less than one-fifth of indoor
paupers; and of those more than three-quarters were relieved
on grounds of sickness or temporary disablement. The
recorded number of adult male paupers relieved because of
'want of employment' in the workhouses of England and
Wales only once exceeded 12,000 in the years between
1886 and 1912.[4] Moreover, the so-called 'stigma of

[1] George Haw, *The Life Story of Will Crooks M.P. From Workhouse to Westminster*, pp. 112–18, 277; Cd. 4499/1909, *RC on the Poor Laws, Minority Report*,
p. 1076.

[2] *Morning Post*, 5 Mar. 1906. Harold Baker, who summarized the reports of
the Royal Commission on the Poor Laws for the Cabinet, suggested that the relaxation of the Poor Law was a reflection of a rising standard of living, which made it
possible to improve workhouse conditions without infringing the principle of
'less eligibility' (Asquith MSS., vol. 78, f. 150). This view was scarcely compatible
with the results of contemporary studies of poverty; but it may have been true of
certain unions.

[3] *Report of A Special Committee of the COS on the Relief of Distress Due to Want of
Employment* (1904), pp. 17–18.

[4] Appendix B, Table 1, p. 373.

pauperism' involved more than the loss of the franchise; and even if the Select Committee's proposal for the abolition of disfranchisement had been accepted it is unlikely that poor relief would ever have been acceptable to the bulk of the unemployed.

Nor was it a desirable remedy from an administrative point of view, since able-bodied paupers were difficult to control and work in the stoneyards inevitably conformed to the standard of the least efficient workmen. In 1894 and 1895, when there was a temporary increase in the number of stoneyards, workhouse officials found that discipline was almost impossible to maintain. In Bermondsey able-bodied paupers terrified the workhouse staff and played leapfrog in the stoneyard; and in Poplar they formed a union and went on strike for a higher scale of relief.[1] According to Beveridge, 1895 was therefore 'the year of the guardians' stoneyards and of their final condemnation as means of dealing with the unemployed'.[2] The labour test became little more than a formality; and in Poplar in 1904 it was remarked that the able-bodied 'have practically no tasks. They sit there hands folded all day long. In the old days they used to do stone-breaking, but stone-breaking was thought to be degrading, and therefore now they do nothing. It is a question whether doing nothing may not be even more degrading.'[3] By 1904 a COS committee of inquiry found that, as a remedy for unemployment, poor relief had 'fallen into desuetude. The able-bodied register their names and apply for employment, which is in fact employment-relief, to the Borough Councils; and the Poor Law has lost its former relation to the problem.'[4]

This account of the situation was strictly unhistorical since, partly owing to the influence of the COS, the Poor Law had had little relevance to the problem of unemployment

[1] 25th Annual Report of the L.G.B. (1895–6), pp. 162–4; C. S. Loch, Methods of Relief Adopted in the Metropolis During the Winter of 1895, p. 4.

[2] W. H. Beveridge, 'Emergency Funds for the Relief of the Unemployed: A Note on their Historical Development', Clare Market Review, 1, no. 3 (May 1906), 74.

[3] Report of a Special Committee of the COS on the Relief of Distress Due to Want of Employment (1904), Evidence of Mr. W. G. Martley, Q. 57.

[4] Ibid., p. 13.

over the previous twenty-five years. But it reflected accurately what many contemporaries believed to be the true state of affairs—that for political reasons the unemployed disdained poor relief, and that the Poor Law had therefore ceased to fulfil its proper function, leaving a vacuum in administration of relief which the borough councils were only very imperfectly able to fill. The problem had been partially eclipsed by the full employment which prevailed in the late 1890s and during the South African war; but the post-war depression once again drew attention to the shortcomings in public provision for the unemployed. As in the early 1890s the guardians, local authorities, and charitable associations were seen to be duplicating their efforts to give relief;[1] and recorded admissions to casual wards were higher than they had ever been before.[2] It was within this context that the first steps were taken to create a more permanent machinery for dealing with the 'respectable' able-bodied workman who was temporarily unemployed.

EXPERIMENTS IN LONDON 1903–5

The trade boom began to slacken at the end of 1902; and early in 1903 the L.C.C. convened two conferences of London administrative bodies, which recommended legislation compelling local authorities to deal with unemployment, and the establishment of a ministry responsible for all labour questions.[3] They also suggested that a national organization with branches in every town should organize labour bureaux, supervise emigration and colonization schemes, provide technical education and lectures on social problems, and promote 'the dissemination of such information as will

---

[1] e.g. in a 'wild warfare of distribution' in West Ham, the unemployed were relieved by the guardians, the borough council, newspaper relief funds, and private charity, and their children were fed by the Education Committee (*Report of a Temporary Colony at Garden City for Unemployed Workmen mainly from West Ham*, organized by the settlement of Trinity College, Oxford, Feb.–Apr. 1905).

[2] 'The Relief of Poverty: Vagrancy and Lunacy', *Charity Organisation Review*, N.S. 15 (1904), 231–6.

[3] London County Council, *Lack of Employment in London*, Minutes of the Proceedings at a Conference, 13 Feb. and 3 Apr. 1903, between representatives of the L.C.C. and of Administrative Authorities in London . . . together with a report of the General Purposes Committee of the L.C.C., adopted by the Council on 27 Oct. 1903, with regard to the recommendations made by the conference, p. 4.

enable the working-class to realise what steps they should themselves take to improve their position'.[1]

These resolutions were adopted by the L.C.C. in October 1903. In November 1903 a study group of Oxford graduates at Toynbee Hall, Balliol House, and Wadham House formed a 'committee on the unemployed', to investigate schemes for labour colonies, labour exchanges, and emigration, and to recommend suitable unemployed workmen to local employers.[2] The secretary of the committee was William Beveridge, and its members included H. R. Maynard and R. H. Tawney, all of whom were to become more closely involved in the administration of unemployment relief. The committee concluded that labour colonies had so far been unsuccessful, but that emigration offered 'a real though limited solution' to the problem of the unemployed.[3]

At the end of November the Lord Mayor reconvened the Mansion House committee on the unemployed which had been in abeyance since 1895. This committee was again dominated by members of the COS; but it also included representatives of the new generation of social reformers, like Beveridge and Percy Alden. The new scheme was confined to Stepney, Poplar, Bethnal Green, and Shoreditch. It was financed by a fund of £4,000 raised by the Lord Mayor, and administered in each borough through local relief committees, representing the guardians, borough councils, and local charities.[4] No attempt was made to start relief works, but the executive committee arranged to employ the unemployed at Hadleigh, and at a private estate on Osea Island in the Thames estuary, which belonged to Frederick Charrington, the temperance reformer.[5] Successful

---

[1] Ibid., p. 32. Delegates to the conference also proposed the equalization of rates, the restriction of alien immigration, and the purchase of land by Poor Law guardians.

[2] Beveridge MSS., A. 6. 104, Minutes of a committee of residents from Toynbee Hall, Balliol House, and Wadham House, 9 Nov. 1903, 1 Dec. 1903, 15 Dec. 1903.

[3] Ibid., Minutes, 15 Dec. 1903.

[4] *Mansion House Committee on the Unemployed 1903–4. Report of the Executive Committee*, pp. 5–6.

[5] Son of the Mile End brewer and founder of the Tower Hamlets Mission. Beveridge eventually withdrew from the committee as he 'objected to being responsible for dealings with Charrington', Beveridge MSS., A1. 100, Box 1, Beveridge's Diary, 3 Feb. 1904.

applicants were selected on the basis of employers' testimonials, evidence of thrift, and regularity of previous
employment. Their families were relieved at home by district visitors, who superintended the expenditure of the
relief and 'guarded against abuse'. Only 467 men were
employed and the quality of their work was poor; but the
committee reported that the scheme 'proves conclusively
that there was, during last winter, a large number of genuine
working men of good industrial character unemployed,
and that these could be discovered and assisted when once
steps had been taken to exclude the chronically unemployed
and the casual labourer.'[1]

The Mansion House scheme of 1903–4 was important
only because its decentralized administrative structure was
copied by subsequent schemes, and because it attracted a
new generation of social reformers to the study of unemployment.[2] The deepening of the depression also aroused concern
for the unemployed outside the rather esoteric circle of the
Mansion House and the university settlements. In September 1904 the T.U.C. debated the problem of unemployment
for the first time for seven years, and instructed its Parliamentary Committee to press for the creation of a special
department of the Board of Trade which would co-ordinate
the efforts of local authorities in combating distress.[3] It
was, however, local pressure that persuaded the Government to take action. In 1903 there had been an outbreak
of the kind of organized demonstrations that had disturbed
London in the 1880s. 'It is all forgotten now,' Walter Long
told the Royal Commission on the Poor Laws four years
later, 'but . . . the methods adopted by the unemployed
towards all the authorities, municipal and Poor Law, were
violent in the extreme.' In London, Leeds, Manchester,
Liverpool, Birmingham, and other great cities, massive
demonstrations of the unemployed urged local authorities

---

[1] *Mansion House Committee on the Unemployed 1903–4. Report of the Executive
Committee*, p. 26, para. 2.

[2] W. H. Beveridge, *Power and Influence*, p. 23: 'I was set to learn about the main
economic problem of those days, not from books, but by interviewing unemployed
applicants for relief, taking up references from former employers, selecting the men
to be helped, and organising the relief work.'

[3] *Report of the Leeds T.U.C.*, 1904, pp. 88, 89–90.

to start public works and guardians to give relief.[1] According to the Chief Commissioner of the Metropolitan Police 'the margin of safety was slight and the strain on the police . . . unduly heavy'[2] Poor Law and municipal authorities in London and the provinces urged the Prime Minister to summon an autumn session of Parliament to deal with the unemployed.[3] Balfour, however, had little sympathy for distressed workmen;[4] and it was not the Prime Minister but Walter Long, the President of the Local Government Board, who was moved to take action on behalf of the unemployed. Long's initiative was rather surprising, since he represented that bucolic section of the Conservative party that was traditionally hostile to increased burdens on the rates.[5] But Long became convinced that the majority of the unemployed were honestly seeking work, and that it was a 'national crime' to turn such men into paupers by refusing to relieve them until they were on the verge of starvation.[6] He also hoped that permanent physical and moral deterioration could be prevented by making better administrative provision for the unemployed.[7]

On 14 October 1904 Long therefore convened a conference of metropolitan guardians,[8] ostensibly to discuss proposals but in fact to outline his own scheme for the unemployed. Long was anxious to find a *via media* which would relieve distress but not commit the Government to a

---

[1] Cd. 5066/1910, Q. 78466.

[2] CAB 37/65/33, Aretas Akers-Douglas on ' "Unemployed" Processions', 22 May 1903.

[3] George Haw, op. cit., pp. 237–40. This demand was endorsed by the T.U.C.'s Parliamentary Committee on 24 Oct. 1904 (*Report of the Hanley T.U.C.*, 1905, p. 60).

[4] About this time he was invited by Sydney Buxton to address a meeting of the unemployed in Tower Hamlets, and begged to be excused the ordeal. 'I am no good in the East End. *Entre nous* I hate the poor—when they struggle!—I like 'em best in workhouses; my sister tells me that in *Prisons* they are *too* delicious—but in Poplar I am sure they are odious. Therefore be merciful to me a sinner and let me off' (Buxton MSS., unsorted, Arthur Balfour to Sydney Buxton, n.d. (probably late 1904/early 1905)).

[5] Walter Long (1854–1924), first Viscount Long of Wraxall; Conservative M.P. 1880–1921; Parliamentary Secretary to L.G.B. 1886–92; President of Board of Agriculture, 1895–1900; President of L.G.B. 1900–5 and 1915–16.

[6] Cd. 5066/1910, Q. 78461, para. 11.

[7] Lord Long of Wraxall, *Memories*, p. 139.

[8] H.L.G. 29/85, vol. 77, ff. 467–8, R. G. Duff to the metropolitan guardians, 6 Oct. 1904.

relaxation of the Poor Law or to any irrevocable obligation
to find work for the unemployed.[1] He was sceptical about
public works and labour colonies as remedies for unemployment, and denied that it was in his power to unify metropolitan administration or to create a single statutory body
responsible for the London unemployed.[2] But he was ready
to make use of existing powers and authorities, and proposed
that throughout London joint committees representing
guardians, borough councils, and charitable associations
should investigate applications for relief and separate the
'respectable, temporary out-of-work men' from the 'ordinary
pauper'.[3] These committees would appoint delegates to a
central committee, which would raise subscriptions and
devise 'a common policy for London'.[4] Employment without
disfranchisement would be available on borough relief works
to temporarily unemployed workmen who had been resident in London for a minimum period of six months. Such
a policy, Long claimed, was 'capable not only of being use-
fully acted on immediately, but also very possibly of considerable extension in the future'.[5]

Long's proposals were a complete reversal of the 'divide
and rule' policy which the Conservatives had previously
pursued in the metropolis;[6] and it was welcomed by the
Liberal leader, Sir Henry Campbell-Bannerman, as a tentative step towards the administrative unification of London.[7]
The first meeting of the new 'Central Committee' was held
at the Guildhall on 25 November 1904. An executive committee was appointed, with sub-committees on public works,
'working colonies', emigration, classification, and finance.[8]

---

[1] Cd. 5066/1910, RC on the Poor Laws, Minutes of Evidence, Q. 78461, para. 8.

[2] H.L.G. 29/85, vol. 77, ff. 475–80, Copy of Mr. Long's concluding speech
at the conference of metropolitan guardians, 14 Oct. 1904, pp. 2–5.

[3] Ibid., pp. 5–6. The structure of the committees and the method of inquiry
were based on a relief scheme in Camberwell during the previous winter (H. R.
Maynard, 'Mr. Long's Proposals', Toynbee Record, Nov. 1904, 25–6).

[4] H.L.G. 29/85, vol. 77, ff. 475–80, p. 8.

[5] Ibid., p. 12.

[6] R. C. K. Ensor, England 1870–1914, pp. 296–7.

[7] Speech at Limehouse, 21 Dec. 1904, on 'Social Reform and Fiscal Policy',
Speeches by Sir Henry Campbell-Bannerman 1899–1908, Selected and Reprinted
from the Times, p. 157.

[8] Report of the Central Executive Committee of the London Unemployed Fund,
1904–5, pp. 11–16.

It was resolved that a uniform policy should be adopted for the whole of London, that assistance should be limited to the provision of work for temporarily unemployed workmen and that 'preference should be given to persons who have established homes with wives and families'.[1] On 12 December the Lord Mayor appealed for funds and raised over £50,000, 'largely owing to Mr. Long's personal efforts'; and a loan of £20,000 at 2 per cent was floated on the stock market in January 1905.[2] Work was provided at Hadleigh and Letchworth Garden City, in the L.C.C. parks and the City of London markets, and in preparing land for a new County Asylum.[3] In October 1904 Joseph Fels offered Long an estate of 1,200 acres at Hollesley Bay in Essex, and this was taken over by the Central Committee in January 1905 with a view to preparing workmen for permanent settlement on the land.[4]

Long had originally proposed that the men relieved should be paid at less than the standard rate, but the L.C.C. objected to this policy and they were therefore employed at standard wages for a reduced number of hours per day.[5] Work places were made available to each borough on the basis of the degree of poverty ascertained by Booth's survey, corrected for recent changes in local population.[6] The Classification Committee planned to divide the applicants into artisans, 'regular unskilled', and casual workmen and to regulate their work accordingly; but in practice it was difficult to impose a standard classification for all areas, since in some boroughs the bulk of the working class were irregularly employed and relief committees were flooded by 'batch after batch . . . of the semi-casual class'.[7]

---

[1] H.L.G. 29/85, vol. 77, f. 484, Statement by the Local Government Board on 'The Unemployed—Mr. Long's Scheme. The Central Committee', 29 Oct. 1904, para. 9 (c).

[2] *Central Executive Committee of the London Unemployed Fund*, 1904–5, Report of Finance sub-committee, pp. 17–23.

[3] Ibid., Report of Works sub-committee, pp. 48–60.

[4] H.L.G. 29/85, vol. 77, f. 487, Joseph Fels to Walter Long, 29 Oct. 1904; f. 148, Walter Long to Joseph Fels, 31 Oct. 1904.

[5] *Central Executive Committee of the London Unemployed Fund*, 1904–5, Report of Works sub-committee, p. 47.

[6] Ibid., Report of Classification sub-committee, p. 28.

[7] Ibid., p. 35.

The Classification Committee tried, moreover, to evolve a method of giving assistance which was wider than the mere mechanical investigation of cases and allocation of employment. Relief committees were encouraged to take account of the domestic situation of the unemployed, to promote 'the virtues of spending wisely and of keeping things clean' and to advise on the choice of employment for children.

It must be the ideal of unemployed administration to see that the offer of employment, or whatever step is recommended, is more than the palliative of the moment; that it leaves men more independent than it found them, their industrial status unimpaired if not improved, and their homes, when these have been characterised by mismanagement, by lack of parental foresight or by acceptance of a low hygienic standard, raised.[1]

With these ends in view, relief was paid directly to the wives of workmen employed in labour colonies rather than to the workmen themselves; and 'wise expenditure' was encouraged by visits from charitable volunteers.[2] A labour exchange was opened in Victoria Street to assist the relief committees in restoring workmen to normal employment.[3] By October 1905 it was estimated that 17,705 persons had been directly relieved by the committees, 17 had been settled in the country, and 215 had been given help in emigration.[4]

[1] *Central Executive Committee of the London Unemployed Fund*, 1904–5, Report of the Classification sub-committee, pp. 44–45.

[2] This system of payment had been initiated by the Mansion House committee of the previous year, which had also graduated relief according to the number of a workman's dependants (*Mansion House Committee on the Unemployed, 1903–4, Report of Executive Committee*, p. 11). The system of direct payment to families was adopted by many distress committees under the Unemployed Workmen Act, which found that it effected a remarkable improvement in the physique of children and home standards (*Report of Stepney Distress Committee*, year ending 30 June 1908, p. 16). A similar method of payment to the wives of servicemen was used during the First World War, and inspired Eleanore Rathbone's campaign for child allowances (Eleanore Rathbone, *Family Allowances* (1949 ed.), pp. 47–9).

[3] *Central Executive Committee of the London Unemployed Fund*, 1904–5, Report of the Classification sub-committee, p. 31.

[4] Ibid., p. 15. These figures included workmen and their dependants; but they did not include those for whom employment was found through the L.U.F.'s Central Employment Exchange, or those employed directly by borough councils on works subsidized by grants in aid from the London Unemployed Fund.

THE UNEMPLOYED WORKMEN ACT 1905

Almost from the start, persons involved in the administration of the London Unemployed Fund were planning to use it as a model for a more ambitious programme. The Lord Mayor, Mr. John Pound, in his plea for subscriptions had announced that 'a question of social and national importance is . . . at stake, for it is permissible to hope that the experience gained by a combined effort on the part of the committee this year may suggest the lines of a more permanent solution of the problem of the unemployed.'[1] It was found that certain aspects of the Central Committee's work—notably the employment exchange and the labour colony at Hollesley Bay—needed a more regular source of income than the spasmodic assistance of charity. Moreover, charitable contributions were more readily available in some areas than others and were in any case confined mainly to periods of emergency. It was therefore desirable to establish the committees on a more permanent basis with a salaried staff of investigators, colony supervisors, and social workers, so that schemes and statistics could be prepared during the summer months, and preparations made in advance for the relief of distress.[2]

Pressure for more permanent measures also came from guardians in distressed areas, from organized labour, from the Liberal opposition, and from London M.P.s. In October and November 1904 several metropolitan boards of guardians 'found themselves unable to cope with the difficulties presented, and from many quarters came allegations that they were face to face with a labour crisis such as had never occurred before'.[3] In December the Parliamentary Committee of the T.U.C., the General Federation of Trade Unions, and the Labour Representation Committee sent to the Prime Minister a report on unemployment, calling for the re-issue of the Chamberlain circular, the prohibition of overtime, the provision of co-operative farms, and invest-

---

[1] *The Times*, 12 Dec. 1904.

[2] H.L.G. 29/85, vol. 77, ff. 577-8, Unsigned typescript memorandum on the Unemployed Workmen Bill, June 1905; G. Haw, op. cit., p. 236.

[3] *The Times*, 21 June 1905, speech by Walter Long.

ment by the Government in 'works of public utility'.[1] On
21 December the Liberal leader, Sir Henry Campbell-
Bannerman, proposed in a speech at Limehouse that financial
responsibility for the unemployed should be made a common charge on the whole of London.[2] And on 11 January
1905 an all-party delegation of metropolitan M.P.s headed
by Sydney Buxton urged Long to convert the Central Committee of the London Unemployed Fund into a statutory
body with power to levy a uniform rate.[3] Buxton also pressed
for the inclusion of boroughs like Tottenham and West
Ham, which were outside the metropolitan boundary but
were too poor to support their own unemployed. Long
indicated that he intended to take action on these lines, and
that similar authorities would be created in other urban
areas, which would be grouped together by the L.G.B. for
the administration of unemployment relief. He hoped that
each unemployment authority would be empowered to levy a
penny rate, though he warned the deputation that many of
his colleagues and the richer London boroughs were opposed to a levy on the rates. Buxton reported to Campbell-
Bannerman several days later that Long's proposals were a
great concession to the campaigns for the equalization of
rates and for the assumption of state responsibility towards
the unemployed.

It is not as yet generally recognised what a tremendous principle is
involved, and being accepted with hardly a murmur; i.e. that it is the
duty of the locality (and therefore logically of the State, which will
certainly be called on to organise and to help, possibly to subsidise, the
localities) to provide *work* . . . for those suffering from want of employment, if they are genuine cases. Personally I am glad: I believe
much distress will be averted, and little or no harm done.[4]

On 24 January Long circulated a memorandum to his
Cabinet colleagues, proposing that the administrative
machinery established under the London Unemployed Fund
should be placed on a permanent footing, and that the London

---

[1] *Report of the Hanley T.U.C.*, 1905, pp. 60–4.
[2] *Speeches by Sir Henry Campbell-Bannerman 1899–1908. Selected and reprinted
from the Times*, pp. 156–7.
[3] Lansbury MSS., vol. 2, section I, f. 95, S. Buxton to G. Lansbury, 12 Jan. 1905.
[4] Add. MS. 41238, ff. 8–12, S. Buxton to Campbell-Bannerman, 16 Jan. 1905.

boroughs should be required to contribute to the scheme out of a uniform penny rate. He admitted that this was tantamount to a measure of redistribution, and that the West End boroughs would probably cavil at having to support the unemployed of the North and East End; but he argued that 'Parliament has already recognised the principle that the wealthier parts of London should assist the poorer, both in Poor Law and sanitary administration, by establishing the Metropolitan Common Poor Fund and by the Equalisation of Rates Act, 1894.' He suggested also that workmen relieved under the new scheme should not be deprived of the right to vote.[1]

A Bill for the 'Organisation of Workmen', creating statutory authorities based on Long's scheme and subsidized from the rates, was drafted in the Local Government Board in February 1905. The measure was to be compulsory in London and other large urban centres, and optional elsewhere. It was proposed that workmen should be paid 'less than that which would be given under ordinary circumstances to an unskilled labourer', and that no one should be employed on relief works for more than sixteen weeks and in more than two successive years.[2] At this stage in the preparation of the Bill, Long definitely envisaged that precepts on local authorities should be levied not only for the management of the scheme but to pay for public employment: '. . . the scheme', he argued, 'would relieve the rates, and hence it does not seem unjust that the ratepayers should make a limited contribution to the necessary expenditure involved.' Long was aware that the use of public money might be open to misinterpretation by advocates of the right to work; but he pointed out that 'if my scheme broke down for lack of funds, as I fear might be the case if the rates are not to be charged, the demand for such state intervention would be greatly strengthened.'[3]

This view was not shared, however, by the richer London

[1] CAB 37/74/17, W. H. Long, 'The Unemployed', 24 Jan. 1905.
[2] H.L.G. 29/85, vol. 77, ff. 493-6, 'The Unemployed; Instructions for Bill', 31 Jan. 1905, ff. 500-3, Second Draft of a Bill to Establish Organisation with a View to the Provision of Employment or Assistance for Unemployed Workmen in Proper Cases, 9 Feb. 1905.
[3] CAB 37/74/31, W. H. Long, 'The Unemployed', 16 Feb. 1905.

boroughs nor by Long's permanent officials nor by his
Cabinet colleagues.[1] 'I am bound to confess that my policy
did not find favour in any quarter' he recalled in his memoirs.
'Among my own friends it was regarded as being too much
akin to socialism, while the more advanced thinkers looked
upon it as incomplete and insufficient'.[2] The Lord Privy
Seal, Lord Salisbury, maintained that if the machinery of
relief was made permanent, every concession to the unemployed evoked by a crisis would become irreversible.

Routine is the servant of precedents, and these precedents will be
progressive in one direction. The variation will always be the other
way, towards enlargement and each enlargement will be immediately
stereotyped as the date point from which the next enlargement will be
conceded.

At present, 'the benefits conferred are still *ex gratia*. The
money is still hard to come by ... no one has yet acquired a
claim.' But once the justification for a penny rate had been
accepted, there would be no logical limit to the liabilities
of the ratepayers and the demands of the unemployed; an
outcome which would be 'disastrous from the point of view
of the working classes, whose besetting temptation is to
believe that their welfare is to be sought in less effort, and
that State intervention should make good the deficiency.'[3]
These criticisms were accepted by the Cabinet; and at the
suggestion of Sir Samuel Provis, the L.G.B.'s Permanent
Secretary, the Bill as introduced into the House of Commons
provided only for rate contributions to the management
expenses of distress committees, the upkeep of labour
colonies, the establishment of labour exchanges, and assistance to migration and emigration.[4]

[1] CAB 41/30/2, Arthur Balfour to the King, 1 Mar. 1905 (report of meeting on
28 Feb.).   [2] Lord Long of Wraxall, *Memories*, p. 139.
 [3] CAB 37/75/44, 'The Unemployed. Mr. Long's Scheme', Cabinet memorandum
by Lord Salisbury, 1 Mar. 1905. (James Edward Gascoyne-Cecil, 4th Marquess
of Salisbury (1861–1947), supporter of Balfour in the tariff-reform controversy,
and subsequently President of the Board of Trade, Mar.–Dec. 1905.)
 [4] H.L.G. 29/85, vol. 77, ff. 526–30, Seventh Draft of Unemployed Workmen
Bill with MS. annotations by S. B. Provis, 29 Mar. 1905, ff. 531–5, Eighth Draft,
30 Mar. 1905.
  Sir Samuel Provis (1865–1926), Permanent Secretary of the L.G.B. 1898–
1910; a member of the Royal Commission on the Poor Laws 1905–6, and of the
Reconstruction Committee of 1918.

The Webbs severely criticized the policy of the Board in giving statutory endorsement to the policy of municipal relief works without first inquiring into the almost universal failure of such relief works over the previous twenty years.[1] This criticism was to a certain extent justified. It was true that the Board made a major miscalculation in framing the Act for an 'élite' of temporarily unemployed workmen—classes 'D' and 'E' of Charles Booth's London survey[2]—when the experience of the London Unemployed Fund and of most municipal relief schemes showed that in practice relief works were flooded with chronically irregular casual labourers. But the administrative structure of the Act was not the invention of the Local Government Board officials; it was rather a formalization of the methods of relief independently developed by local, Poor Law, and charitable authorities. The Board's aim was not so much to endorse those methods nor to create new statutory powers as to repair the chaos into which the system of tripartite responsibility had fallen, and to establish it 'on more stable lines'.[3] As Long had told the metropolitan guardians in 1904:

We want ... no overlapping ... no visitation of the same house by two different representatives of different institutions, so that charity shall not undo one moment the work which a local governing body has tried to do in another.[4]

The Board's officials appear to have had no ambition to extend their responsibility for the relief of unemployment and several years earlier had implied that problems of the labour market should be dealt with by the Board of Trade.[5]

The aim of the Bill is, by enlisting the co-operation of all local governing bodies, to assist the more deserving cases among the unemployed and to prevent those cases from coming upon the poor law,

---

[1] S. and B. Webb, *English Poor Law History*, II. ii. 652.

[2] Cd. 5066/1910, Q. 77738: evidence of Gerald Balfour. (Class D = 'regular workers at low wages'; Class E = 'regular workers at standard rates of payment'.)

[3] H.L.G. 29/85, vol. 77, f. 587, Typescript notes on amendments, n.d.; Cd. 4499/1909, *RC on the Poor Laws Majority Report*, Part VI, para. 451.

[4] H.L.G. 29/85, vol. 77, ff. 475-80, Copy of Mr. Long's concluding speech at the Conference of Metropolitan guardians, 14 Oct. 1904, p. 3.

[5] H. L. G. 29/69, vol. 63, f. 60, Unsigned memorandum on the Labour Bureaux (London) Bill, Feb. 1902. See also below, p. 284.

wrote Provis in his brief for Gerald Balfour on the second
reading of the Bill.

> It is hoped that the various local bodies . . . will by reason of their
> partaking in it, be stimulated to arrange as far as may be the sequence of
> the works which they are themselves called upon to undertake, so that
> such of them as can be performed when other work is slack may be done
> then. Also that by a more complete system of labour exchange, em-
> ployers and employed may be more readily brought together. In this
> way employers in the country may be brought into touch with men in
> towns and some may be assisted to obtain work in the Colonies.[1]

The Bill was introduced into the House of Commons in
April 1905 by Gerald Balfour, who had succeeded Long as
President of the Local Government Board a few weeks
before.[2] It was subsequently denounced by conservatives
and welcomed by radicals and labour representatives as a
recognition of the 'right to work'—which was exactly what
Long in framing the scheme had hoped to avoid.[3] On 19
May, however, a joint meeting of the Labour Representation
Committee, the Parliamentary Committee of the T.U.C.,
and the General Federation of Trade Unions decided that
organized labour could not support the Bill unless it was
made compulsory in all areas and paid for out of the national
exchequer, and 'unless all limitations upon rates of wages
paid, which are a serious menace to trade unions, are removed'.[4] These conditions were submitted to Gerald Balfour
on 25 May by a deputation headed by James Sexton, Isaac
Mitchell, Ramsay Macdonald, and Arthur Henderson. Balfour refused to make the scheme a national charge but pro-

[1] H.L.G. 29/85, vol. 77, f. 588, Typescript memorandum on Unemployed
Workmen Bill, June 1905, almost certainly written by Provis. He conceded that
'the Bill is, in some sense, an experiment . . . and like most experiments, it is open
to the criticism that it is subversive of the principle upon which existing arrangements have proceeded . . .' (ibid., f. 587).

[2] Gerald Balfour (1853–1945), brother of the Prime Minister; Conservative M.P.
or East Leeds 1885–1905; Chief Secretary for Ireland 1895–1900; President of the
Board of Trade 1900–5; President of the Local Government Board, Mar.–Dec.
1905.

[3] *The Times*, 8 Aug. 1905, Speeches by F. Banbury and D. Lloyd George.

[4] *Report of the Hanley T.U.C.*, 1905, pp. 64–5. According to John Burns, who
thought the Bill 'barbaric', it was favoured only by Keir Hardie and by delegates
looking for parliamentary seats (Add. MS. 64323, Burns Diary, 19 and 30 May
1905).

mised to consider amendments for extending the Bill compulsorily to all large towns. He pointed out that standard hourly wages were going to be paid, but that workmen would be employed for less than the full working day in order to maintain the principle of financial deterrence.[1]

On the second reading Labour members supported the Bill, although they continued to criticize its 'less eligibility' aspects, and Keir Hardie urged that the scheme for the metropolis should be extended to outlying areas such as West Ham.[2] The Liberal party welcomed the Bill, although Sydney Buxton pressed for the payment of wages out of the rates, Herbert Samuel for greater powers of compulsory purchase, and John Williams Benn for an exchequer contribution.[3] Nevertheless, 'so far as any Liberal opposition was concerned the Bill was never for a moment in danger of being lost'.[4] The most serious attack came from the Government's own supporters. H. Lawson, newly elected on a xenophobia ticket for Mile End, declared that the Bill 'would tend to decrease rather than increase employment. All attacks on capital discouraged private enterprise, and he attributed much of the shortage of employment in London to municipal socialism, which made people hesitate about investing their money.'[5] Sir George Bartley deplored the creation of a 'new spending authority of the state for the purpose of providing relief' and the omission of specific provision for the prior investigation of the moral character of those relieved.[6] He thought that 'constructively though not textually the Bill gave the right of employment to

[1] *Report of the Hanley T.U.C.*, 1905, pp. 65–6.

[2] West Ham was still outside the metropolitan boundary, and therefore outside the scope of the London Unemployed Fund of 1904/5 and outside the area compulsorily covered by the Act. In 1904 the unemployed of West Ham had been relieved by charitable subscriptions raised by the *Daily Telegraph* and *Daily News* (Cd. 5066/1910, QQ. 78681–4).

[3] H.L.G. 29/85, vol. 77, ff. 585, 590–9, Typescript notes on Amendments to Unemployed Workmen Bill.

[4] *Liberal Magazine*, Sept. 1905, p. 493.

[5] *The Times*, 21 June 1905.

[6] 'Sir George Bartley is afraid that the Bill will undermine the national welfare and the independence of the workmen', commented an L.G.B. official. 'He will probably quote the evils which preceded the introduction of the Poor Law Act of 1834 and will claim that the Measure is the first step on a dangerous course which leads towards socialism' (H.L.G. 29/85, vol. 77, Notes on amendments, f. 587).

those who could not find employment'. But Bartley's amendment condemning relief outside the Poor Law was defeated by 228 votes to 11, and the Bill passed its second reading with only thirteen dissenters on 20 June 1905.[1]

Conservative ministers had been frightened, however, by the radical interpretation that the measure had been given by social reformers, and for nearly three months the Government prevaricated over the final stages of its own Bill. A massive demonstration in favour of the Bill was convened in Hyde Park on 9 July,[2] and Opposition members urged the Government to proceed to the committee stage and third reading in order to 'prevent the breaking of the public peace'.[3] The Prime Minister replied that the crisis in the Scottish Church[4] had legislative priority over a chronic problem like unemployment. Balfour's attitude convinced Labour representatives of the bad faith of Conservative ministers,[5] and played into the hands of Opposition leaders, who were forced into the paradoxical position of defending the Bill against the Government which had introduced it. Sir Charles Dilke warned Balfour that a 'revolutionary situation' would ensue if the Government proved incapable of passing its own measures of social reform;[6] and strong pressure from Crooks and Keir Hardie, Buxton and Campbell-Bannerman appeared to dissuade the Prime Minister from entirely abandoning the Bill.[7] Early in August he announced that its operation would be limited to three years, pending the inquiries of a Royal

[1] *The Times*, 21 June 1905.

[2] *Hansard*, 4th series, vol. 149, col. 853. According to George Lansbury, the demonstrations in favour of the Bill were financed by Joseph Fels (Lansbury MSS., vol. 7, f. 162, G. Lansbury to M. Fels, 30 June 1914).

[3] *Hansard*, 4th series, vol. 149, cols. 410–12; vol. 148, cols. 475–6, 791–2.

[4] i.e. over the division of property and educational endowments between the Free Church and United Free Church of Scotland.

[5] *Hansard*, 4th series, vol. 150, cols. 355–7. 'If a crisis is necessary,' remarked Keir Hardie, 'I can promise the right hon. Gentlemen that there will be one on this question in the winter.'

[6] Ibid., col. 972.

[7] Ibid., cols. 961, 984–7, 1013–18. Lansbury recalled that it was a speech by Chamberlain that tipped the scales in persuading Balfour to forge ahead with the Bill (Lansbury MSS., vol. 7, f. 162, G. Lansbury to M. Fels, 30 June 1914). But Chamberlain made no speech at this stage of the Bill. Lansbury may have been referring to Churchill, whose last speech as a Conservative was a slashing attack on the Government's negative record of social reform (*Hansard*, 4th series, vol. 150, cols. 996–1001).

Commission on the Poor Laws, which would report on 'everything which appertains to . . . the problem of the poor, whether poor by their own fault or by temporary lack of employment'.[1] The Bill was committed on 4 August, and amended to provide for the inclusion on local distress committees of women and 'persons experienced in the relief of distress'. Provision was made for the keeping of separate accounts for contributions from the rates and for charitable subscriptions.[2] The Bill was hustled through the third reading and the House of Lords with little more opposition before the recess of August 1905.[3]

MACHINERY AND FINANCE

As a system of social administration the representative committees created by the Unemployed Workmen Act were doomed to failure from the outset by internal and external political factors and by the statutory limitations on their powers.[4] The Act was initially passed as an interim measure for three years only, and before its machinery was under way the Royal Commission on the Poor Laws had been appointed. The Conservative government which framed the Act had little positive enthusiasm for its aims. Walter Long and Gerald Balfour saw it as a useful safeguard against more sweeping reforms while the Poor Law was being overhauled; but almost immediately the Government was under pressure to extend or amend its provisions. At the Trades Union Congress in September 1905 the President, James Sexton attacked the Act as an 'abortion born of political expediency and desperation', liable to depress wages and to promote the recruitment of blackleg labour.[5] In October a deputation of the unemployed from Poplar was received by King Edward at the ceremonial opening of the Aldwych–Kingsway junction;[6] but the Prime Minister refused to hold an autumn session of Parliament to introduce

---

[1] *Hansard*, 4th series, vol. 150, col. 13148.
[2] *The Times*, 5 Aug. 1905.   [3] Ibid., 10 and 11 Aug. 1905.
[4] Cd. 4499/1909, *RC on the Poor Laws*, Memorandum by Mr. T. Hancock Nunn in regard to Unemployment, p. 712.
[5] *Report of the Hanley T.U.C.*, 1905, pp. 48–9.
[6] *Daily News*, 10 Oct. 1905.

new legislation on the unemployed.[1] Lansbury's request that a delegation of unemployed women should be received by the Queen was denied as 'contrary to all custom and quite impossible'.[2] But on 6 November over 3,000 working women marched from Walworth and Poplar to Westminster in silent demonstration of the plight of the unemployed.[3]

These demonstrations were a marked contrast to the disturbances which had terrified the West End twenty years before and troubled the East End guardians in 1903. The organizers were careful to refrain from 'anti-loyal explosions' and 'anything of antagonism and menace'.[4] The Reverend Herbert Stead, who had negotiated with the Royal Family on Lansbury's behalf, warned him to avoid any action that would alienate the churches and the middle and upper classes. 'The moment there is any sign of the unemployed being exploited by extremists as a means of propagating Republican or Socialist opinions there will be a shutting of the heart against the unemployed.'[5] The women's demonstration was by no means ineffective.[6] The Prime Minister refused to contemplate an extension of public contributions to the Act; but he urged his supporters to contribute generously to the fund opened in the name of Queen Alexandra.[7] 'He pledged himself, as head of the Government to do his utmost to see that requisite funds were forthcoming from private charity' wrote Stead triumphantly to George Lansbury.

The logic of the situation compels him to the same result. If the Mansion House Fund is a 'fizzle', the absurdity of the Unemployed Workmen's Act as it now stands will become too painfully transparent. I admit that herein your tactics have been successful. By insisting on

---

[1] Lansbury MSS., vol. 2, f. 118, A. J. Balfour to the clerk of Poplar Borough Council, 20 Oct. 1905.

[2] Ibid., f. 120, M. Chalmers to G. Lansbury, 21 Oct. 1905.

[3] The Times, 7 Nov. 1905.

[4] Lansbury MSS., vol. 2, ff. 122–4, Revd. F. H. Stead to G. Lansbury, 24 Oct. 1905. Francis Herbert Stead (1857–1928), a Congregationalist minister, religious journalist, and brother of W. T. Stead; Warden of the Browning Settlement in Walworth Road; a pacifist, and promoter of the National Committee on Old Age Pensions.

[5] Ibid., f. 139, Revd. F. H. Stead to G. Lansbury, 8 Nov. 1905.

[6] Ibid., vol. 29, ff. 4–5, Revd. Russell Wakefield to G. Lansbury, n.d.

[7] The Times, 7 Nov. 1905.

State help, you have made him more than ever anxious that private charity should be copious.[1]

Sydney Buxton, the spokesman of the London radicals, observed that Balfour's statement was 'really a . . . Treasury speech against his own Bill. The step of an Unemployed Act having been taken, the further step of having some public funds available cannot be resisted for long.'[2]

Nevertheless, although radicals welcomed the measure as an admission of principle and as a useful channel of temporary relief, the Act commanded little support or enthusiasm in the Liberal government which took office at the end of the year. 'The idea of having to administer the unemployed act is a nightmare, enough to kill one' commented Edmond Fitzmaurice to the new Prime Minister, Sir Henry Campbell-Bannerman;[3] and John Burns, the new President of the Local Government Board, was actively hostile to all schemes for 'artificial' employment.[4] Moreover, since the end of 1904 Liberal politicians had been deliberating about the relief of unemployment along lines very different from those laid down by the Unemployed Workmen Act;[5] and several members of the new Cabinet, notably Buxton and Lord Ripon, were definitely committed to its amendment or repeal.[6]

Almost from the start of its operations, therefore, the Act was known to be impermanent and to lack the positive support of the central government. But persons involved in the actual administration of the Act were scarcely more enthusiastic than the Liberal cabinet about the new system of 'employment relief'. John Burns recorded that the chairman of the newly appointed Central (Unemployed) Body for London 'shares my view of the act—cause and consequence—apprehends *now* its ultimate failure . . .';[7] and

[1] Lansbury MSS., vol. 2, f. 139, Revd. F. H. Stead to G. Lansbury, 8 Nov. 1905.

[2] Buxton MSS., unsorted, Sydney Buxton to Charles Buxton, 7 Nov. 1905.

[3] Add. MS. 41214, ff. 205–6, Edmond Fitzmaurice to H. Campbell-Bannerman, 13 Dec. 1905.

[4] Add. MS. 46324, Burns Diary, 12 May 1906.   [5] Below, pp. 219–24.

[6] Buxton MSS., unsorted, Lord Ripon to Sidney Buxton, 28 May 1906; Add. MS. 46299, f. 39, Lord Ripon to John Burns, 29 Jan. 1906.

[7] Add. MS. 46324, Burns Diary, 2 June 1906.

administrators in London were pressing for an amendment of the Act as early as the spring of 1906.[1] Labour representatives on distress committees were primarily interested in demonstrating the shortcomings of the Act and using it as a stepping-stone to more radical reforms. Members of charitable organizations on the other hand were concerned to prevent the 'dangerous' potential of the Act from being fulfilled; whilst individuals like Beveridge and Lansbury saw the machinery of the Act mainly as a useful arena for the furtherance of their own ideas of reform. Nobody was prepared to defend the Act as an end in itself, as a permanent feature of social administration;[2] and it is within this context that the working of the Act must be discussed, from the point of view of organization, public finance, and policy formation.

The administrative machinery created by the Unemployed Workmen Act was based on two existing networks of local administration—borough councils and boards of guardians—and in this respect differed significantly from subsequent experiments in social legislation for the unemployed, which circumvented local authorities and worked through specially constituted central and regional authorities. Even so, the administration of the Act was not directly representative; it foreshadowed the view expressed by the Minority Report of the Royal Commission on the Poor Laws that authorities immediately subject to popular control could not be trusted to administer social services with rigorous impartiality.[3]

The Act laid down that 'distress committees' should be established in all metropolitan boroughs, and in all provincial boroughs and urban districts with a population of not less than fifty thousand. Elsewhere distress committees could be established at the discretion of the L.G.B.[4] In the prov-

---

[1] C.U.B. Minutes, i. 108, 4 May 1906.

[2] Hansard, 4th series, vol. 161, col. 426. The Act was also condemned by the Central Poor Law Conference and the Municipal Corporations Association.

[3] 5 Edw. 7, c. 18, Sections 1 (1) and 2 (1).

[4] Ibid., Section 2 (2). County councils and county borough councils could also execute the Act in the absence of a distress committee (Section 2 (3)).

inces distress committees were to be responsible for both the registration of applicants and the provision of work; but in London distress committees would merely register and investigate cases of distress, and the actual administration of relief would be the responsibility of a 'central body', which would also supervise and co-ordinate the work of distress committees in the twenty-nine London boroughs.[1] The Act was primarily designed to meet the situation in the metropolitan labour market;[2] and since it was in London that most experiments and most administrative developments took place, it is the experience of the working of the Act in London that will mainly be considered here.

The administration for London was based on that of the Mansion House scheme of 1903–4 and the London Unemployed Fund of 1904–5. 'Distress committees' were set up in every London borough in September and October 1905, consisting of nominees of local guardians, borough councils, and charitable organizations; and in November a Central (Unemployed) Body was convened, with two representatives from every distress committee, nominees from the L.C.C. and the Local Government Board, and seven co-opted members.[3] The Reverend Russell Wakefield, the Christian Socialist vicar of St. Mary's, Bryanston Square, was elected chairman.[4] The change of government at the beginning of December 1905 caused some delay in the appointment of government representatives, but on December 8th the L.G.B. nominated Alderman Alliston of the L.C.C., James Ramsay Macdonald of the Labour Representation Committee, C. Waley Cohen of the Jewish Board of Guardians, and Mrs. May Tennant, a retired factory inspector and sister-in-law of Mr. Asquith, the new

[1] Ibid., Section 1 (4, 5, 6).

[2] Twenty-nine distress committees were established in London and eighty-five elsewhere in 1905–6. Ten committees were operating in Scotland and thirty-six in Ireland during 1907–8.

[3] Meetings were held in the Guildhall, at first weekly and then fortnightly. The minutes and reports of the C.U.B. are in Beveridge MSS. (Coll. B), vols. ix, xi–xiii, xviii–xx.

[4] Henry Russell Wakefield (1854–1933), rector of St. Mary's 1894–1909, Bishop of Birmingham 1911–24; a member of the RC on the Poor Laws 1905–9; President of the National Council of Public Morals, and subsequently President of the Christian Counter-Communist Crusade.

Chancellor of the Exchequer.[1] Sub-committees were set
up to deal with Finance, Classification, Emigration, Works,
Working Colonies, Employment Exchanges, and Women's
Workrooms.[2]

The Body suffered from all the weaknesses of a 'federal'
organization with none of the compensating advantages. The
attempt to make it representative of all different shades of
social opinion meant that it was too large and too diverse
to be either efficient or single-minded. But at the same time
the members of the Body were so remote from popular
control that their disagreements did not necessarily reflect
any substantial conflict of opinion in the localities they were
supposed to represent. Working men and trade unionists
complained that they were under-represented on distress committees and that the administration of the Act was 'swamped'
with clergymen.[3] Middle-class members of the Central (Unemployed) Body complained on the other hand that they
were 'bullied' by representatives of Poplar and Woolwich.[4]

This weakness was reinforced by the shortage of full-
time professional assistance on distress committees and on
the C.U.B. The C.U.B. inherited a staff of clerks and a
secretary, H. R. Maynard, from the London Unemployed
Fund, but their services were retained on a merely temporary
basis after December 1905.[5] In the middle of 1906 a
special committee of the C.U.B. reported that staffing
arrangements were quite inadequate, and that the pressure
of work made specialization between committees and the
devolution of responsibilities quite impossible.[6] The absence

---

[1] *C.U.B. Minutes*, i. 11, 8 Dec. 1905. Charles Booth and the Bishop of Stepney
were added to the L.G.B.'s nominees early in 1906. The L.C.C. nominated J.
Williams Benn, W. C. Steadman, Edmund Harvey, and Major W. Houghton-
Gastrell. Members co-opted by the C.U.B. included Beveridge, Will Crooks, and
Leonard Cohen.

[2] Ibid., 1 Dec. 1905, unpaginated. The chairman was *ex officio* a member
of all sub-committees; and any member of the C.U.B. could attend the meetings
of a sub-committee of which he was not a voting member.

[3] *Report of the Bath T.U.C.*, 1907, p. 141. Add. MS. 46300, f. 204, Philip Snow-
den to John Burns, 23 Mar. 1909.

[4] Add. MS. 46324, Burns Diary, 13 and 15 Oct. 1906.

[5] *C.U.B. Minutes*, i. 2, 23 Nov. 1905; ibid., p. 22, 22 Dec. 1905.

[6] 'Report of a Special Committee of Chairmen of Standing Committees upon
Office Staff and Organisation', submitted to the C.U.B., June 1906, *C.U.B.
Minutes*, i. 168–76, 15 June 1906.

of expert advice on technical matters was hindering the execution of relief schemes, and practical questions which should have been decided on the spot by the superintendents of relief works were being constantly referred back to the committees of the C.U.B.[1] Moreover, 'the understaffing of the Classification Department made it impossible to provide that close and constant pressure upon the district committees by which alone the vacancies could have been filled after the works were started.' The special committee therefore recommended that a permanent staff on a progressive salary scale should be appointed. The work of this staff was to be divided into specialist departments, corresponding to the different sub-committees of the C.U.B., and supplementary junior staff would be engaged during busy periods.[2] These proposals, if adopted in the summer of 1906, might have streamlined the work of the C.U.B. and given it a more permanent role in metropolitan social administration. But a detailed plan for a rational division of labour within the C.U.B. and the creation of a career structure for its employees was not finally ratified until July 1907, when discontent with the Unemployed Workmen Act was almost universal and the possibility that its administrative machinery might be used to carry out more extensive measures of assistance had virtually disappeared.[3]

Moreover, the relationship between the C.U.B. and the local distress committees was nebulous and in some local areas it was difficult to establish central control. Members of the C.U.B. were not merely delegates of local committees. They were supposed to exercise the supervisory powers and to impose the administrative norms prescribed in the regula-

[1] Beveridge in his evidence to the RC on the Poor Laws, 14 Oct. 1907, deplored the fact that members of the C.U.B. and of distress committees were chosen 'on the ground of their knowledge of poverty' and had little or no experience of problems of the labour market and the management of contracts (Cd. 5066/1910, Q. 77832, para. 44). See also Cd. 4944/1909, *Replies by Distress Committees on the Subject of the Unemployed Workmen Act*, 1905, p. 9 [15], statement of George Lansbury.

[2] *C.U.B. Minutes*, i. 169–75, 15 June 1906.

[3] Ibid., ii. 274–7, Report of the Finance Committee, 19 July 1907. This plan provided for a staff of thirty-six full-time officials, at an aggregate cost of £4,623. 16s. 0d. per annum. The chief official would be the clerk of the C.U.B. at a salary of £350 per annum. Even so, posts could not be made permanent because of the uncertain future of the Act.

tions issued by the L.G.B.[1] But the distress committees
came into operation before the Central (Unemployed)
Body;[2] and in several boroughs joint committees of councillors and guardians had been relieving the unemployed for at
least two years before the Unemployed Workmen Act
gave them a statutory function.[3] By the middle of 1907 a
majority of distress committees were opposed to the continuation of the Act without drastic amendments;[4] and the
committees of Poplar, St. Pancras, Islington, and Woolwich exerted continuous pressure for the transfer of responsibility to the central government and a statutory recognition
of the 'right to work'.[5]

In the day-to-day administration of the Act, local representatives fretted against the authority of the C.U.B., and
urged that the L.G.B. regulations should be modified to
enable distress committees to give 'more immediate relief
and work to the unemployed'.[6] Often they had evolved
methods of giving assistance which were incompatible with
the principles laid down by the C.U.B. and the L.G.B.
The C.U.B. decided that preference on relief works should
be given to the most 'deserving' workmen, or those who
could show evidence of regular employment; but nearly all

---

[1] Statutory Rules and Orders, 1905, No. 1071 (10 Oct. 1905), *The Regulations*
(*Organisation for the Unemployed*). These rules prescribed that all applicants for
relief should be visited in their homes; that 'good character' should be verified by
reference to previous employers; that applicants should not have received Poor
Relief in the previous twelve months, nor relief under the Unemployed Workmen
Act in more than two successive years; and that their resources should be insufficient to maintain themselves and their dependants.

[2] *Preliminary Report of the C.U.B.*, 12 May 1906, pp. 4, 14–15.

[3] e.g. Camberwell had had such a committee since 1903 (*Report of Borough of
Camberwell Unemployed Central Committee*, 1903–4). Bethnal Green, Finsbury,
and Poplar had established 'joint committees' in 1904.

[4] *Views of distress committees as to whether the Unemployed Workmen Act 'has . . .
been so far effective as to justify its renewal'.*

Distress Committees	Yes	No	Only if Amended	Non-Committal	Total
London	5	8	13	2	28
Provinces	23	15	28	23	89
Total	28	23	41	25	117

Based on replies to questionnaire issued by *RC on Poor Laws*, 10 May 1907 (Cd.
4944/1909), pp. 1–7, 27–32).

[5] *C.U.B. Minutes*, i. 145, 1 June 1906.   [6] Ibid., p. 50, 2 Feb. 1906.

distress committees gave priority to workmen with the largest number of dependants or to most urgent cases of need.[1] Moreover, the term 'distress committee' was itself a misnomer, since the terms of the Act applied only to regular workmen temporarily unemployed, and the Treasury and L.G.B. insisted that it was designed to relieve 'unemployment' and not chronic 'distress'.[2] But in areas where casual labour predominated this rule proved almost impossible to apply; and employment on relief works became virtually indistinguishable from the other forms of assistance with which the casual labouring class eked out its precarious way of life.[3]

Within the uneasy hierarchy laid down by the Act the actual process of relief in the metropolitan area was extremely cumbersome.[4] The principles of investigation, scales of remuneration, and the allocation of funds were decided by the Classification committee, in accordance with the regulations issued by the L.G.B. Places on relief works were allotted to each borough in the proportions adopted for the London Unemployed Fund, subject to periodical adjustment as local needs became more apparent.[5] The initial investigation of the circumstances of applicants was made by the distress committees under the supervision of the

[1] *Report on the Work and Proceedings of Distress Committees in London from their Constitution to 30 June 1906*; prepared by the Classification Committee of the C.U.B., Jan. 1907; *3rd Report of the C.U.B.* (1907–9), Report of the Classification Committee, p. 19.

[2] T. 1/10740A/22327, Sir George Murray to H. H. Asquith and W. Runciman, 21 Dec. 1907. This interpretation of the Act caused much misunderstanding and resentment in Ireland, where unemployed distress was predominantly both casual and chronic (ibid., A. R. Barlas (Secretary of the Irish L.G.B.) to the Under-Secretary, Dublin Castle, 17 Dec. 1907, and Sir George Murray's annotations).

[3] *Report of Stepney Distress Committee*, 1906–7. The committee remarked that it could never find enough men of the class for whom the Act was intended to fill its share of places on the C.U.B.'s relief works; and the distress committee therefore tended to become merely another centre of casual employment. In the previous year the Stepney committee had reported that its clients were drawn from 'drunkards, wife-beaters, thieves and burglars'. The Fulham distress committee reported that the distinction between 'free labour' and 'pauperism' was artificial and that the Act merely accelerated the process of pauperization (*Report of the Fulham District Committee*, 1906–7).

[4] Cd. 5066/1910, Q. 77832, para. 42.

[5] *Preliminary Report of the C.U.B.*, to 12 May 1906, pp. 15–16, Report of the Classification Committee. Allotments ranged from 6·5 per cent for Islington, Poplar, and Stepney to 0·5 per cent for Stoke Newington and the City of London.

Classification committee of the C.U.B.; and successful candidates were then referred to the Works Committee or Working Colonies Committee for temporary employment.[1]

The procedure of investigation generated much antagonism between representatives of organized labour and members of the COS. The method of inquiry laid down by the Local Government Board was directly based on the 'casework' system which had been developed by organized charity;[2] and the COS was accused of dominating the machinery of the Unemployed Workmen Act much as it had dominated earlier measures of unemployment relief.[3] Nevertheless, the COS had initially condemned the Act as a 'new pseudo-industrial system of remuneration', which concealed the economic dependence of the workmen relieved.[4] In a series of letters to *The Times*, C. S. Loch deplored the granting of relief without disfranchisement, the payment of standard hourly wages, and the limited levy on local rates, all of which were likely to be used as a precedent for extracting further concessions from 'the social party now in power'. He was sceptical also about the 'alleged safeguard' of preliminary casework, which he thought local distress committees would have neither the desire nor the experience nor the courage to enforce.[5]

In December 1905 an 'unemployed' deputation persuaded Burns and Campbell-Bannerman to modify the 'inquisitorial'

[1] *Preliminary Report of the C.U.B.*, to 12 May 1906, pp. 3–4.

[2] On 7 Aug. 1905 Lloyd George had described the machinery of the Act to the House of Commons as 'a statutory Charity Organisation Committee' (*The Times*, 8 Aug. 1908). Keir Hardie complained that 'every line [of the L.G.B. regulations] has COS stamped across its face' (J. Keir Hardie, *John Bull and His Unemployed*, I.L.P. pamphlet 1905, p. 10).

[3] Add. MS. 46299, f. 344, Margaret Moore to John Burns, 7 Oct. 1907. This influence was probably exaggerated, however, since I have not been able to ascertain that more than 13 of the 82 members of the C.U.B. in 1905–6 were members of the COS. Details of COS membership of distress committees under the C.U.B. do not exist, but COS membership of distress committees under the London Unemployed Fund of 1904–5 was as follows: Chelsea 5, Hampstead 5, Lambeth 4, Finsbury 3, Kensington 2, Woolwich 2, Paddington 1, Islington 1, Poplar 2, elsewhere 0 ('Last Year's Unemployed', *Charity Organisation Review*, N.S. 19 (1906), 61–84). Most distress committees contained 20 to 40 members.

[4] *The Times*, 17 May 1905.

[5] Ibid., 9 and 17 May 1905; C. S. Loch, *Employment Relief* (COS Occasional Paper, No. 23, Fourth Series), p. 5.

system of investigation prescribed under Gerald Balfour;[1] but, even so, Loch's misgivings proved to be well founded, and many local distress committees neglected to conduct exhaustive casework or to enforce the regulations prescribed by the L.G.B.[2] The Stepney distress committee declared that the 'follow-up' of cases relieved was 'undesirable' and impracticable;[3] and the distress committee of Kensington came to the conclusion that thorough casework was pointless, since it merely raised false expectations in the minds of the unemployed.[4] Even in Chelsea, where the influence of the COS was unusually strong, the committee reported that they could 'only regard their operations as a waste of time, work and money', from which the 'beneficial results' were 'ridiculously small'.[5] The first report of the C.U.B. in May 1906 showed that, in a sample of cases over which correspondence with distress committees had arisen, 15 per cent had received poor relief in the previous year, 4 per cent had not been resident in London for twelve months, and in 55 per cent of cases insufficient inquiry had been made.[6] Cyril Jackson and J. C. Pringle reported to the Royal Commission on the Poor Laws that many distress committees were anxious to conduct proper inquiries but did not know how to do so;[7] and certainly the rigour of investigation varied enormously from borough to borough, the proportion of cases accepted by distress committees that were subsequently rejected by the C.U.B. ranging from 1 per cent

[1] Add. MS. 46323, Burns Diary, 13 Dec. 1905; Add. MS. 46299, ff. 25–34, draft of a letter from John Burns to the Editor of *The Times*, 24 Jan. 1906; *Report of the Liverpool T.U.C.*, 1906, pp. 63–6. The main alteration brought about by this deputation was the partial depersonalization of the 'inquiry papers' used by distress committees. Reference was henceforth omitted to arrears of rent, membership of provident institutions, and details of employment during the previous five years. The disqualification of applicants who had received poor relief or who had been relieved by distress committees in two successive years was not lifted until 1908 (L.G.B. Order No. 53056, 17 Nov. 1908).

[2] Cd. 4499/1909, *RC on the Poor Laws, Majority Report*, Part VI, paras. 444–5; Cd. 5066/1910, *RC on the Poor Laws, Minutes of Evidence*, Q. 78704, para. 6.

[3] *C.U.B. Minutes*, i. 146, 1 June 1906.   [4] Ibid. ii. 356, 15 Nov. 1907.

[5] *Report of the Chelsea Distress Committee*, 1906–7.

[6] *Preliminary Report of the C.U.B.*, to 12 May 1906, Report of the Classification Committee, p. 24.

[7] Cd. 4795/1909, *Report on the Effects of Employment or Assistance given to the 'Unemployed' since 1886 as a means of Relieving Distress outside the Poor Law*, pp. 67–8.

in Battersea to 33 per cent in Poplar and 75 per cent in Stepney.[1] Bad casework of this kind was, according to the COS, 'worse than useless because it causes needless irritation, and because it has a sort of plausibility which disguises the real issues'.[2] Not until 1909 could the C.U.B. report that the principles and practice of casework had been standardized throughout the metropolis;[3] and even so the labour and expense of casework was out of all proportion to the number of workmen relieved.[4] 'I am horribly disgusted at the local expenditure all over London on the purposes of the Act' wrote Russell Wakefield to Burns in November 1906.

In 28 districts an average of £300 per Borough is being spent in enquiries, classification etc., and then only a few get work. This is *not* the fault of the people, it is the fault of the Act. If all this work were done from *one* centre, the expense would be half and the methods uniform, whereas now it is muddle. . . .[5]

One of the motives behind the Unemployed Workmen Act had been to place the planning and execution of relief schemes on a sound financial footing; but Walter Long's original intention to subsidize relief works out of the rates had been frustrated by his Cabinet colleagues. The Act provided, however, for the levy of a halfpenny or—with special permission from the L.G.B.—a penny rate for labour exchanges, emigration, management expenses, and the aquisition of land. In November 1905 a national unemployed fund to raise charitable subscriptions was opened in the name of Queen Alexandra, and in the following winter £154,000 was raised of which £45,000 was allocated to the Central (Unemployed) Body for London.[6]

---

[1] Cd. 5066/1910, *RC on the Poor Laws, Minutes of Evidence*, Q. 77832, para. 43.

[2] Ibid., Q. 78704, para. 6, Statement of William Bailward, Chairman of the Bethnal Green Committee of the COS.

[3] *3rd Report of the C.U.B.* (1907–9), Report of the Classification Committee, pp. 19–20.

[4] The average cost of investigation incurred by distress committees was £2. 8s. 6d. for every place allotted on relief works (*C.U.B. Minutes*, ii. table between pp. 270 and 271, 5 July 1907).

[5] Add. MS. 46299, f. 131, Russell Wakefield to John Burns, 11 Nov. 1906.

[6] W. H. Beveridge, 'Emergency Funds for Relief of the Unemployed: A Note on their Historical Development', *Clare Market Review*, i, no. 3 (May 1906), 77.

Outside London, local distress committees were responsible both for the arrangement of relief works and for the management of charitable funds.[1] But in London income and expenditure under the Act were supervised by the Finance Committee of the C.U.B., under the chairmanship of Sir Edward Brabrook, the ex-Registrar of Friendly Societies and a member of the Council of the COS.[2] This committee allocated funds between distress committees, controlled the C.U.B.'s spending departments and kept two separate accounts for voluntary subscriptions and contributions out of the rates.[3]

The process by which subsidies were given to relief works was complicated and expensive. After investigation by the Classification Committee, approved workmen were referred to the Works Committee to be employed on public works. But the Works Committee had no power to give employment directly; it merely negotiated with bodies like the L.C.C., the borough councils, and the Office of Works to employ workmen on its behalf.[4] The C.U.B. was officially the employer of such workmen and paid their wages; and the Finance Committee was supposed to recover from the authority which actually gave the employment the value of the work done. This policy caused much friction between the employing bodies and the C.U.B., because no prior arrangement was made for an independent assessment of the value of the work. In January 1906 the C.U.B. agreed not to demand 'full recoupment',[5] but it was difficult to exact even token repayment for work done by the unemployed. The Finance Committee estimated that workmen employed

---

[1] Eighty-five distress committees had been established in the provinces by Mar. 1906, although no 'central bodies' were set up outside London (H. of C. 392/1906, *Return of the Proceedings of Distress Committees*, up to 31 Mar. 1906). In ten county boroughs—Bath, Blackpool, Canterbury, Chester, Exeter, Gloucester, Lincoln, Oxford, Southport, Worcester—where there was no distress committee, 'special committees' were set up under Section 2 (3) of the Unemployed Workmen Act; these committees were mainly concerned with the organization of labour exchanges (H. of C. 173/1908, *Return of the Proceedings of Distress Committees under the Unemployed Workmen Act*, during the year ending 1 Mar. 1908).

[2] Edward William Brabrook (1839–1930) Chief Registrar of Friendly Societies (1891–1904) and a leading opponent of national Old Age Pensions.

[3] *Preliminary Report of the C.U.B.*, to 12 May 1906, Report of the Finance Committee, pp. 11–12.

[4] Ibid., pp. 27–33.   [5] *C.U.B. Minutes*, i. 47, 19 Jan. 1906.

on relief works were 75 per cent as efficient as the average unskilled labourer; but the L.C.C. claimed that work performed by the unemployed was worth only one-fifth of that done by ordinary workmen, and the Office of Works would pay nothing at all for the unemployed.[1] The C.U.B. was in a bad position to bargain over recoupment, since it had great difficulty in finding employment which was suitable for all kinds of labourer and which conformed to the regulations under the Unemployed Workmen Act.[2] 'The present position is dangerous largely because wage and work are not fitting in one with the other,' reported Wakefield in 1907, '. . . the difficulty is less one of money than of work.'[3] A further problem arose from the disparity of rateable values in the different London boroughs. The creation of a central rate-contribution fund, and its allocation on a basis of need rather than population, was intended as a measure of financial redistribution. But even so, the halfpenny rate bore far more heavily on some boroughs than others; and some of the poorer boroughs, notably Poplar, St. Pancras, Islington, and Woolwich, continually pressed to have this charge transferred to the national exchequer.[4]

The decision in July 1906 to make a parliamentary grant of £200,000 for the administration of the Act was largely a political manœuvre, designed to enable the Liberal government to postpone the fulfilment of its promise to amend the Act until after the Royal Commission on the Poor Laws had reported.[5] This grant was to be allocated by the Treasury between the three Local Government Boards of England and Wales, Scotland, and Ireland, and the details of expenditure were subject to Treasury approval.[6] In each of the

[1] Beveridge MSS., Coll. B, vol. iv, item 35, Extracts from the 2nd Report of the C.U.B., with comments by Russell Wakefield, May 1908, para. 3.

[2] This problem was increased by the decision of the C.U.B. not to give financial assistance to local authorities merely to bring forward works which would be paid for out of the rates in the normal course of events (*C.U.B. Minutes*, i. 29, 5 Jan. 1906). [3] *2nd Report of the C.U.B.* (1906–7), p. 62, paras. 2 and 10.

[4] *C.U.B. Minutes*, i. 67, 6 March 1906; p. 145, 1 June 1906; p. 304, 19 Oct. 1906.

[5] CAB 41/30/69, 13 July 1906.

[6] This control was only rigorously exercised in Scotland and Ireland, where the Treasury suspected that the grant was being used for purposes not authorized by the Unemployed Workmen Act (T. 1/10560/22606/19330, E. W. Hamilton to the Secretary of the L.G.B., 17 Nov. 1906; T. 1/11149/24652/5556, C. Hobhouse to A. Birrell, 8 Mar. 1909).

five years for which the grant was made, the English L.G.B. got a disproportionately large share of the total grant[1] and London and West Ham got a disproportionately large share of the English grant.[2] It was, however, as difficult to find suitable openings for the expenditure of the parliamentary grant as for voluntary funds; and in April 1907 John Burns handed back to the Treasury over half of the English share of the grant.[3] Burns's action was widely condemned by labour leaders, who were demanding a parliamentary subsidy for a national system of public works.[4] But such a system could not be developed within the framework of regulations imposed under the Act. The parliamentary grant was supposed to supplement the system of charitable contributions, and to be allocated to distress committees in proportion to the funds that had been locally raised. But subscriptions to the Queen's Unemployed Fund were dwindling soon after the initial publicity given to the Act had subsided. 'Indeed, the day of such funds is probably over—for better or for worse,' commented William Beveridge in the *Clare Market Review* in May 1906.[5] Once a parliamentary grant had been issued, the charitable public declined to subscribe voluntarily to a scheme for which they were being compulsorily taxed.[6] Moreover, many local authorities which received a share of the parliamentary grant neglected or refused to levy the local rate. In July 1907 the borough council at Bethnal Green decided to

---

[1] Allocation of the parliamentary grant:

	England £	Scotland £	Ireland £
1906–7	120,000	10,000	11,000
1907–8	140,000	19,448	4,500
1908–9	227,000	47,253	13,750

(T. 1/11149/14652/5588 and 19493).

[2] T. 1/11149/24652, J. Paterson (clerk to the Glasgow distress committee) to the Local Government Board, 10 Dec. 1909.

[3] *Hansard*, 4th series, vol. 179, cols. 1833–4.

[4] *C.U.B. Minutes*, ii. 160, 19 Apr. 1907, letter from the Town Clerk of Woolwich; *Report of the Bath T.U.C.*, 1907, pp. 140–1.

[5] W. H. Beveridge, 'Emergency Funds for the Relief of the Unemployed: A Note on their Historical Development', *Clare Market Review*, 1, no. 3 (May 1906), 78.

[6] *3rd Report of the C.U.B.* (1907–9), Report of the Finance Committee, p. 10. Only £210. 9s. 6d. had been voluntarily subscribed during the two years under review.

discontinue payments out of the rates as a protest against 'the futility of the Central Body's powers and operations';[1] and between 1907 and 1909 less than £70,000 was contributed out of the rates in the whole of London.[2] In many cases the grant became not merely a subsidy to charitable subscriptions, but a disguised form of imperial relief to local taxation.

SOME POLICY ALTERNATIVES 1905–1908

The most serious conflicts within the Central Body arose, however, not over casework or over methods of raising money but over policies for the provision of employment. Under the terms of the Act, workmen referred to the Central Body could be employed on relief works financed out of charitable funds; they could be transferred to rural labour colonies while their families received domestic maintenance; they could be given financial assistance for migration and emigration; or they could be given advice on finding normal employment through a rate-assisted labour exchange.

In theory all these functions were complementary. Labour colonies were supposed to train men for emigration; labour exchanges to minimize the period of time during which workmen were dependent on 'employment relief'. But in practice they competed with each other for financial assistance and political support; and this was particularly important at a time when the Royal Commission on the Poor Laws and a Cabinet committee on unemployment were examining experiments with a view to framing recommendations on future national policy.[3]

The policy of relief works was discredited almost from the start. Nobody with any experience of the management of relief works expected them to be either commercially successful or to effect any permanent improvement in the situation of the unemployed. Initially the Works Committee of the Central (Unemployed) Body under the Reverend J. Anderson tried to find work which would not compete

---

[1] *C.U.B. Minutes*, ii. 253, 5 July 1907.
[2] *3rd Report of the C.U.B.* (1907–9), p. 6.
[3] Below, pp. 233, 248–64.

with private enterprise nor deprive regular workmen of employment;[1] and works undertaken by public authorities were supposed to be confined to 'exceptional' projects, which would not be financed out of the rates in the normal course of events.[2] Even so, the Poor Law Commissioners found several cases in which the employment of subsidized unemployed labour on relief works had persuaded public bodies to reduce their regular staff.[3] Work was arranged in the parks of the L.C.C. and on the renovation of Alexandra Palace; new cemeteries were laid out in Bermondsey, Fulham, Islington, and Shoreditch, and recreation grounds in Wandsworth, Battersea, and Camberwell.[4] In 1906 and 1907 a number of workmen were employed in private labour colonies; but in 1908 the L.G.B. decided that the Act authorized the employment of workmen out of the rates by public authorities only.[5] It proved, however, virtually impossible to find sufficient suitable employment on which to spend the available funds. 'It is *work* under the ordinary conditions of labour that is wanted', reported the C.U.B. in 1909. '[We] are unable to offer either the proper sort of work or the proper amount of it.'[6] Moreover, the bottleneck in the work of the Classification Committee meant that there was often a delay in filling vacant places on relief works, even when a long queue of applicants was waiting for employment. Local committees fretted against their inability to start relief works, and against the delays incurred in referring cases to the Central Body.[7] The works undertaken suffered from all the disadvantages of previous

[1] *Preliminary Report of the C.U.B.*, to 12 May 1906, p. 27.

[2] Ibid., pp. 32-3.

[3] Cd. 4795/1909, *Report on the Effects of Employment or Assistance given to the 'Unemployed' since 1886 as a means of Relieving Distress outside the Poor Law*, p. 101 [117].

[4] *Preliminary Report of the C.U.B.*, to 12 May 1906, Report of the Works Committee, pp. 29-34.

[5] *C.U.B. Minutes*, iii, 3 Apr. 1908, Report of the Works Committee, p. 150. The L.G.B. refused to sanction an arrangement with Joseph Fels's Vacant Land Cultivation Society because it was not a 'public body' as defined by Article V of the Unemployed Workmen Act regulations.

[6] *3rd Report of the C.U.B.* (1907-9), p. 84.

[7] *C.U.B. Minutes*, i. 50-1, 2 Feb. 1906, Resolutions from the City of London Distress Committee and the St. Pancras Borough Council, complaining about delays and lack of powers.

relief-work schemes. Professional supervision was inadequate and, since payment was by the hour rather than by the piece, the standard of efficiency tended to conform to that of the least skilled or least industrious workmen.[1] The works committee tried to find work on which skilled workmen could be employed at a higher rate of wages, and a few workmen of this kind were employed at Alexandra Palace;[2] but in most cases artisans and casuals, indoor and outdoor labourers worked together on the same tasks of digging, levelling, and demolition, and at the same hourly rates.

Very little attempt was made to evaluate relief works on a social rather than a commercial basis; their futility was in fact prejudged by twenty years of previous failure. Foremen, social workers, and local officials agreed that they were useless to the community and demoralizing to the unemployed.[3] John Burns in 1906 and 1907 toured relief works in London and the provinces, and recorded in his diary his disgust with the results. 'Went to Wanstead Park, where 146 unemployed were playing at work,' reads a typical entry, dated 28 March 1907,

. . . resting, talking, smoking, between the intervals of work. Looked on, walked about and chatted with foreman who confirmed views. From Park to East Ham where 100 were similarly engaged, foreman confirmed here the view I have, that 30 percent is all they do . . . but perhaps it is a sign of improvement that the shirker has been shamed into the pretence of work.[4]

Burns was by no means an impartial observer. He combined a ministerial prejudice against wasteful expenditure with an artisan's contempt for inferior workmanship. Moreover, his denunciations were rather unrealistic, since the L.G.B.'s refusal to sanction 'relief works' that were part of a local authority's normal programme of development meant that such works were almost necessarily uneconomic, especially if carried out with inexperienced labour at an unfavourable

1 *4th Annual Report of the Manchester Distress Committee*, p. 11.
2 *3rd Report of the C.U.B.* (1907–9), pp. 15–16.
3 Add. MS. 46324, Burns Diary, 4 Oct. 1906. Add. MS. 46327, Burns Diary, 7 Jan. 1909.
4 Add. MS. 46325, Burns Diary, 28 Mar. 1907.

season of the year.[1] However, his dissatisfaction with the relief-works policy was echoed by many local authorities and distress committees, some of whom called for a reversion to *ad hoc* charity,[2] others for a programme of national rather than local public works.[3] Russell Wakefield observed in May 1908 that most of the men who applied for work were technically inefficient and drawn from the chronically under-employed class.

When out of employment (they) automatically turn to the State or to the Municipal Authority to help them through. Their work when thus provided is never profitable either to the authority for whom they work or for themselves. There is no incentive to finish the work quickly; its completion only means to them a return to unemployment and the loss of the regular wage.[4]

The workmen relieved under the Act conspicuously failed to show signs of the 'permanent improvement' that was the explicit aim and justification of charitable administration;[5] and in November 1908 Burns recorded that he had at last converted Wakefield to the view that 'the endowment of a professional recurring lazaroni . . . means a serious and chronic centre of disturbance, a burden civil, social and financial.'[6]

Even Burns admitted, however, that not all the works authorized under the Unemployed Workmen Act were entirely wasteful and unproductive. The men employed were improved in health and physique, if not in industrial status;[7] and during the depression of 1907–8, when distress committees began for the first time to attract a large number of skilled workmen, some of the work performed was of unexpectedly high quality.[8] The Act was an important source

[1] The C.U.B., which had approved this policy in Jan. 1906, asked for its reversal in Oct. 1908 (*3rd Report of the C.U.B.* (1907–9), pp. 7–8).

[2] *Report of the Manchester Distress Committee*, 1908–9, p. 11.

[3] *C.U.B. Minutes*, i. 68, 16 Feb. 1906.

[4] Beveridge MSS., Coll. B, vol. iv, item 35, Extracts from 2nd Report of the C.U.B., with comments by Russell Wakefield, May 1908, para. 13.

[5] W. H. Beveridge, 'Labour Exchanges and the Unemployed', *Economic Journal*, 17 (Mar. 1907), 69–70.

[6] Add. MS. 46326, Burns Diary, 21 Nov. 1908.

[7] Beveridge MSS., Coll. B, vol. iv, item 35; Extracts from 2nd Report of the C.U.B. with comments by Russell Wakefield, May 1908, para. 1.

[8] *3rd Report of the C.U.B.* (1907–9), pp. 15–16.

of short-term funds during the credit crisis of 1907–8; and between 1905 and 1910 over £1,000,000 was spent by local distress committees in England and Wales—the largest programme of 'artificial' public works since the Lancashire cotton famine.[1] The most serious objection to relief works, however, was that they were merely a temporary solution to what was shown conclusively by the administration of the Unemployed Workmen Act to be a permanent problem. The majority of workmen investigated by distress committees were always in precarious employment; and after a maximum of sixteen weeks a year on relief works they returned to a casual style of industrial and domestic life. They were 'irregular workmen, normally in or on the verge of distress' reported Beveridge after the Act had been in operation for two and a half years. 'Their case is rather chronic than acute.'[2] By 1908, therefore, it was clear that relief works had to be established on a more permanent basis—which meant national workshops and a recognition of the 'right to work'—or they had to be replaced by some other form of assistance for the unemployed.

The alternative policies authorized by the Unemployed Workmen Act were employment exchanges, labour colonies, and assisted emigration. Emigration as a remedy for unemployment had attracted widespread support since the early nineteenth century from reformers who believed that the congestion of the labour market was caused by over-

[1] Appendix B, Table 6, p. 377. This was in addition to the much larger sums, amounting to over £100,000,000, raised by local authorities in England and Wales in the form of loans for public services between 1906 and 1910 (B. Mitchell and Phyllis Deane, *Abstract of British Historical Statistics*, p. 420). Burns spent the summer of 1908 touring the country and urging local authorities to combat unemployment by increasing their normal public works expenditure; and in Oct. 1908 Asquith resisted the Labour demand for an increased rate-contribution to relief works with the arguments that the borrowing capacity of municipal authorities far exceeded the yield of a 1*d*. rate (*Hansard*, 4th series, vol. 194, cols. 1161–71). Burns's preference for municipal loans rather than imperial grants to increase employment was not merely a reflection of a departmental point of view, since he had taken the same attitude on the Select Committee on Distress from Want of Employment in 1895 (H. of C. 111/1895, QQ. 977, 1019).

[2] Beveridge MSS. i. b. 356; 'Unemployment in Utopia', Address by W. H. Beveridge to the students' union of the London School of Economics, 1907/8. Casual or general labourers accounted for 51·5 per cent of workmen relieved under the Act in 1905–6; 52·2 per cent in 1906–7; 53·3 per cent in 1907–8; 47·4 per cent in 1908–9; and 47·0 per cent in 1909–10.

population or by the alienation of urban workmen from the land.[1] Socialists and trade unionists had been consistently sceptical about organized emigration as a means of relieving unemployment, suspecting that it was mainly designed to depress the wages of colonial labourers;[2] but in May 1906 the C.U.B. reported that there were many openings in Canada for 'the general labouring class . . . of which there is a superabundance in London'.[3] The Emigration Committee under the chairmanship of Walter Hazell decided to concentrate primarily—though not exclusively—on sending emigrants to agricultural employment. In each case the character, physique, and industrial experience of intending emigrants were carefully scrutinized; and preference was given to families rather than individual emigrants, and to workmen under 45 years old.[4]

At first the committee decided to work through existing emigrant organizations, and arrangements for co-operation were made with the East End, Self Help, and British Women's Emigration Societies and with the Church and Salvation Armies.[5] This policy gave rise to conflict within the C.U.B., however, because the COS were accused of carving out a monopoly on behalf of their own subsidiary organization, the East End Emigration Society.[6] Moreover, the 'triangular correspondence' between distress committees, C.U.B., and emigration societies caused much administrative delay; and some members of the C.U.B. were doubtful about the 'desirability of expending public funds through the agency of religious and charitable organisations'.[7] In 1908 it was therefore decided that emigration

[1] For a summary of emigration and colonization schemes see Cd. 2978/1906, *Report of the Departmental Committee appointed to consider Mr. H. Rider Haggard's Report on Agricultural Settlements in British Colonies*, paras. 4–30.

[2] Lansbury MSS., vol. 1, ff. 38–47, Papers and correspondence on emigration, 1886, ibid., vol. 28, ff. 20–5, G. Lansbury to Walt Sewell (written from Brisbane), 1 Mar. 1885.

[3] *Preliminary Report of the C.U.B.*, to 12 May 1906, Report of the Emigration Committee, p. 47.

[4] Ibid., pp. 35–6.   [5] *C.U.B. Minutes*, i. 18, 15 Dec. 1905.

[6] Add. MS. 46299, f. 344, Margaret Moore to John Burns, 7 Oct. 1907. Mrs. Moore claimed that out of £35,156 spent by the C.U.B. on emigration, £15,220 had gone to the East End Emigration Fund and only £14,400 had been spent on 'direct emigration or migration'.

[7] *3rd Report of the C.U.B.* (1907–9), p. 56.

should be organized directly by the staff of the Central
(Unemployed) Body.[1] By the middle of 1909 nearly 2,000
families had been helped to emigrate to Canada; fares were
advanced in the form of a loan, and families were provided
with suitable clothing.[2] A booklet giving 'Advice to Applicants for Emigration to Canada' was issued, which urged
emigrants to avoid large towns, to join benefit societies,
and to repay their loans as quickly as possible in order to
retain their independence.[3] Early in 1907 three members of
the Emigration Committee visited Canada to examine the
prospects for potential British settlers. They issued glowing
reports on the high level of success among those already
emigrated by the Central Body.[4] By the middle of 1909,
however, the Emigration Committee was less optimistic.
Many emigrants had found the assistance of the Central
Body to be more of a liability than an asset; and 'however
deserving the men may be, the fact that they are practically
labelled "London unemployed" gives them a bad start'.
Four per cent had been deported from Canada under the
stricter emigration laws passed in 1906; and the international
depression of 1908 had reduced the opportunities for foreign
settlers. The Committee had advised that all emigrants
should receive preliminary agricultural training; but the
L.G.B. refused to sanction expenditure for this purpose out
of the rates or the parliamentary grant.[5]

The third report of the Central Body concluded that one
of its most useful functions had been the 'provision of
adequate machinery' for emigration.[6] But outside the Emigration Committee there was little positive enthusiasm for
emigration as a remedy for unemployment. Russell Wakefield thought that it was depriving the country of its most
efficient workmen and leaving distress committees to deal
with problem families and the unemployable.[7] Lansbury,

[1] Beveridge, *Unemployment* (1930 ed.), p. 182.

[2] *3rd Report of the C.U.B.* (1907–9), p. 55; *Preliminary Report of the C.U.B.*,
to 12 May 1906, p. 37.

[3] *3rd Report of the C.U.B.* (1907–9), p. 62.

[4] *2nd Report of the C.U.B.* (1906–7), pp. 49–55.

[5] *3rd Report of the C.U.B.* (1907–9), Report of the Emigration Committee,
pp. 57–9.   [6] *3rd Report of the C.U.B.* (1907–9), p. 85.

[7] Beveridge MSS., Coll. B, vol. iv, item 35, Extracts from the 2nd Report of
the C.U.B., with comments by Russell Wakefield, May 1908, para. 7.

who had himself been an unsuccessful emigrant in the 1880s, declared that no workman should be forced by economic pressure to go to the colonies whilst thousands of acres of English land remained untilled.[1] And by the end of 1907 Beveridge had come to the conclusion that the over-population thesis which inspired emigration had been falsified by the evidence of economic growth.[2]

The activities that attracted most attention from social and administrative reformers outside the Central Body were, however, employment exchanges and labour colonies; of which the chief protagonists were, respectively, William Beveridge and George Lansbury. 'Two of us, George Lansbury and I, each wanted something quite different' Beveridge recalled, many years later. 'We were the two wild young men of the C.U.B., he urging "back to the land" and I urging Labour Exchanges; the Minister in charge of us at the time—John Burns of Battersea—thought us both equally foolish.'[3] The success of Beveridge's plans and the failure of Lansbury's were to have a significant influence on the subsequent shaping of government policies. But in 1905 the eclipse of labour colonies by labour organization was by no means a foregone conclusion—in fact, rather the reverse. When the Central Body was created, labour colonies were far more familiar to the social reforming world than 'labour bureaux' or employment exchanges. It has been shown that since the 1880s they had been the subject of a great deal of publicity and of practical experiment;[4] and the schemes of the Salvation Army, the Church Army, the Christian Social Service Union, and the Home Colonisation Society had aroused interest and support far outside their own membership. The report of the Departmental Committee on

[1] Lansbury MSS., vol. 29, ff. 53–4, MS. speech on unemployment, Oct.–Nov. 1906.
[2] Cd. 5066/1910, RC on the Poor Laws, Minutes of Evidence, QQ. 77834–7. He thought, however, that under the prevailing 'bad system' of industry, decasualization would leave a redundant residuum who should be trained in 'recuperative or convalescent' colonies as a preliminary to emigration (ibid., Q. 78296).
[3] W. H. Beveridge, 'The Birth of Labour Exchanges', Minlabour, 14, no. 1 (Jan. 1960), 2–3.
[4] Above, Chapter III.

Vagrancy had awakened a widespread expectation that statutory labour colonies would be set up for penal and reformatory purposes. George Lansbury, the leader of the labour colony movement in London, was appointed to the Royal Commission on the Poor Laws in 1905. A conference on labour colonies and unemployment convened by the Christian Social Service Union in the summer of 1905 showed how widespread was the interest among the upper classes in a form of social organization originally promoted by people who were regarded as social and religious eccentrics.[1] And on 22 December 1905 Campbell-Bannerman's speech at the Albert Hall included a general reference to the 'colonisation of underdeveloped home estates' as a central part of the Liberal election programme.[2]

Why then did 'labour colonies' fail? This failure arose partly from the inherent difficulties involved in the management of a subsidized colony of inferior workmen; partly from the active hostility of the Local Government Board; and partly from the public discrediting of the policy of the Poplar guardians, who under the leadership of George Lansbury had pioneered the use of labour colonies as an alternative to the workhouse and the stoneyard. Lansbury was elected chairman of the Working Colonies committee of the Central Body in December 1905,[3] and initially found no lack of support for 'home colonization' among his colleagues—particularly from May Tennant and from C. H. Grinling, the Chairman of the Woolwich Distress Committee, who had been influential in persuading the executors of the London Unemployed Fund to take over the Hollesley Bay estate in 1904.[4] Russell Wakefield and the COS members were overtly sceptical about the capacity of

---

[1] *The Problem of the Unemployed*, Notice of a Conference under the auspices of the Christian Social Service Union, on 'Labour and Training Colonies'.

[2] 'The Liberal Government's Programme', 22 Dec. 1905, *Speeches by Sir Henry Campbell-Bannerman, 1899–1908. Selected and reprinted from the Times*, p. 182.

[3] *C.U.B. Minutes*, i. 12, 8 Dec. 1905.

[4] Lansbury MSS., vol. 7, f. 162, G. Lansbury to Mary Fels, 30 June 1914. A convert from the COS to the I.L.P., Grinling was founder of the Woolwich Labour Representation Association, the Woolwich Dispensary, the Invalid Children's Aid Committee, and a promoter of the 'Woolwich Pioneer' (P. Thompson, *Socialists, Liberals and Labour*, pp. 23, 257–62).

the London unemployed for life on the land;[1] but even so
the Preliminary Report of the Central Body remarked that
'the agricultural training colony is . . . the scheme of employment which . . . offers most prospect of permanent usefulness' to the unemployed.[2] Beveridge in particular was interested in the possibility of a labour colony as part of the
machinery of decasualization.[3] He was warned against
Lansbury, and against the principles that the Poplar guardian was supposed to represent, by Mrs. Rose Dunn
Gardner, the COS representative of the Chelsea distress
committee.[4] But, he recalled, 'in practice, George Lansbury
and I made a deal; he was ready to support my Employment
Exchanges, so long as I supported his Farm colonies for the
unemployed, like Hollesley Bay.'[5]

This was an uneasy alliance, of more benefit to Beveridge
than to Lansbury. There seems to have been no personal
antagonism between them; but Beveridge was an archetype
of the 'settlement' school of social reformer, whose influence Lansbury so much disliked.[6] Moreover, Beveridge's
idea of a labour colony was very different from Lansbury's.
The Poplar guardian hoped to create a community in which
individuals would be trained for a new life of permanent
settlement on the land, and which would provide a model
for the substitution of co-operative for capitalist production.[7]
Beveridge, on the other hand, saw labour colonies as peripheral institutions for the misfits of the economic system,
to which the 'unemployable' would be expelled from the
industrial system with loss of civil rights.[8] He thought,
moreover, that since material 'less-eligibility' was becoming
increasingly unacceptable to scientific social reformers,

[1] Beveridge MSS., Coll. B, vol. iv, item 35, Extracts from the 2nd Report of
the C.U.B., with comments by Russell Wakefield, May 1908, para. 5. Wakefield
remarked that training on the land was physically, and perhaps morally, but not
economically, useful.

[2] *Preliminary Report of the C.U.B.*, p. 29.

[3] W. H. Beveridge, 'The Problem of the Unemployed', *Sociological Papers*,
3 (1906), 331.

[4] W. H. Beveridge, *Power and Influence*, p. 47.

[5] Beveridge MSS., D. 047, Draft of speech for the Labour Exchange jubilee,
Jan. 1959.   [6] G. Lansbury, *My Life*, pp. 130-1.

[7] Lansbury MSS., vol. 29, ff. 43-55, MS. copy of a speech on unemployment,
Oct.-Nov. 1906.

[8] W. H. Beveridge, 'The Problem of the Unemployed', loc. cit., p. 327.

labour colonies would have to be based on principles of
'moral' restraint.[1] Both Beveridge and Mrs. Dunn Gardner,
however, became members of the Working Colonies committee and took part in the deployment of workmen to
rural estates.[2] In November 1905 the Central Body took
over the management, though not the ownership, of the
colony at Hollesley Bay;[3] and during the winter of 1905–6
arrangements were made to employ workmen in the colonies
at Lingfield and Osea Island, at Letchworth Garden City,
and on the reclamation of floodland at Fambridge in Essex.[4]
But temporary employment of this kind was merely an
institutionalized rural version of the relief works in the
metropolis, of which Lansbury was as contemptuous as any
member of the COS.[5] He was anxious to experiment on a
more permanent basis, and in March 1906 the Working
Colonies committee recommended that the Central Body
should take over the ownership and liabilities of the Hollesley
Bay estate from the trustees of the London Unemployed
Fund.[6]

From the start the decision to acquire Hollesley Bay was
fraught with difficulties. The original aim of the colony,
as approved by Long and Gerald Balfour, was not merely to
train men but to provide them with a permanent settlement
on the land.[7] The new President of the L.G.B., however,
was known to be opposed to subsidized agricultural enterprises. By the spring of 1906 little progress had been made
in this direction. The colony was in danger of folding up
through lack of funds and the colonists complained of the
inefficient management of the superintendent, Bolton Smart.[8]
However, the Central Body approved the purchase, and
permission for the transfer of the property was sought from

---

[1] *Morning Post,* 5 Mar. 1906.
[2] W. H. Beveridge, *Power and Influence,* pp. 47–8.
[3] *Preliminary Report of the C.U.B.,* p. 29.   [4] Ibid., pp. 29–34.
[5] See Lansbury's comments on the futility of works authorized by the Chamberlain circular (Lansbury MSS., vol. 29, f. 47, MS. text of a speech on unemployment,
Oct.–Nov. 1906).   [6] *C.U.B. Minutes,* i. 70–1, 2 Mar. 1906.
[7] Cd. 2561/1905, *Preliminary Statement on the Work of the London Unemployed
Fund,* May 1905, pp. 35–7.
[8] Lansbury MSS., vol. 2, f. 236, Hubert Hammond to G. Lansbury, 13 Feb.
1906. Smart had previously acted as secretary to the Mansion House unemployed
funds of 1892–4.

the L.G.B.[1] Before it was granted Burns and his officials visited the colony; and Burns's comments suggest that he was not at this time wholeheartedly opposed to the scheme. He approved of Bolton Smart, and thought that

the whole place [was] fit for a doubtful experiment. Given a scheme, granted an experiment, this is the ideal place to try it. Much will depend on personnel and administration, but we have to pay for learning 20 years' hence what the verdict will be. I will however consent to the transfer.[2]

Formal permission for the acquisition of the property by the Central Body was granted on 30 May, subject to the condition that 'the scheme must be regarded as one of an experimental character and that it should be carefully watched both from the point of view of cost and administration'.[3] Lansbury and Fels tried to stipulate as a condition of the transfer that the estate should be used for the development of smallholdings as well as for the temporary testing and training of the unemployed. But some members of the Central Body were reluctant to acquire land subject to restrictive covenants; and this attitude was endorsed by Burns against the wish of Russell Wakefield and the Working Colonies committee in June 1906.[4]

Having authorized the acquisition of a colony, however, Burns's opposition, both to home colonization and to Lansbury personally, appear to have increased. This hostility to labour colonies, and to other innovations in unemployment policy, was ascribed by contemporary critics to Burns's indoctrination by his departmental officials.[5] But this view exaggerated the degree of sympathy that prevailed between Burns and the staff of the L.G.B.[6] Some at least of

[1] C.U.B. Minutes, i. 71, 2 Mar. 1906.

[2] Add. MS. 46324, Burns Diary, 18 Apr. 1906.

[3] C.U.B. Minutes, i. 146, 1 June 1906.

[4] Ibid., p. 36, 19 Jan. 1906; p. 146, 1 June 1906. Add. MS. 46324, Burns Diary, 12 June 1906.

[5] B. Webb, Our Partnership, p. 393. On the conservative influence of the L.G.B. permanent officials see also Charles Masterman to H. H. Asquith, 15 Jan. 1909, quoted in Lucy Masterman, C. F. G. Masterman, pp. 121–2.

[6] Add. MS. 46325, Burns Diary, 19 Apr. 1907; Add. MS. 46326, Burns Diary, 11 Jan. 1908. On Burns's curious relationship to his Permanent Secretary see W. H. Beveridge, Power and Influence, pp. 144–5.

the permanent officials at the L.G.B. were in favour of labour colonies—of the kind prescribed by the Vagrancy Committee if not by the guardians of Poplar.[1] Colonel Lockwood, the metropolitan Poor Law inspector—whom Burns regarded as a weak and impractical 'drifter'[2]—thought that colonies of the kind founded by the Poplar guardians at Laindon might be turned into compulsory detention centres where the able-bodied would be persuaded to work by a 'scientifically contrived subsistence diet'.[3] Burns, on the other hand, had condemned labour colonies long before his appointment to the L.G.B.[4] On the Select Committee of 1895 he had attacked the Salvation Army colony at Hadleigh for promoting unfair competition and unnatural vice.[5] And in 1905 he had condemned farm colonies in the *Daily Chronicle* as 'the pauperisation of the really one decent industry left to us'.[6] His attitude can best be explained by a skilled workman's contempt for inefficient workmanship; by a trade unionist's prejudice against cut-price labour and the marketing of charitably subsidized products; by a puritan aversion to personal economic dependence; and by his dislike of the colonization movement's failure to recognize the essential 'urbanity' of contemporary industrial society.[7] These prejudices were reinforced but not created by his contact with a department that was responsible for preserving the poor from demoralization and for preventing unauthorized local expenditure on behalf of the unemployed.

Burns's opposition to the labour-colony movement hardened during the spring and summer of 1906, when in an 'investigating fever', he toured the labour colonies of England and the Continent, and interviewed the leading authorities on different types of home colonization.[8] The

[1] *27th Annual Report of the L.G.B.* (1897–8), p. 66.
[2] Add. MS. 46324, Burns Diary, 23 Jan. 1906.
[3] *36th Annual Report of the L.G.B.* (1906–7), pp. 284–5.
[4] John Burns, *The Unemployed* (Fabian Tract No. 47, Nov. 1893), p. 17.
[5] H. of C. 111/1895, QQ. 1939–40; H. of C. 365/1895, Q. 5005.
[6] Add. MS. 46323, Burns Diary, 29 Nov. 1905.
[7] Cf. Burns's comments on the Smallholdings and Allotments Act (Add. MS. 46325, Burns Diary, 16 Aug. 1907).
[8] Beveridge MSS., L. ii. 218a, J. Burns to W. H. Beveridge 1906; Lansbury MSS., vol. 2, f. 269, J. Fels to G. Lansbury, 29 June 1906; Add. MS. 46324, Burns Diary, 12 Sept. and 1 Oct 1906.

Poplar colony at Laindon, the colony opened by the West Ham distress committee at Ockenden, the private colony at Osea Island, and the land-reclamation scheme at Fambridge in the Thames Estuary, were all condemned as demoralizing and uneconomic.[1] Most scathing of all were Burns's views of the semi-co-operative community established at Letchworth Garden City, where the Working Colonies committee had arranged employment for several hundred of the London unemployed. 'A picturesque aviary of cranks, foreman called them "monkey nuts and macaroni" type. Sandals. No hats. Liberty ties etc. . . . 10 years hence it will be the ordinary town with persons of marked eccentricity of manner and character.'[2]

The events, however, that confirmed Burns in his determination to prevent Hollesley Bay from becoming a permanent establishment, and alienated support from Lansbury both inside and outside the Central Body, were the 'East End scandals' which broke upon metropolitan Poor Law administration in 1906-7. Special inquiries were conducted into the affairs of the guardians in Hammersmith, Edmonton, Mile End, Poplar, and West Ham. In each case the guardians were found guilty of negligence or extravagance, and in West Ham five guardians and four Poor Law officials were imprisoned for corruption.[3] But the case that attracted most public attention was the Poplar inquiry, opened by one of the L.G.B.'s assistant secretaries, J. S. Davy, in March 1906.[4] For the previous twelve years the Poplar guardians had moved towards a policy, often advocated by socialists and anarchists in the 1880s and 1890s,[5] of exploiting all the legal loopholes in the Poor Law in order to use their relieving powers to the fullest possible extent. Davy found that in the depression of 1904-5 the discretionary power to give

---

[1] Add. MS. 46324, Burns Diary, 3 Mar. 1906, 5 June 1906, 9 Oct. 1906.

[2] Ibid. 11 Oct. 1906.

[3] 37th Annual Report of the L.G.B. (1907-8), pp. 289-90.

[4] The best account of the administrative crisis in Poplar is Brian Keith-Lucas, 'Poplarism', Public Law, 1962, pp. 52-80, which deals mainly with the Poplar revolt of 1920s; see, however, pp. 52-5. On the political background to the case and the challenge of the Poplar Municipal Alliance see George Haw, op. cit., pp. 271-95.

[5] e.g. Henry B. Samuels, What's To Be Done? The Unemployed Question Considered (1892), pp. 4-5.

assistance without prior investigation in 'cases of sudden
or urgent necessity' had been used on behalf of nearly every
able-bodied applicant for relief;[1] and the administration of
workhouse contracts was both extravagant and corrupt.[2]
At the same time the Poplar distress committee had been
referring applicants *en masse* to the guardians rather than
to the Central (Unemployed) Body;[3] and pauperism in
Poplar had doubled between 1894 and 1906.[4]

The results of Davy's investigation were laid before
Parliament in July, and his full report and notes on the case
were published in October 1906.[5] The full significance of the
Poplar revelations cannot be discussed here; but Davy's
condemnation of the regime of the Poplar guardians also
extended to the colony at Laindon and to Poplar's country
workhouse at Forest Gate. It was found that, out of 653
unemployed workmen admitted to the colony, only one had
been found suitable for promotion to the permanent staff.[6]
Twenty-five had been helped to emigrate, of whom all but
three had disappeared without trace.[7] Genuine unemployed
workmen were demoralized and 'contaminated' by contact
with the pauper class.[8] The average cost of maintaining a

---

[1] Cd. 3240/1906, *Report to the President of the L.G.B. on the Poplar Union*,
by J. S. Davy, pp. 21–3. A condition of this discretionary power was that all such
cases should be reported retrospectively to the L.G.B., which the Poplar guardians
had failed to do. The fact that such relief had fallen by 50 per cent since Feb. 1906,
when the inquiry was first announced was interpreted by Davy as an acknowledgement of the guilt of the guardians (pp. 23–4). Poplar sympathizers complained that
the results of the inquiry were a foregone conclusion, since Davy had made up his
mind to condemn the Poplar guardians in advance (Ensor MSS., T. Edmund
Harvey to R. C. K. Ensor, 17 May 1906).

[2] B. Webb, *Our Partnership*, p. 337.

[3] Lansbury MSS., vol. 2, f. 270, Sydney Buxton to George Lansbury, 30
July 1906.

[4] Cd. 3240/1906, *Report on the Poplar Union*, pp. 3, 7, 19–20. The mean number
of indoor paupers had risen from 2,623 in 1894 to 3,465 in 1904. Outdoor paupers
per 1,000 of population had trebled. The cost of indoor and outdoor relief had
doubled, rateable value increased, population remained stable.

[5] Cd. 3240/1906, *Report to the President of the L.G.B. on the Poplar Union*,
by J. S. Davy, C.B., Chief Inspector to the Board; and Cd. 3274/1906, *Transcript
of the Shorthand Notes taken at the Public Inquiry held by J. S. Davy . . . into the
General Conditions of the Poplar Union, its Pauperism, and the Administration of the
Guardians and their Officers.*

[6] Cd. 3274/1906, Evidence of Mr. J. Clarke, Superintendent of the Laindon
Branch Workhouse, p. 158.

[7] Ibid., p. 261.   [8] Ibid., p. 259.

colonist and his family was 25s. a week, compared with an average wage of 17s. among Essex farm labourers.[1] Residents of Laindon and the local police complained that the colonists were drunken and violent, and that they abused and terrified the local female population.[2] Railway tickets with which they had been issued to search for work were used to spend long week-ends with their families; and in November 1905 many of the colonists had received free tickets from the guardians to attend the West End demonstrations of the Poplar unemployed.[3]

Davy's report damped the enthusiasm of many who had previously been sympathetic to the colony movement; and the analogy with Hollesley Bay was overwhelming.[4] Burns regarded the report as a confirmation of his views; and when the parliamentary grant was voted in October 1906 he declined to allow any of London's share to be used for consolidating the colony movement along the lines suggested by Lansbury.[5] The C.U.B. was refused permission to build cottages or to create smallholdings at Hollesley Bay, on the ground that this was a 'departure from the original purpose' of the colony, not authorized by the Unemployed Workmen Act which provided only for temporary relief.[6]

This decision evoked an indignant protest from the Working Colonies committee, which stated that the aim of permanent settlement had been clearly laid down by the London Unemployed Fund. In April 1906 the Central Body itself had informed the L.G.B. that its aim was to create

[1] Ibid., pp. 156–8.

[2] Ibid., Evidence of Police Superintendent A. Marden and Police Constable G. Reeve, pp. 150–7.

[3] Ibid., Evidence of J. Clarke, pp. 158, 161.

[4] Add. MS. 46299, ff. 128–32, Russell Wakefield to John Burns, 11 Nov. 1906: 'This report is *very* important to our Central Body . . . the colonies we have—especially Hollesley—are very little more justifiable than Laindon . . . the Poplar report has revolutionised the whole matter . . .'

[5] *36th Annual Report of the L.G.B.* (1906–7), pp. 277–8, Circular letter from Sir Samuel Provis to the Clerks of certain Distress Committees, 12 Oct. 1906. The circular stated that the parliamentary grant could be used only to assist in 'the provision of temporary work', thereby implicitly excluding permanent settlement on the land.

[6] *C.U.B. Minutes*, i, 31 Oct. 1906, Report of Working Colonies Committee, p. 334.

an 'ultimate settlement in some form of permanent occupation'; and fifty families had been kept at Hollesley Bay
throughout the summer in anticipation of the fulfilment of
this policy. The Central Body therefore approved a resolution asking Burns to reconsider his decision.[1] Unbeknown
to the Central Body, however, its chairman Russell Wakefield was himself encouraging Burns's policy of restraint.
On 11 November 1906 he wrote to Burns that Hollesley
Bay was 'little more justifiable than Laindon', even though
the workmen employed there were of a superior class to the
paupers of Poplar. But he pointed out that both Laindon
and Hollesley Bay had been promoted by the previous holders
of Burns's own office, and that it was difficult to discard
unsound schemes once they had been authorized and set in
motion.

> The question now is *what are we to do?* I should like to get from you
> the absolute 'thus far and no farther' of Hollesley Bay—and I promise
> to see that the limit is put definitely. Only how can I, with Lansbury
> as Chairman of a Committee pledged to a certain course of action?
> It is to ask an enthusiastic father to kill his child. Is there to be *any*
> market gardening? If *yes*, there will be competing with other people
> in the business. If there is *not*—most of the past expenditure is useless
> and it is difficult to see how we are to employ the 350 men.[2]

Lansbury's influence on the Working Colonies committee had been much weakened by the Poplar inquiry.
His colleague Will Crooks, the chairman of the Poplar
guardians, resigned after the publication of Davy's report.[3]
And the conservative reaction in the municipal elections
of November 1906 eliminated many of Lansbury's supporters, including C. F. Grinling, among borough council
representatives on the C.U.B.[4] For a while Lansbury hoped
that Burns might be persuaded to change his mind about
Hollesley Bay, or that an estate suitable for smallholdings

---

[1] *C.U.B. Minutes*, i, 31 Oct. 1906, Report of Working Colonies Committee,
pp. 335–8.

[2] Add. MS. 46299, ff. 130–1, Russell Wakefield to John Burns, 11 Nov. 1906.

[3] *C.U.B. Minutes*, ii. 25, 21 Dec. 1906. Crooks had had a nervous breakdown
in 1904 (G. Haw, op. cit., p. 242), and was brought near to collapse by the strain
of the Poplar inquiry (Add. MS. 46324, Burns Diary, 16 Nov. 1906).

[4] Beveridge MSS., L. ii. 218a, G. Lansbury to W. H. Beveridge, 1 Feb. 1907.
Lansbury MSS., vol. 3, f. 21, May Tennant to G. Lansbury, 1 Jan. 1907.

might be granted to the Central Body out of Crown lands.[1] But early in 1907 he admitted to Beveridge that his case appeared to be lost on the Central Body.

You see, the men with real brains on the committee are against me either in principle or in detail and a Chairman can never run a committee properly when such is the case. After Grinling left, there was no one to put my point of view; for me to do it in face of Mumford, Bailward, yourself and Mrs. Dunn Gardner was out of the question. On top of that was the whole question of the management of Hollesley. . . .[2]

Lansbury tendered his resignation as chairman of the Working Colonies committee on 1 February 1907, but was persuaded to remain on the Central Body.[3] Henceforward, however, the activities of the Working Colonies committee were very circumscribed. In April 1907 Burns concluded that Hollesley Bay was 'an expensive and demoralising toy. A costly piece of political bribery . . . a holiday for 250 men from London who deteriorate and get soft by a process of coddling that unfits them for emigration and is useless for migration.'[4] He therefore refused an offer from Joseph Fels of another estate to be used exclusively for smallholdings; and also refused to authorize a colony for unemployed women.[5] Schemes for sending groups of unemployed workmen to other parts of the United Kingdom were frustrated by the hostility of local labourers and by the reluctance of the L.G.B. to finance such removal.[6] The L.G.B. urged the Central Body to concentrate on employing more workmen for a shorter period rather than fewer for a longer period;[7] and Hollesley Bay therefore remained little more than a complicated and expensive form of temporary relief for the unemployed. When the Webbs visited the estate in January 1908 they found

three hundred 'unemployed' living in the settlement . . . unintelligent and unhappy looking, angry with the cold wind and unaccustomed

---

1 *C.U.B. Minutes*, ii. 33, 21 Dec. 1906.
2 Beveridge MSS., L. ii. 218a, G. Lansbury to W. H. Beveridge, 1 Feb. 1907.
3 *C.U.B. Minutes*, ii. 74, 1 Feb. 1907.
4 Add. MS. 46325, Burns Diary, 4 and 13 Apr. 1907.
5 Add. MS. 46327, Burns Diary, 16 Jan. 1909; *3rd Report of the C.U.B.* (1907–9), p. 83.
6 *C.U.B. Minutes*, ii. 299, 4 Oct. 1907.   7 Ibid. i. 355, 16 Nov. 1906.

work and longing for wife and child or the Public House of the London slums. They are all married men and are supposed to be training for conscription as land settlers, but the great bulk of them are simply destitute men who accept the 14 weeks residence as one job like another, but rather an unpleasant one as they have only 6*d*. a week to spend on beer [and] tobacco and nowhere to spend their leisure.[1]

Beatrice Webb thought, however, that the colony was 'quite well worth watching', and that it might be improved by 'better organisation of the men's leisure'. Shortly afterwards an attempt was made to improve the efficiency of Hollesley Bay by employing a larger proportion of ordinary workmen and by placing the different departments of the colony—farming, gardening, and marketing—under a single superintendent.[2] But the sale of produce on the open market evoked protests of 'unfair competition' from local farmers.[3] Lansbury hoped that the Smallholdings and Allotments Act of 1907 might enable the Central Body to acquire land under the supervision of the Board of Agriculture:[4] but in fact this was not the case and the Central Body was unable to do more than exert pressure on the L.C.C. and other local authorities to use their powers under the Act.[5]

The administrative straitjacket imposed by the L.G.B. by no means put an end to labour-colony experiments. Although Lansbury finally resigned in October 1908, both the expenditure and income of Hollesley Bay were undiminished until 1914;[6] and outside London, the distress committees in Glasgow, Manchester, and West Ham opened labour colonies which provided work in accordance with the L.G.B. regulations.[7] A co-operative colony for smallholders and their families was opened by Joseph Fels at Maylands in

---

[1] Passfield MSS., ii, 4, d, item 1, B. Webb to Mary Playne, 1 Jan. 1908.
[2] *C.U.B. Minutes*, iii. 53–5, 17 Jan. 1908.
[3] Ibid., p. 256, 17 July 1908.
[4] Ibid., p. 27, 20 Dec. 1907.
[5] Ibid. iv. 195–6, 19 Feb. 1909.
[6] Income from the work of the colonists rose from £6,280 in 1907–8 to £8,723 in 1913–14. Annual expenditure on the colony was maintained at about £25,000 from 1906 to 1914 (*Annual Distress Committee Returns*).
[7] CAB 37/93/64, Cabinet memorandum by John Sinclair on 'Administration in Scotland of the Unemployed Workmen Act, 1905', 22 May 1908. *37th Annual Report of the L.G.B.* (1907–8), p. clxxvii.

Essex.[1] The Royal Commission on the Poor Laws heard evidence on no less than nineteen labour colonies of different kinds in the United Kingdom.[2] Politicians of all parties continued to show an interest in penal colonies for the work-shy and unemployable, although as the metropolitan Poor Law inspector remarked in his report for 1907 'the House of Commons is yet to come which would legislate in the direction indicated'.[3] After 1906, however, there was very little chance of co-operative labour colonies being adopted as part of national unemployment policy. This was not because they had been given a fair trial and been found wanting. But the manifest inefficiency of Hollesley Bay and the pre-judice aroused by the Laindon inquiry tended to dissipate the support of members of the Central Body and of poli-ticians and public administrators. Moreover, there was much confusion about the economic aspects of labour colonies. One of the most potent arguments of their critics was the fact that the net cost of employing an unemployed workman in a labour colony and supporting his family was greater than the cost of giving unconditional outdoor relief.[4] Lansbury tried to contravert this argument, not by empha-sizing the social value of employment and training, but by claiming that agriculture was about to experience a great international revival and that co-operation would ultimately prove more efficient than capitalist production.[5] These argu-ments, whether true or false, were inherently implausible to most of his contemporaries.

Labour Exchanges on the other hand had received com-paratively little attention from social and charitable ad-ministrators before 1905. Several public and private bureaux had been in spasmodic operation since the 1880s; and

[1] Lansbury MSS., vol. 7, ff. 163–4, G. Lansbury to M. Fels, 30 June 1914.

[2] At Chingford, Fambridge, Hempstead, Hadleigh, Hollesley Bay, Laindon, Leicester, Letchworth Garden City, Lingfield, Newdigate, Osea Island, Starn-thwaite, Walsingham, Cumbernauld, Dumfries, Edinburgh, Glasgow, Murieston, Palacerigg.   [3] *36th Annual Report of the L.G.B.* (1906–7), p. 285.

[4] Cd. 3274/1906, *Transcript of Shorthand Notes taken at the Public Inquiry held by J. S. Davy . . . into the General Conditions of the Poplar Union, its Pauperism and the Administration of the Guardians and their Officers,* p. 158.

[5] Beveridge MSS., L. ii. 218a, G. Lansbury to W. H. Beveridge, 1 Feb. 1907.

several witnesses to the Royal Commission on Labour had suggested that a national system of exchanges should be created by the Board of Trade.[1] In 1902 the Labour Bureaux (London) Act had authorized the maintenance of municipal labour bureaux in the metropolitan area out of the rates.[2] In 1905 there were ten municipal bureaux established under this Act, mostly 'passive' or 'moribund', and also a 'Central Employment Exchange' set up by the London Unemployed Fund.[3] At the time of the passing of the Unemployed Workmen Act, however, there was no significant body of support for labour exchanges as there was for labour colonies. Contrary to the impression given in his autobiography, Beveridge did not enter the C.U.B. in December 1905 as a fully fledged protagonist of a labour-exchange system.[4] His interest in labour organization had been aroused by the Toynbee Hall study group of 1903; and in February 1904 he was introduced by the Webbs to 'many people unashamed of talking L.C.C. and Unemployed'.[5] In 1905 he acted for a brief period as secretary to a COS inquiry into problems of unskilled labour.[6] But he was at this stage primarily interested in persuading private employers to regularize casual employment; and since the Unemployed Workmen Act was not supposed to apply to casual workmen there was no *prima facie* reason why the machinery of the Act should realize this aim. Although he drafted the plan for metropolitan exchanges in February 1906 he continued to treat them as peripheral to the process of labour organization, and as useful mainly for elderly workmen.[7] By September 1906 he had come to see them as 'one of the most thorough-going and indispensable measures of reform falling within the practical politics of this century';[8] but at the end of the year he still

---

[1] Below, pp. 279–81.   [2] 2 Edw. VII, c. 13.
[3] H. of C. 86/1906, *Labour Bureaux*, Report made to the President of the L.G.B. by Arthur Lowry, Nov. 1905, pp. 3–12.
[4] W. H. Beveridge, *Power and Influence*, pp. 44–5.
[5] Beveridge MSS., A1. 100, Box 1, Beveridge's Diary, 10 Feb. 1904.
[6] *Report of a COS Special Committee on Unskilled Labour*, June 1908.
[7] Beveridge MSS., Coll. B, vol. xvi, item 2, Draft memorandum on 'Employment Exchanges', by W. H. Beveridge, 14 Feb. 1906.
[8] W. H. Beveridge, 'Labour Bureaux', *Economic Journal*, 16, no. 63 (Sept 1906), 437. He thought, however, that their usefulness would lie mainly in organizing casual workmen (ibid., p. 439).

thought that skilled workmen could provide their own information service, and that 'the most important object' of public exchanges was 'dealing with the semiskilled or unskilled labourer'.[1] It was not until the beginning of 1907 that he came to the conclusion that 'the universal application of the principle of the Labour Exchange' was a prerequisite of decasualization, of the forestalling of depressions, and of the maintenance of a 'maximum of mobility' throughout the economy.[2]

The Central (Unemployed) Body decided in March 1906 to take over existing metropolitan exchanges, including that of the London Unemployed Fund, and to spend £10,725 on establishing exchanges in boroughs where they did not exist under the Act of 1902.[3] A committee under the chairmanship of Beveridge was elected to direct the policy and management of labour exchanges in April 1906.[4] By the beginning of 1907 there were twenty-five 'employment exchanges' in London, registering over 10,000 applicants a month.[5] Registration was free and workmen seeking employment were required to attend weekly.[6] Special provision was made for the registration of women and children, either in separate waiting-rooms or during separate hours of the day. Women were eligible for registration only if they could prove that they were family breadwinners or dependent on their own earnings; and no provision was made for domestic servants, who were adequately covered by private and charitable registries; advice to school-leavers was given in conjunction with voluntary Apprenticeship and Skilled Employ-

[1] Beveridge MSS., Coll. B, vol. xvi, item 10; 'Memorandum on the Relation of Employment Exchanges to Trade Unions', by W. H. Beveridge, 17 Dec. 1906, p. 3.

[2] W. H. Beveridge, 'Labour Exchanges and the Unemployed', *Economic Journal*, 17, no. 65 (Mar. 1907), 76.

[3] *C.U.B. Minutes*, i. 91–2, 23 Mar. 1906. C.U.B. Classification Committee, *Interim Report and Recommendations on Employment Exchanges* (7 Mar. 1906), revised by the meeting of the C.U.B., 21 Mar. 1906.

[4] *C.U.B. Minutes*, i. 96, 6 Apr. 1906; 115, 4 May 1906. The committee included Wakefield, Lansbury, Mrs. Dunn-Gardner, and the founder of the first free labour exchange, Nathaniel Cohen (see below, p. 279).

[5] W. H. Beveridge, 'Labour Exchanges and the Unemployed', *Economic Journal*, 17 (Mar. 1907), 67.

[6] *Preliminary Report of the C.U.B.*, to 12 May 1906, Report of Employment Exchanges Committee, p. 42. Employed workmen who wished merely to change their jobs were charged a 6d. fee.

ment committees and with the education department of the L.C.C. The character and industrial capacity of each applicant was supposed to be carefully verified, so as to 'offer to the employer a better sort of man than he would be able to get in the ordinary way'.[1]

From the start, however, the 'labour exchange' principle came into conflict both with the other functions of the Central (Unemployed) Body and with other commercial and industrial interests. A report by Arthur Lowry, an assistant inspector to the Local Government Board, in November 1905, had emphasized that association with charity was fatal to the efficient working of exchanges and that a 'Bureau should occupy itself solely with the normal labour market'.[2] Where exchanges were actually managed by distress committees, as they were in many provincial areas, strict adherence to this principle was manifestly impossible; but the federal system of exchanges created in London was administratively quite separate from distress committees and was supposed to act as a 'neutral medium' of commercial intercourse between employers and employed.[3] In practice, however, it proved difficult to convince employers that this was the case, particularly as many of the applicants for work were passed on to exchanges by local distress committees.[4] Moreover, contrary to the regulations, local employment exchange officials persisted in giving priority to applicants according to social need rather than industrial capacity; and one local exchange hired out unemployed workmen as political agitators.[5] For the first two years, therefore, the exchanges had very limited success in winning the confidence of employers. In 1907 the exchanges registered an unsatisfied demand for highly skilled workmen,[6] but the

[1] Beveridge MSS., Coll. B, vol. xvi, item 2, Draft memorandum on 'Employment Exchanges', by W. H. Beveridge, 14 Feb. 1906.

[2] H. of C. 86/1906, *Labour Bureaux*, Report made to the President of the L.G.B. by Arthur Lowry, Nov. 1905, p. 19.

[3] *2nd Report of the C.U.B.* (1906–7), pp. 56–7.

[4] Beveridge MSS., Coll. B, vol. iv, item 35, Extracts from the 2nd Report of the C.U.B., with comments by Russell Wakefield, May 1908, para. 10; vol. xvi, item 42, Report by the Organising Superintendent of Employment Exchanges on Various Matters, 4 May 1908.

[5] *C.U.B. Minutes*, ii. 115, 1 Mar. 1907; 144, 15 Mar 1907.

[6] *2nd Report of the C.U.B.* (1906–7), p. 63.

workmen who passed through the exchanges were rarely of that class of labour; and even in 1908–9, when an office for canvassing was opened in the City, situations were found for only 26,386 workmen out of 177,969 applications for work.[1]

Moreover, workmen in trade unions were suspicious of the policy of the exchanges towards trade disputes and standard rates.[2] In December 1906 Beveridge laid down that employment exchanges should be 'markets for labour in time of peace rather than time of war'; and exchange officials were instructed to avoid the registration of vacancies caused by strikes.[3] But he refused to concede that exchanges should only advertise situations which paid either trade-union rates or the local standard wage.[4] Beveridge's attitude was endorsed by the Local Government Board;[5] but the fears of trade unionists were confirmed in November 1907 when the railway companies were thought to be recruiting labour through the exchanges in anticipation of a strike.[6] The clash of principle between exchanges and unions was partially relieved by the creation of Local Advisory Committees, representing employers and workmen, to advertise and promote the interest of each exchange;[7] and by allowing unions to deposit their own 'vacant books' at an exchange, so that they could use its facilities and at the same time retain their own rules and system of information.[8] But trade unionists did not support labour exchanges to any great

---

[1] 3rd Report of the C.U.B. (1907–9), pp. 74, 233–5.

[2] C.U.B. Minutes, i. 209, 20 July 1906; ii. 221, 7 June 1907.

[3] Beveridge MSS., Coll. B, vol. xvi, item 10, 'Memorandum on the Relation of Employment Exchanges to Trade Unions', Dec. 1906.

[4] Ibid. 14, item 14, Report of the Trade Unions Sub-Committee of the Employment Exchanges Committee of the C.U.B., 14 Feb. 1907.

[5] Ibid., item 11, Frederick Johnson (clerk to C.U.B.) to W. H. Beveridge, 4 Sept. 1907, enclosing copy of a letter from the L.G.B., 31 Aug. 1907.

[6] Ibid., item 36, Frederick Johnson to labour exchange superintendents, 6 Nov. 1907.

[7] Ibid., item 41, 'Report and Recommendations on Local Advisory Committees', by the Employment Exchanges Committee of the C.U.B.; and item 47, 'Scheme for the Constitution of Local Advisory Committees'. Local Advisory Committees were to be appointed annually, five members being local borough councillors and four members nominated by the C.U.B. from local representatives of employers and workmen. They were to meet at least once a month, supervise local exchanges, and fill staff vacancies.

[8] C.U.B. Minutes, iii, 3 July 1908, Report of Employment Exchange Committee, p. 232.

extent until they were reinforced by a system of insurance in 1912.[1]

A third problem in the running of exchanges under the Central (Unemployed) Body arose over the appointment of suitable staff. The registrar of the pioneering voluntary labour exchange at Egham had stated that the manager of a labour exchange should be 'competent, earnest and impartial . . . possessing business aptitude and a kindly interest in the success of . . . those entered on his books'.[2] It was not at all clear, however, whether this ideal type should be drawn from the managerial or working class, and whether his experience should be clerical or industrial. Some proponents of labour exchanges suggested that it was necessary for managers to have personal experience of unemployment; but Arthur Lowry thought that 'there are certainly men in whom imagination can supply the place of actual experience'.[3] In March 1906 the Central (Unemployed) Body decided that each exchange needed a minimum staff of a superintendent and one clerk; but since the whole scheme was authorized only for an experimental three-year period such officials could be given no security of tenure nor prospects of future advancement.[4] In 1907 the staff quota of each exchange was increased, salaries were incremented, and plans were made for unifying local labour exchanges into a single service for the whole metropolis;[5] but even so, the precarious future of exchanges meant that it was difficult to recruit officials with 'business' capacity, who could command the confidence of both employers and workmen.[6]

[1] Below, p. 354.

[2] 8th Annual Report of the Egham Free Registry for the Unemployed, 1892.

[3] H. of C. 86/1906, Labour Bureaux, Report to President of L.G.B. by Arthur Lowry, Nov. 1905, p. 19.   [4] C.U.B. Minutes, i. 92, 23 Mar. 1906.

[5] 3rd Report of the C.U.B. (1907-9), pp. 74-7. Under the new system all metropolitan exchanges were made 'district branches' of a single, co-ordinated system, staffed by eighteen 'district managers', at £140 rising to £164 per annum, eighteen assistant managers, at £78 rising to £104 per annum, and office boys at £26 per annum. A female superintendent was attached to the central exchange. These arrangements were, however, incomplete by June 1909 (ibid., p. 75).

[6] Applicants for posts as exchange superintendents included a school attendance officer, traffic manager, journeyman compositor, librarian, warehouse foreman, political agent, trade union organizer, works manager, accountant, laundry manager, several kinds of clerk, and persons employed by private labour bureaux (Beveridge MSS., Coll. B, vol. xvi, items 28 and 29).

Finally, it was found that makeshift exchanges located in municipal buildings were 'with very few exceptions . . . calculated to repel rather than attract employers of labour'; and they were often remote from centres of employment. The Employment Exchanges committee therefore came to the conclusion that, in London at least, the local borough was an inappropriate unit for the organization of labour, and that exchanges should henceforth be located according to the distribution of industry and commerce.[1]

The 'employment exchange' aspect of the Unemployed Workmen Act was therefore little more successful than labour colonies or relief works in actually improving the situation of the unemployed. But Beveridge was able to convince people both inside and outside the Central (Unemployed) Body that the administrative deficiencies of labour exchanges could be overcome by the application of business principles and by national rather than local organization. 'Labour exchanges need to be recognised, industrialised, nationalised', he wrote in a memorandum to John Burns in June 1907, '. . . as instruments of industrial organisation they need industrial management.' He suggested three possible lines of future policy in order of preference. Firstly, a national network of local exchanges, directly established and supervised by the Board of Trade. Secondly, the creation of a special department by the Board of Trade to encourage and subsidize exchanges set up by local initiative. And thirdly, 'as second best, except the L.G.B. in place of the Board of Trade . . . but it has to be remembered that it is the Board of Trade which is in touch with Trade Unions and employers.'[2]

In September, Beveridge visited the German labour exchanges, which had been noted as an 'interesting experiment' in a Board of Trade inquiry in 1904.[3] He found that there were over 4,000 exchanges of various kinds in the German empire, filling over one and a quarter million

---

[1] *C.U.B. Minutes*, iii. 240, 3 July 1908.

[2] Beveridge MSS., Coll. B, vol. xiv, item 20, 'Memorandum as to Future of Labour Exchanges', by W. H. Beveridge (June 1907).

[3] Cd. 2304/1905, *Report to the Board of Trade on Agencies and Methods for Dealing with the Unemployed in Certain Foreign Countries*, by David Schloss, pp. 51–105.

vacancies a year. Of these the most important were public exchanges, organized and financed by municipalities, and voluntary exchanges, which received a municipal subsidy. The German exchanges were universally accepted by employers and trade unions, and in many cases were managed by joint committees of employers, workmen, and public officials. During trade disputes some German exchanges closed down, some carried on business as usual; but the most popular practice was for exchanges to stay open and to inform applicants which vacancies had been specifically caused by strikes. The most highly developed German exchange was at Strasbourg, where the 'labour office' was used for paying subsidies to trade-union insurance, 'testing' the destitution of able-bodied paupers, and supplying workmen to municipal contractors. The most successful exchange in relation to population was that of Freiburg, which was used by workmen as a regular means of changing their jobs without necessarily incurring interim unemployment.[1]

Beveridge's scheme for a national system of labour exchanges was submitted to the Royal Commission on the Poor Laws in October 1907. He proposed that such a system should be centrally controlled, based on the distribution of industry rather than local-government areas, and managed by persons experienced in business rather than charity. The management of exchanges should be assisted by local advisory bodies of masters and workmen; and strict neutrality should be observed in trade disputes. He claimed that a national system of labour exchanges, besides reducing to a minimum the unemployment caused by fluidity within the labour market, would also provide the necessary machinery for the treatment of all other aspects of the problem; for the collection of statistics, which would help to predict and stabilize trade depressions; for the verification of authentic unemployment as a necessary condition for the payment of relief or insurance benefits; and for the arrangement of employers' schemes to regularize employment and to dovetail casual occupations.[2]

[1] W. H. Beveridge, 'Public Labour Exchanges in Germany', *Economic Journal*, 18, no. 69 (Mar. 1908), 1–18.
[2] Cd. 5066/1910, *RC on the Poor Laws, Minutes of Evidence*, Q. 77832, paras. 52–76.

Beveridge's scheme was undoubtedly the most carefully planned and comprehensive plan for the treatment of unemployment that had so far been advanced; but more important than its administrative thoroughness was its political plausibility—a quality that the schemes of General Booth and George Lansbury conspicuously lacked. It was consistent with the realization of many different kinds of social and political principle—with socialism and individualism, trade unionism and charitable organization, the laws of the 'free market' and the inculcation of self-help.[1] Hence its appeal was widespread. The Webbs adopted the labour-exchange principle as a stage in the compulsory organization of the labour market; and in January 1908 Sidney Webb gave evidence to the Royal Commission suggesting that failure to register at a labour exchange on the part of a destitute unemployed workman should be made a serious criminal offence.[2] Even John Burns, to whom Beveridge had submitted his memorandum on 'The Future of Labour Exchanges' in June 1907, subsequently inspected the continental labour-exchange system.[3] In the autumn of 1907 Sir Hubert Llewellyn Smith, the new Permanent Secretary to the Board of Trade, entrusted Beveridge with the preparation of the Board's evidence to the Royal Commission on the Poor Laws, fully nine months before Beveridge, at the invitation of Winston Churchill, became a salaried employee of the Board of Trade.[4]

On the Central (Unemployed) Body, there was some criticism of Beveridge's scheme. George Lansbury objected that labour exchanges could do nothing to increase the total volume of employment,[5] and Russell Wakefield implied that

---

[1] W. H. Beveridge, 'Labour Exchanges and the Unemployed', *Economic Journal*, 17, no. 65 (Mar. 1907), 80–1.

[2] Cd. 5068/1910, *RC on the Poor Laws, Minutes of Evidence*, Q. 93031, Statement handed in by S. Webb, p. 189.

[3] Add. MS. 46299, f. 328, Sir Henry Fairfax Lucy to John Burns, 26 Aug. 1908. From this visit, Beveridge recorded, Burns drew 'the opposite moral to myself. To him as an old trade union organiser the great waiting-room of the Berlin exchange with its rows of men waiting for a call was a repellent sight—strike-breaking fodder; he waxed eloquent to me on this, and would not listen to my argument that it was better for the men to be sitting comfortably there than to be wearing out shoe-leather in blind search of jobs . . .' (*Power and Influence*, p. 92).

[4] Below, pp. 284–5.

[5] Beveridge MSS., L. ii. 218 a, G. Lansbury to W. H. Beveridge, 1 Feb. 1907.

all chairmen of committees had a vested interest in enlarging their own sphere of responsibility.[1] But a majority of the Central (Unemployed) Body voted in favour of a nationalized scheme under the Board of Trade in February 1908,[2] and in April their view was endorsed by a 'consensus' of the London advisory committees.[3] This conversion was highly significant, since it meant that those concerned with the management of exchanges at a local and voluntary level formed a pressure group for rather than against the planning and introduction of a national scheme.

Historians have differed widely about the significance of the Unemployed Workmen Act, some seeing it as an official extension of the policy of the Chamberlain circular,[4] others as the affirmation of an entirely new kind of responsibility for the unemployed.[5] Both views tend to overrate the positive intentions behind the Act, which was designed in the first instance merely to improve and assist the administration of certain forms of charitable relief. It was not originally intended as a temporary measure; but after the decision to hold an inquiry into the Poor Laws its operation was limited to three years, by which time it was hoped that the principles of Poor Law administration would have been restated or redefined. Before its machinery was properly under way, the Government that framed the Act was superseded by a Liberal administration committed to its amendment or repeal.

On the other hand it would be wrong to underestimate the practical significance of the Unemployed Workmen Act. It was a testing-ground for many of the theories of unemployment relief that had been worked out over the previous twenty years; and it gave scope for public experiments in relief works, labour colonies, labour exchanges, assisted emigration, and home visitation, at a time when more extensive measures on these lines were being widely discussed and canvassed among social reforming groups.

The importance of the Act was, however, more than

---

[1] Cd. 5066/1910, RC on the Poor Laws, Minutes of Evidence, Q. 78347.
[2] C.U.B. Minutes, iii. 110, 28 Feb. 1908.   [3] Ibid., p. 153, 3 Apr. 1908.
[4] S. and B. Webb, English Poor Law History, II. ii. 650–6.
[5] Kar de Schweinitz, England's Road to Social Security, pp. 179–80.

merely experimental. It also enabled a much more accurate picture to be gained of the nature and extent of unemployment and of destitution among the unemployed. 'Before the passing of the Unemployed Workmen Act . . . it was still possible to regard the problem of unemployment as mainly a question either of misfortune or changes of particular industries or of exceptional trade depression', reported Beveridge to an audience at the London School of Economics in 1907. 'The administration of the Act during the past two-and-a-half years has dispelled that view once and for all. . . . The great bulk (of applicants) are irregular workmen, normally in or on the verge of distress.'[1] Even John Burns, the sternest critic of the Act, admitted that it had 'certainly enabled a better judgment to be formed than was previously possible of the comparative prevalence of unemployment and of the quality of the men who formed the unemployed.'[2] The Act also performed a useful function in providing employment during the most acute commercial crisis since the 1870s, and hence to ward off panic measures of relief or legislation; and it supplied invaluable personal experience in social administration to those who were later to be responsible for more permanent schemes.[3]

From the point of view of future policy, however, the most important aspects of the administration of the Unemployed Workmen Act were threefold. Firstly, it discredited remedies that were external to the normal commercial and industrial system. Secondly, it suggested that charitable casework was useless as a basis for unemployment relief, partly because the unemployed were too numerous[4] and partly because few of them had an industrial or domestic 'character' which would stand up to the kind of inquiry prescribed by the COS. Such an investigation was moreover, increasingly unacceptable to many distress committees and working-class representatives on political grounds.

[1] Beveridge MSS., I. b. 356, 'Unemployment in Utopia', Address by W. H. Beveridge to L.S.E. Students' Union, 1907/8.
[2] CAB 37/91/33, 'The Unemployed Workmen Bill', by John Burns, 9 Mar. 1908.
[3] Cd. 5066/1910, *RC on the Poor Laws, Minutes of Evidence*, Q. 79619, para. 9, statement of the Bishop of Stepney.
[4] R. C. Davison, *The Unemployed*, p. 39.

Thirdly, the administrators of the Act came to the conclusion that 'it is impossible to deal adequately with unemployment by Local Authorities'; and in February 1908 the members of the Central (Unemployed) Body therefore passed a resolution demanding that unemployment should henceforth be treated as a national problem by a central government department, which would establish labour exchanges throughout the country, register the unemployed, locate available employment, and predict and anticipate the onset of depressions of trade.[1]

---

[1] *C.U.B. Minutes*, iii. 110, 28 Feb. 1908.

V

THE DEMAND FOR A NATIONAL
POLICY 1903–1908

By 1908 it was clear that Mr. Long's attempt to rationalize the local provision of relief had failed; and this failure was interpreted by many people, including a majority of the Central (Unemployed) Body, as proof that the problem of unemployment was inherently incapable of solution at a local level.

Since the 1880s, however, it had been evident to many people involved in the problem that isolated local attempts to give assistance or employment to the unemployed were a hindrance to labour mobility and positively attracted unemployed workmen into distressed areas. Moreover, the adequacy of local resources often varied inversely with the needs of the local unemployed. These difficulties had been pointed out by Keir Hardie and Percy Alden to the Select Committee of 1895; and since the early 1900s several groups of social reformers had been trying to work out a policy for the unemployed that would be national in application—either by the direct intervention of the central government or by the imposition of a compulsory and universal responsibility on local authorities.

The demand for a 'national policy' came primarily from three sources; from radical groups within the Liberal party, from the organized labour movement, and finally and most emphatically from the Royal Commission on the Poor Laws and Relief of Distress. These groups will be considered separately, although, as will be shown, there was a certain amount of interaction between them, and the formation of remedial policies was ultimately influenced by all three sources.

LIBERALISM AND UNEMPLOYMENT

The formation of new Liberal ideas on unemployment policy was merely an aspect of a much wider revolution in the Liberal attitude to social administration which occurred during the 1900s; a revolution in which many Liberals were consciously forced to abandon 'those principles of 1834 which have hitherto been the sheet-anchor of our social economics'.[1] This change in the Liberal approach to social problems had many origins, both intellectual and pragmatic. Theoretically, it was made possible by certain shifts of emphasis in orthodox economics, particularly the teaching of Alfred Marshall that gratuitous payments to persons in need did not necessarily depress wages, nor discourage thrift, nor act as an incentive to reckless procreation; they would instead 'raise wages, because the increased wealth of the working-classes would lead to better living, more vigorous and better educated people, with greater earning power, and so wages would rise. . . .'[2] This new teaching was a necessary prelude to the adoption of more positive attitudes towards social distress, and the abandonment of a policy based on 'deterrence'; and it was reflected in the writings of many of the new generation of enthusiasts for social reform.[3] Economic doctrines were, however, probably less important than the real or apparent constraints of democratic politics and the new definition imposed on urban and rural problems by empirical research.

During the late 1890s the discussion of domestic policy within the Liberal party had been eclipsed by controversies over party leadership and the problem of the Empire.[4] But at the end of the Boer war social questions gradually crept back into political discussion. At the end of 1901

[1] Phelps MSS., unsorted, James Bryce to Lancelot Phelps, 11 Mar. 1909.

[2] Marshall's evidence (1893) to the Royal Commission on the Aged Poor, quoted in Alfred Marshall, *Official Papers* (ed. J. M. Keynes), p. 249. Marshall's view is here cited as an indication of a trend rather than the actual cause of a new way of thinking. On the persistence of the opposite point of view see Add. MS. 48675, f. 90, Sir Edward Hamilton's Diary, 8 Nov. 1899.

[3] e.g. W. H. Beveridge, 'The Feeding of Schoolchildren; its Effect on Wages', *Morning Post*, 11 Apr. 1906.

[4] P. Stansky, *Ambitions and Strategies. The Struggle for the Leadership of the Liberal Party in the 1890s*, p. 298.

Lord Rosebery in a speech at Chesterfield echoed Sidney Webb's demand for a policy of 'national efficiency' when he advised the Liberals to replace the obsolete radicalism of the Newcastle programme with social and administrative reform.[1] In the same year a group of 'university settlement' Liberals headed by Charles Masterman called for a new approach to urban social problems through the development of public housing, public transport, and town planning;[2] and Philip Whitwell Wilson, the future Liberal M.P. for South St. Pancras, proposed that the casual labour market should be dissolved by the compulsory decentralization of London's industry and commerce.[3]

The most crucial factor in forcing a redefinition of Liberal policy was, however, the open declaration of Joseph Chamberlain in favour of an imperial tariff in May 1903.[4] The tariff-reform crisis influenced Liberal ideas on domestic policy in three ways. Firstly, by dividing the Conservative party it opened up the probability that the Liberals would win the next election; and from June 1903 onwards, the calculations of the Liberal leader, Campbell-Bannerman, and the Chief Whip, Herbert Gladstone, were framed with the immediate prospect of a general election constantly in mind.[5] Secondly, the tariff reformers claimed that protection would reduce unemployment and finance social reform;[6] and

---

[1] On the political and intellectual background to Rosebery's speech see G. R. Searle, 'The Development of the Concept of "National Efficiency" and its Relations to Politics and Government 1900–1910', Cambridge Ph.D. thesis, 1965, esp. pp. 296–303. The speech was welcomed by 'Liberal Imperialists' and, less enthusiastically, by pro-Boers and centre Liberals as a move towards party unity; but both groups were more interested in his South African proposals than in his rather indefinite suggestions for social reform (Bryce MSS., Box E. 15, H. Gladstone to James Bryce, 18 Dec. 1901).

[2] C. F. G. Masterman et al., The Heart of the Empire (1901). On the background to this work see Lucy Masterman, C. F. G. Masterman. A Biography, pp. 40–3.

[3] P. Whitwell Wilson, 'The Distribution of Industry', in The Heart of the Empire, pp. 111–235.

[4] On the circumstances of the tariff-reform crisis, which led to the resignation of both Chamberlain and the 'free fooders' from the Conservative cabinet in Sept. 1903, see P. Fraser, Joseph Chamberlain: Radicalism and Empire 1868–1914, pp. 221–51.

[5] Add. MS. 45988, ff. 45–6, Unsigned memorandum from H. Gladstone to H. Campbell-Bannerman, 24 June 1903.

[6] 'The Question of Employment', Speech at Newcastle by Joseph Chamberlain, 28 Oct. 1903, in Imperial Union and Tariff Reform, pp. 150–61.

whether or not this was a genuine claim, the Liberals were frightened into believing that this might be the case. 'The only conceivable lasting destroyer of the policy of Mr. Chamberlain is an alert and determined policy of Social Reform' commented the *Daily News* in July 1905.

It is mere folly to imagine that the continual promises of work in a monopolised home market which this great demagogue is continually offering will not sooner or later be accepted in desperation, if the only alternative is the maintenance of the status quo . . .[1]

Thirdly the Liberals, who had traditionally regarded themselves as the party of change, were forced by the challenge of protection into an ambiguous and compromising apologia for the existing social, economic, and fiscal system.[2] Since the permanent staffs of the Treasury and Board of Trade were almost unanimously in favour of free trade,[3] it was not difficult to cite official statistics that proved that Great Britain was commercially more prosperous than protectionist countries;[4] but the Liberals found themselves in a much less secure position when denouncing the proposed taxation of imported food. In June 1903 Campbell-Bannerman quoted the estimates of Booth and Rowntree to show that many working-class budgets contained no margin for additional taxation;[5] and he was thereby trapped into the admission that nearly a third of the population was living in poverty under the system of free trade. It was of little use politically to argue that 'if food had been taxed the condition of matters would have been infinitely worse'.[6] Similarly, by attacking indirect taxation in general as a regressive imposition on lower income-groups, radicals were

1 *Daily News*, 5 July 1905.
2 *Nation*, 7 Mar. 1908, Winston Churchill to the Editor, pp. 812–13.
3 Add. MS. 48681, ff. 11–12 and 113–14, Sir Edward Hamilton's Diary, 18 June 1903 and 14 Jan. 1904; Add. MS. 41237, ff. 171–2, F. Mowatt to H. Campbell-Bannerman, 27 Sept. 1903. See also Walter Layton, *Dorothy*, pp. 29–30.
4 e.g. Cd. 1761/1903 and Cd. 1337/1905, *First and Second Series of Memoranda, Statistical Tables, and Charts prepared in the Board of Trade with Reference to Various Matters bearing upon British and Foreign Trade and Industrial Conditions.*
5 Speech at Perth, 6 June 1903, on 'Mr. Chamberlain's Fiscal Proposals', printed in *Speeches by Sir Henry Campbell-Bannerman. Selected and Reprinted from the Times*, pp. 101–11.
6 Add. MS. 41237, f. 194, H. Campbell-Bannerman to J. Holmes Wood, 30 Oct. 1903.

forced to concede that the main instrument of direct taxation—the tax on annual incomes of £150 and over—had
become a permanent feature of the fiscal system.[1] In practice
this had been the case for the previous fifty years; but the
recognition in principle of the permanence of the income-
tax was a major departure from Gladstonian finance, and of
revolutionary significance for the future pattern of welfare
legislation.[2]

In order to retain their 'progressive' identity the Liberals
were therefore gradually compelled to redefine their attitude
to economic and social policy, and to counter Chamberlain's
argument that the protection of home manufactures would
provide work for the unemployed. The Liberal leaders were,
however, rather slow to respond to this aspect of the protectionist attack. Campbell-Bannerman was reluctant to
exploit working-class fear of unemployment;[3] and when in
1903 the Chief Whip, Herbert Gladstone, compiled a list
of topics on which prospective Liberal candidates should
be prepared to comment he included 'Poor Law reform'
but made no specific reference to the unemployed.[4] Moreover, Liberal sources of information suggested that protection was evoking little response from the working-class
voter. In October 1903 it was feared that London working
men were opting for protection;[5] but in the suburban by-
elections at the end of the year it was the business-men and
'clerk and villa' voters, rather than workmen liable to
irregular employment, who seemed to have been seduced
by the tariff-reform campaign.[6]

[1] Sydney Buxton, *Political Questions of the Day* (11th ed., 1903), p. 302, section
on 'The Graduation of the Income Tax'. Buxton remarked that 'In the earliest
editions of this book, this section stood under the heading of "Direct versus Indirect
Taxation", and the question of the retention or non-retention of the Income Tax
was argued out. But since the book first appeared, the retention of the Income
Tax is no longer open to question.'

[2] On the traditional view that the income-tax was a crisis tax that ought ideally
to be abolished in peace time see Add. MS. 48654, f. 56, Sir Edward Hamilton's
Diary, 23 Nov. 1890.

[3] Add. MS. 41214, ff. 131–2, Campbell-Bannerman to John Ellis, 10 Nov. 1903.

[4] Add. MS. 45988, ff. 45–6, Unsigned memorandum by Herbert Gladstone
24 June 1903.

[5] Add. MS. 41225, f. 15, Lord Ripon to Campbell-Bannerman, 18 Oct. 1903.

[6] Add. MS. 41211, ff. 254–7, James Bryce to Campbell-Bannerman, 28 Dec.
1903.

As unemployment increased in the winter of 1903–4, however, protectionist propaganda concentrated increasingly on the displacement of workmen by foreign competition.[1] The theoretical 'free trade' rejoinder—that tariffs would increase unemployment by reducing the purchasing power of foreign customers[2]—did not translate easily into terms that were comprehensible to the voting public;[3] and in December 1903 Herbert Gladstone came to the conclusion that it was politically essential to devise a more positive alternative to the threat of tariff reform.[4]

Liberal remedies for unemployment between 1903 and 1906 fell into two broad categories; those that were extensions of conventional radical policies of the 1890s, such as land revival and the equalization of rates; and those that prescribed a new kind of state intervention in education and industry. Typical of the former were the views of London radicals like Sydney Buxton, who thought that unemployment could best be relieved by a redistribution of the financial burdens of local authorities;[5] and of Lord Carrington and Sir Walter Foster, who ascribed the congestion of urban labour markets to agricultural decline and rural depopulation. 'The Land question, including as it does the question of Rating or Taxing Land Values, Housing and Royalties, with more easy access to the land for the peasant, must be the main alternative [to protection],' wrote Sir Walter Foster to Campbell-Bannerman in the autumn of 1903. 'Railway Rates too may become part of a new scheme for helping trade.'[6]

Pressure for a more dynamic policy came initially from a group of 'business' Liberals, who favoured such 'Germanic'

---

[1] Joseph Chamberlain, 'The Question of Employment', Speech at Liverpool, 28 Oct. 1903, printed in *Imperial Union and Tariff Reform*, pp. 150–61.

[2] PRO 30/60/44, Paper on 'The Fiscal Policy of International Trade' by Alfred Marshall, 31 Aug. 1903.

[3] The difficulty of popularizing the free-trade position on unemployment was pointed out by J. A. Hobson, 'A Tariff as a cure for Unemployment', *Nation*, 28 Mar. 1908, p. 933.

[4] Add. MS. 41217, ff. 42–3, H. Gladstone to Campbell-Bannerman, 5 Dec. 1903.

[5] Add. MS. 41238, ff. 8–11, S. Buxton to Campbell-Bannerman, 16 Jan. 1905.

[6] Add. MS. 41237, f. 214, Sir Walter Foster to Campbell-Bannerman, 28 Nov. 1903.

innovations as public investment in scientific research and the creation of a 'ministry of commerce'.[1] Since the early 1890s Sir John Brunner, the Germanophile chemical manufacturer and Liberal M.P. for Northwich, had been warning Parliament that the only way to ward off foreign competition was by improving British technical and commercial education;[2] and in November 1903 he urged Campbell-Bannerman to abandon *laissez-faire* in favour of public investment in technical education and the nationalization of canals.[3] There is no evidence to suggest that these proposals were specifically directed towards the reduction of unemployment among the working class; Brunner was far more interested in providing the commercial community with a counter-attraction to tariff reform. But Herbert Gladstone suggested that technical and commercial development might be welded together with a scheme for 'the reorganisation of Poor Relief'.[4] Campbell-Bannerman was unimpressed: 'Brunner is vague and wild,' he protested to Gladstone, '. . . I cannot at a moment's notice be expected to develop a new policy. . . . I can't propose social purchase in a casual way in a party speech.'[5]

In May 1904, however, Brunner presented Campbell-Bannerman with a memorandum, signed by a group of Liberal backbenchers, which amplified his proposals of the previous year. He argued that, while the protectionists greatly exaggerated Britain's commercial decline, it was nevertheless essential to counter their arguments with a positive

[1] The Associated Chambers of Commerce had been pressing for greater public investment since the early 1890s (Add. MS. 48661, f. 105, Sir Edward Hamilton's Diary, 28 Oct. 1893, report of a conversation with Sir Courtenay Boyle, Permanent Secretary at the Board of Trade: 'I asked him what the bent was of recent resolutions which had been received at the B. of T. from Chambers of Commerce. Was it fair trade? bimetallism? or what? It was, he said, state-aid. They wanted the state to build harbours and lighthouses, to do this and to do that for them. In fact, the Chambers were all on the track to socialism . . .').

[2] John Tomlinson Brunner (1842–1919), Liberal M.P. for Northwich 1885–6 and 1887–1909; Chairman of Brunner Mond & Co., the alkali manufacturers; pioneer of the eight-hours day (above, p. 69) and member of the Royal Commission on Canals and Inland Waterways.

[3] Add. MS. 41237, ff. 203–4 and 205, John Brunner to Campbell-Bannerman, 15 and 16 Nov. 1903.

[4] Add. MS. 41217, ff. 42–3, H. Gladstone to Campbell-Bannerman, 5 Dec. 1903.

[5] Ibid., f. 37, Campbell-Bannerman to H. Gladstone, 16 Nov. 1903.

policy of commercial and social 'betterment' which would keep industrial interests in the Liberal ranks.

It is the duty of the Liberal party . . . to advocate and strenuously to take in hand the development of the internal resources of the United Kingdom; for it is by a vigorous home policy, rather than by questionable schemes and extravagant expenditure abroad . . . that the discontent will be appeased and the welfare of the country best promoted.

Brunner's proposals centred on the internal and external development of all forms of transport and communication. Inside the country, he envisaged a nation-wide network of canals, the creation of County Highways Boards, and the nationalization and subsidization of trunk roads, which would stimulate domestic trade and regional development and at the same time 'smash Joe'. He cited the examples of Belgium, France, Germany, Austria-Hungary, and the Netherlands, where such a policy had diffused 'life, activity and energy . . . into all parts of the country'. To encourage foreign trade, he urged that ports and harbours should be modernized, and that the Board of Trade consular service, which collected commercial intelligence in foreign countries, should be 'manned by officials specially trained and selected for their work after the manner of our foreign competitors'. And finally, he proposed that public funds should be invested in scientific research and higher education—a policy that had been pursued with great commercial benefit in Germany, Holland, Switzerland, and the United States.[1]

Brunner's memorandum sounded a note that was becoming increasingly recurrent in discussions of social and economic reform—the inefficiency of the United Kingdom in comparison with foreign countries.[2] Campbell-Bannerman, whose 'little Englandism' embraced domestic as well as foreign policy, was slow to respond to these ideas; but he discussed them with Brunner during a visit to Manchester at the end of November. He admitted that a 'network of well-appointed canals would be an immense blessing'. But

[1] Add. MS. 45988, ff. 96–102, Memorandum to Campbell-Bannerman from John Brunner, Francis Evans, William Holland, Christopher Furness, D. A. Thomas, James Kitson, Charles McLaslen, M. Foster, 6 May 1904.

[2] See e.g. Ernest Williams, 'Made in Germany' (1896); Arthur Shadwell, Industrial Efficiency (1905).

he feared that this would involve not merely state investment in canals—'which would be a very new departure'—but the nationalization of railways, which would antagonize 'all the quiet people who live on railway dividends' and mean a great extension in the area of government employment.[1] Herbert Gladstone thought that Brunner was 'on the right track' and that the business aspects of 'canalization' should be referred to a Royal Commission. He proposed, however, that the question be deferred until the General Election, lest the Government should 'say, certainly—and then . . . claim all the credit'.[2]

Nevertheless, it was Herbert Gladstone who clearly brought social reform and particularly unemployment to the forefront of Liberal policy discussions. Gladstone's attitude at this time is rather surprising, in view of the rather negative attitude to labour questions that he displayed as Home Secretary after 1906. His interest in unemployment may be partly explained by the fact that as First Commissioner of Works he had been personally involved in promoting some of the schemes authorized by the Chamberlain circular in 1894–5.[3] But probably the most crucial factor was that the Conservative government showed signs of taking the problem seriously through the promotion of the London Unemployed Fund of 1904–5.[4]

Gladstone proposed initially that a committee should be set up to consider the affairs of the Local Government Board 'because I believe it to be of the utmost importance and questions connected with it are likely to come to the forefront at once'.[5] This was implicitly a radical departure from the traditional view that local government and Poor Law problems were purely administrative and outside the sphere

[1] Add. MS. 45988, ff. 139–41, Campbell-Bannerman to Herbert Gladstone, 2 Jan. 1905.

[2] Add. MS. 41217, f. 153, H. Gladstone to Campbell-Bannerman, 30 Dec. 1904. This compromise was accepted by the Brunnerites early in 1905 (Add. MS 41217, ff. 171–3, H. Gladstone to Campbell-Bannerman, 4 Jan. 1905).

[3] Add. MS. 46118, H. Gladstone's MS. and typescript autobiography, ff. 35–6.

[4] Above, pp. 153–4. It is possible also that Gladstone was influenced by the fact that the L.C.C. and his electoral allies the L.R.C. were advocating policies of counter-depressive public works.

[5] Add. MS. 41217, ff. 139–40 Herbert Gladstone to Campbell-Bannerman, 27 Nov. 1904.

of 'high policy'.[1] He also urged Campbell-Bannerman that 'informal sub-committees' should be set up to prepare a Liberal brief on problems of special importance; and the party leader agreed that Ireland, Education, School-feeding, Licensing, and Unemployment were subjects that 'require some more definite decision than we have yet given them'.[2] These 'small and confidential committees' were to 'consider practical methods for dealing with certain questions and where necessary to acquire special information'.[3] These proposals were endorsed by Asquith;[4] but a problem arose over the personnel of such committees, since Gladstone wanted to include Brunner and Sir Francis Mowatt in the unemployment committee and to use John Burns and Sidney Webb as 'consulting doctors'.[5] Campbell-Bannerman was reluctant, however, to commit himself to future appointments by formally consulting anyone who had not held office in a previous Liberal administration; even though, as he himself admitted, 'we sadly want new blood and fresh views from all corners of the party'.[6] He therefore let the matter slip, and there is no evidence to suggest that any of these committees had more than a hypothetical existence; but the correspondence about them was indicative of a new consciousness of social problems, not merely among radical backbenchers but in the Liberal party machine.

Public pressure by and on behalf of the unemployed, however, made a more definite statement of policy imperative. During his visit to Manchester at the end of November

[1] e.g. CAB 37/67/79, Memorandum on 'Alien Immigration' by Walter Long, 30 Nov. 1903. Long remarked that questions of state intervention were matters of 'high policy', and 'in this my Department is not specially concerned'.

[2] Add. MS. 45988, f. 129, Campbell-Bannerman to Herbert Gladstone, 23 Nov. 1904.

[3] Add. MS. 41217, f. 145, H. Gladstone to Campbell-Bannerman, 10 Dec. 1904.

[4] Add. MS. 41210, ff. 237–8, H. H. Asquith to Campbell-Bannerman, 6 Dec. 1904.

[5] Add. MS. 41217, ff. 139–40, H. Gladstone to Campbell-Bannerman, 27 Nov. 1904.

[6] Add. MS. 45988, ff. 132–3, Campbell-Bannerman to Herbert Gladstone, 5 Dec. 1904: 'If at this time we pick out people for confidential consultation it comes precious near (in their eyes) a rehearsal of a cast for a new Government. Would it not be safer to keep it among ourselves and let the members of the Committees beat about for opinions and advice among outsiders . . .'

1904 Campbell-Bannerman was lobbied by representatives of the local unemployed, who urged him to press for the instant recall of Parliament, the abolition of disfranchisement for unemployed paupers, and the transfer of responsibility for relief from local to central government. The Liberal leader was caught unprepared. He equivocated by conceding that the temporarily unemployed workman 'who had always kept his head straight' ought not to 'lose his citizenship'; and that 'while the central authority ought to give large latitude to local authorities, the central authority ought to have large powers of compelling the local authorities to do their duty'.[1] This reply was so vague as to be meaningless. But the incident was useful in awakening Campbell-Bannerman to the sheer embarrassment of having no preconceived policy on such a pressing problem; for, as he later confided to Herbert Gladstone, he 'did not know what the mischief to say'.[2]

Gladstone was himself challenged on the subject at a Liberal rally in Leeds on 5 and 6 December; and three days later at a meeting in West Ham he 'found the place alive with unemployed men in conjunction with the socialists, in particular the S.D.F. men and Hyndmanites generally'. The East End was 'red hot' with the discontent of the unemployed, of which not only socialists but tariff reformers were seeking to take advantage.[3] Chamberlain in a speech at Limehouse on 15 December compared the standard of living of the English workman unfavourably with that of his American and colonial counterparts, and blamed free imports and the competition of alien labour. He promised that the result of protection and immigration restriction would be 'more remunerative employment for those who have to gain the subsistence of themselves and their families

---

[1] *Liberal Magazine*, Dec. 1904, pp. 705–6.

[2] Add. MS. 45988, f. 133, Campbell-Bannerman to Herbert Gladstone, 5 Dec. 1904.

[3] Add. MS. 41217, ff. 143–4, H. Gladstone to Campbell-Bannerman, 10 Dec. 1904. Gladstone also warned Campbell-Bannerman that the unemployed were complaining about 'some vote which *you* gave some time ago on I think the Labour Comⁿ'; this was possibly a reference to the Select Committee of 1895, when Campbell-Bannerman had voted against the proposed abolition of disfranchisement for deserving unemployed paupers (H. of C. 111/1895, *SC on Distress from Want of Employment, Proceedings*, p. xv. Discussion of para. 8).

by the work of their hands'.[1] The situation in Leeds and the East End persuaded Herbert Gladstone of the need to crystalize his shadowy ideas on unemployment; and he became convinced that 'we have got to open this question and personally I am quite clear that we ought to make a new departure'.[2]

He therefore drew up a memorandum designed to 'draw fire' from his colleagues, which outlined a new Liberal policy on unemployment.[3] He pointed out that at the best of times there were between two and three hundred thousand unemployed workmen in the United Kingdom. In periods of depression, as many as 16 per cent of workmen in skilled trades might be out of work, and an immeasurably larger proportion of unskilled and casual workmen: '. . . this condition of things continually recurs at intervals of a few years; and . . . is growing more serious and intolerable.'[4] Moreover, the impact of unemployment was concentrated in certain areas, which were often those least able to cope with the problem.

West Ham, for example, is absolutely incompetent to deal with the mass of men out of work—men who work in national industries to the benefit of the nation. The state of things there is a disgrace to the community and the Poor Law is wholly inefficient to deal with it. Why should charitable people bear the burden of giving relief when it is no concern of theirs? While the vast majority of well-to-do people never lift a finger? Genuine workmen in thousands are demoralised by being forced to depend upon this private charity or the workhouse.[5]

Gladstone therefore proposed that the central as well as the local authorities should be responsible for the unemployed; that the State should undertake public works which would in the long run be both useful and profitable, but where the returns to investment were too slow to attract private capital; and that the operation of these works should be regulated to

1 *Liberal Magazine*, Jan. 1905, p. 727.
2 Add. MS. 41217, ff. 143–4, H. Gladstone to Campbell-Bannerman, 10 Dec. 1904.
3 No copy of this memorandum exists in the Gladstone, Campbell-Bannerman, Bryce, or Asquith papers. Its contents are here summarized from the subsequent correspondence in which Gladstone's proposals were criticized and defended.
4 Bryce MSS., Box E. 15, H. Gladstone to James Bryce, 17 Dec. 1904.
5 Add. MS. 41217, f. 165, H. Gladstone to H. Fowler, 1 Jan. 1905.

absorb temporarily unemployed workmen without removing them permanently from the open labour market.[1] He urged Campbell-Bannerman to make a public statement on unemployment and to 'press the government to make a searching examination into the possibilities of a national scheme'.[2]

Gladstone's analysis of the problem was not basically different from that of the London radicals who saw unemployed distress mainly as a projection of the problem of local rates. But his remedy was closer to the ideas that the I.L.P. and collectivist Liberals like Percy Alden had been advancing since the 1890s.[3] The view that public works should be 'reproductive' was clearly influenced by the Brunnerites. In principle, Gladstone's proposals lay halfway between the Chamberlain circular which suggested merely that normal public works schemes should be synchronized with depressions, and the Minority Report of the Poor Law Commission which advocated ten-year programmes of public works deliberately designed to eliminate fluctuations of trade.[4]

This memorandum was circulated by Campbell-Bannerman to Asquith, Bryce, Spencer, Fowler, and John Morley in December 1904.[5] Significantly, the criticisms that it evoked were mostly political rather than economic. Fowler, who had been more involved than other Liberal politicians in Poor Law questions, was reluctant to abandon the local basis of relief.[6] He thought that, whatever the faults in the administration of the Poor Law, its principles were 'sound and ... impregnable', and he recalled that 'Lancashire as a *county* dealt with the terrible distress of the cotton famine'. He believed that the permanent establishment of centralized public works would be tantamount to national workshops—

---

[1] Add. MS. 41217, ff. 165–6, H. Gladstone to H. Fowler, 1 Jan. 1905.

[2] Bryce MSS., Box E. 15, Herbert Gladstone to James Bryce, 17 Dec. 1904.

[3] Below, pp. 227, 235.

[4] Cd. 4499/1909, *RC on the Poor Laws, Minority Report*, p. 1197.

[5] Add. MS. 41210, ff. 241–2; Add. MS. 41217, ff. 164–6; Add. MS. 41223, f. 138; Add. MS. 46019, ff. 84–5.

[6] Fowler's biographer claimed, however, that he was anxious to give legislative priority to Poor Law reform, non-contributory Old Age pensions and national expenditure on 'legislation to improve the condition of the people' (E. Fowler, *Life of H. H. Fowler, 1st Viscount Wolverhampton*, p. 139).

'a remedy from which the country would shrink'.[1] Fowler's criticisms were echoed by Bryce, who discussed the proposals at length with Campbell-Bannerman.[2] He thought that there could be no valid objection to 'reproductive works' designed to coincide with periods of temporary unemployment;[3] but he warned the Liberal leader to tread with extreme caution, since 'we are exposed just now not only to Tory criticism, but to the risk of being asked to fulfil promises, so it is more than usually necessary to guard against misconstruction'.[4] And to Gladstone Bryce wrote that it was essential to avoid any admission 'that it is the duty of the State to provide work—a doctrine which would cause general alarm. . . . The enemy would lay hold of this at once—and indeed it would be a dangerous admission.'[5]

The most encouraging reaction to Herbert Gladstone's scheme came from Asquith, who agreed that it was necessary to plan public works in advance, and that 'the burden should fall, not on the charitable, nor on the specially afflicted districts, but on the whole community'. He was, however, opposed to projects that could not be easily adjusted to trade fluctuations, or that involved 'the use of expensive plant which cannot be allowed to be idle'; and he suggested that a 'large well-thought out scheme of afforestation' would be 'the most practical plan'. Such a scheme would require no special preliminary inquiry and 'is a matter which any Government ought to be prepared to handle and make the subject of proposals'.[6]

Nevertheless Campbell-Bannerman's response to the whole discussion of a public-works policy seems to have been

---

[1] Add. MS. 41214, ff. 258–60, H. Fowler to Campbell-Bannerman, 26 Dec. 1904.

[2] James Bryce (1838–1922), historian, lawyer, and Gladstonian Liberal; President of the Board of Trade 1894–5; Chairman of the RC on Secondary Education 1894–5; Chief Secretary for Ireland 1905–7; British Ambassador to Washington 1907–13; Viscount 1914.

[3] Add. MS. 46019, ff. 84–5, James Bryce to H. Gladstone, 14 Dec. 1904.

[4] Add. MS. 41211, ff. 290–1, James Bryce to Campbell-Bannerman, 19 Dec. 1904.

[5] Add. MS. 46019, ff. 84–5, James Bryce to H. Gladstone, 14 Dec. 1904; ff. 86–7, 19 Dec. 1904.

[6] Add. MS. 41210, ff. 241–2, H. H. Asquith to Campbell-Bannerman, 1 Jan. 1905.

little more than superficial. His speech at Limehouse on
20th December reflected Bryce's warning about the danger
of the 'right to work'. He urged public and private employers
to exercise greater foresight in planning their work and to
bring forward 'necessary public works which were not
necessarily urgent'. He rejected the idea of state works 'for
the mere purpose of giving employment', but approved of
works 'which could be justified on their merits as likely to
prove reproductive'. The main part of his speech was,
however, devoted to an attack on Chamberlain's promises
of increased employment through tariff reform. He prophesied that London workmen would be particularly vulnerable to tariffs, since so many of them were engaged
directly or indirectly in the import and export trades; and
he evoked the shadow of corruption that had characterized
the old protectionist system. 'Employment? Yes, yes; more
employment to the customs officer and to the smuggler, and
to another class, the class of trust promoters and monopoly-
mongers and tariff touts.' For the immediate relief of distress
he proposed instead the removal of rating inequalities and
the unification of London government. It was 'wrong in
principle and disastrous in practice to constitute a system of
administration for London in watertight compartments'.
And he praised the organization of Long's scheme as a
recognition of this evil, although he was sceptical about the
usefulness of 'forms and machinery' to the actual relief of
the unemployed.[1]

Campbell-Bannerman's Limehouse speech did not really
signify a redirection of Liberal unemployment policy. It was
welcomed as a 'judicious' new departure by his front-bench
colleagues, but they were less impressed by its originality
than relieved by its moderation;[2] and a far more radical
speech that Campbell-Bannerman made a month later,
recommending the abolition of poverty and the raising of
domestic consumption as remedies for unemployment, went

[1] Speech at Limehouse, 21 Dec. 1904, on 'Social Reform and Fiscal Policy',
*Speeches by Sir Henry Campbell-Bannerman, 1899–1908. Selected and Reprinted
from the Times*, pp. 156–65.
[2] Add. MS. 41217, ff. 147–8, H. Gladstone to Campbell-Bannerman, 25 Dec.
1904; Add. MS. 41214, f. 258, H. Fowler to Campbell-Bannerman, 26 Dec.
1904.

virtually unnoticed by his colleagues.[1] In fact, the discussion
raised by Gladstone in the autumn of 1904 seems to have
petered out in the early spring of 1905. In December 1904
Campbell-Bannerman had mentioned the possibility of
setting up an unemployment committee under Henry
Fowler, but this never materialized.[2] The eclipse of Gladstone's scheme can be explained partly by the triumph of
more cautious counsels; partly by the revival of trade; and
partly by the announcement of the President of the Local
Government Board that he intended to introduce legislation
which promised to fulfil some if not all of the conditions
laid down by the Gladstone memorandum.[3]

Throughout 1905 it is difficult to discern any clear or
unanimous line of approach to the problem in Liberal discussions. Unemployment figured as a secondary item in the
Liberal attack on the Government's 'Chinese labour' policy,
when it was argued rather implausibly that indentured
coolies on the Rand were taking employment that had been
designed for British workmen.[4] In January Sydney Buxton
made it clear to Campbell-Bannerman that he welcomed the
Unemployed Workmen Bill as a recognition of the duty of
the State to provide employment;[5] and during the second
reading of the Bill Lloyd George, the leader of the Welsh

[1] Speech at Stirling, 17 Jan. 1905, printed in the *Liberal Magazine*, Feb. 1905,
p. 42; 'It could not be too often repeated and enforced that the way . . . to organise
their home market was not the crude and unequal and exploded method of setting
up tariffs. It was to raise the standard of living, abolishing those centres of stagnant
misery which were a disgrace to our name, and when once their home market was so
organised the demand for labour would be larger and more sustained and more
capable of ensuring itself against fluctuation. The worst obstruction to a steady and
growing home market were these very same bad conditions of life which created
bad workmen and bad customers.'

[2] Add. MS. 45988, ff. 134–5, Campbell-Bannerman to H. Gladstone, 23 Dec.
1904.

[3] Above, pp. 158–9.

[4] Thomas Burt, *The Chinese Labour Question*, Liberal Leaflet No. 2067/1905.
Lord Ripon had remarked to Campbell-Bannerman in 1904 that 'it is curious to
remember how one of Rhodes's great arguments in favour of his policy used to be
that the Transvaal and Rhodesia would afford a wide field for the employment of
British labour, which is now the one thing which the Mine Magnates will not hear
of!' (Add. MS. 41225, f. 22, Lord Ripon to Campbell-Bannerman, 18 Jan. 1904.)

[5] Add. MS. 41238, ff. 8–11, S. Buxton to Campbell-Bannerman, 16 Jan. 1905.
The Liberal leader agreed that Long's proposals appeared to be 'satisfactory' and
'we should strike no discordant note' (Buxton MSS., unsorted, Campbell-Bannerman
to S. Buxton, 24 Jan. 1905).

radicals, proclaimed to the House of Commons that it was an admission of the 'right to work'.[1] The Liberal Publication department, which by 1905 was a good deal to the 'left' of the party leaders on domestic issues, warmly recommended Percy Alden's book on *The Unemployed: a National Question*, which called for the creation of a Minister of Commerce and Industry, and the provision of labour colonies, state subsidies to trade-union insurance, and counter-depressive public works.[2] But when Samuel Barnett and William Beveridge addressed a meeting at the House of Commons early in March they found that Liberal M.P.s in general were lamentably ignorant about current proposals for dealing with the unemployed.[3] It was in fact the enthusiasm of individuals and of Liberal writers and publicists that created the impression that the Liberal party had a policy on unemployment. Bryce on 7 November 1905 wrote to Herbert Gladstone in some alarm about 'the reckless way in which some treat the unemployed question. When these bills, drawn in a light heart, come to be paid there will be trouble. Could anything be done to whisper words of caution?'[4] John Morley in a speech at Walthamstow on 20 November claimed that there was no simple explanation for unemployment and that 'it was the mark of a narrow and untrained mind to believe ... that you could give the question a single answer'.[5] Morley's speech infuriated all those Liberals who had been striving to find a remedy,[6] and forced Campbell-Bannerman to clarify the party's attitude to the problem of the unemployed. In a speech at Partick on 28 November he defined unemployment in terms of an imbalance between urban and rural population and resources, which he proposed to remedy by 'colonising our own Country';[7] and in the following month the policies pre-

[1] *The Times*, 8 Aug. 1905.
[2] *Liberal Magazine*, Feb. 1905, p. 63; P. Alden, *The Unemployed. A National Question*, pp. 137–43.
[3] Beveridge MSS., L. i. 203, W. H. Beveridge to Annette Beveridge, 4 Mar. 1905; H. Barnett, *Canon Barnett. His Life, Work and Friends*, ii. 244.
[4] Add. MS. 46019, f. 101, James Bryce to H. Gladstone, 7 Nov. 1905.
[5] *Liberal Magazine*, Dec. 1905, pp. 670–1.
[6] Add. MS. 41238, ff. 97–9, Sydney Buxton to Campbell-Bannerman, 29 Nov. 1905.
[7] Ibid., ff. 114–15, Sir Walter Foster to Campbell-Bannerman, 30 Nov. 1905.

scribed in this speech were adopted as part of the Liberal party's official programme in the election campaign.[1]

The promises of the Liberal party in the election of 1906 have been subject to a certain amount of confusion among historians. In view of the Liberal government's subsequent record of social legislation it was often assumed by writers involved in these events that the Liberals had come to power with a commitment to, and a mandate for, social reform.[2] More recent historians have suggested that this was not entirely true, and that very little of the social legislation of the next eight years was foreshadowed at the time of the election.[3] Nor is there any reason to suppose that anticipation of social change was a very potent factor in gaining electoral support. For reasons that have never been fully explained, 'social reform' was notoriously lacking in electoral appeal, even after the extension of the franchise.[4] A French political scientist, Émile Boutmy, analysing the political character of the English in the early 1900s, marvelled at the complaisance of the poor in the face of flagrant social inequality; and astutely surmised that such discontent as there was arose mainly from a comparison with more prosperous members of their own class rather than with the classes above them.[5] In January 1906 the electoral turnout

[1] Speech at the Albert Hall, 22 Dec. 1905, 'The Liberal Government's Programme', *Speeches by Sir Henry Campbell-Bannerman 1899–1908. Selected and Reprinted from the Times*, pp. 174–83.

[2] R. C. K. Ensor, *England 1870–1914*, pp. 384, 391, stated that in 1906 Campbell-Bannerman's 'name was . . . the watchword of the radicals and the young . . . Radicalism and socialism alike, released from the suppressions of two decades were radiant with sudden hopes of a new heaven and a new earth . . .'.

[3] Bentley G. Gilbert, *The Evolution of National Insurance in Great Britain*, p. 202, states that 'the Liberals . . . showed no interest in social reform legislation during the general election of January 1906'. This is true only if one adopts a very narrow definition of 'social reform', since although the election was primarily fought on the tariff issue, Liberal ministers also committed themselves to home colonization, taxation of site values, Poor Law reform, development of public transport, and the protection of trade unions from liability to damages. This was, however, a very different social programme from that enacted by the Liberals several years later; moreover, the fact that Campbell-Bannerman also promised massive public retrenchment suggests that neither he nor his colleagues fully understood the financial implications of social welfare policies.

[4] For an interesting discussion of this point see H. Pelling, 'The Working Class and the Origins of the Welfare State', in *Popular Politics and Society in late Victorian Britain*, pp. 1–18.

[5] Émile Boutmy, *The English People. A Study of their Political Psychology*, pp. 133–4.

was high and widespread excitement seemed to pervade the voting public. For the first time Labour appeared as a significant political force. But, as John Morley observed to Lord Minto

the wonder is . . . that it did not come sooner, considering that town workmen have had votes for forty years and rurals for twenty. There will be some wild-cat talk, but I represented workmen in Newcastle for a dozen years and I always felt that British workmen are essentially bourgeois, without a bit of the French Red and the Phrygian cap about them.[1]

Herbert Gladstone, summarizing the reasons for the Liberal victory in order of importance, placed the electoral pact with Labour, free trade, the critique of the South African war, and the desire for change for its own sake before 'arrears of industrial and social legislation'.[2] Analysing the composition of the new Liberal majority, he told Campbell-Bannerman that 'the most striking thing about it is the preponderance of the "centre" Liberals. There is no sign of any *violent* forward movement in opinion. . . . The dangerous element does not amount to a dozen.'[3]

Nevertheless, it was not true that the Liberal party was not committed to certain social improvements, or that they were not looked upon by contemporaries as a party of reform.[4] During the years in opposition, the constraints of

[1] John Morley, *Recollections*, ii. 157. Cf. Add. MS. 46287, ff. 325–6, G. B. Shaw to John Burns, 11 Sept. 1903.

[2] Add. MS. 46118, H. Gladstone's MS. and typescript autobiography, f. 102. The Liberals won 401 seats, the L.R.C. 29, Unionists 157, and Irish Nationalists 83. The Government therefore had an absolute majority of 132 over all possible combinations of opposition parties.

[3] Add. MS. 41217, f. 294, H. Gladstone to Campbell-Bannerman, 21 Jan. 1906. Political opponents claimed that the Liberal majority contained a disproportionate number of 'eccentric, dissident or disreputable MPs', with little experience of public life (Earl Winterton, *Orders of the Day*, p. 47). But in fact it was largely composed of middle-aged members of the commercial and professional middle class; many of them necessarily lacked parliamentary experience, but a very high proportion had previously taken an active part in extra-parliamentary public affairs (J. A. Thomas, *The House of Commons, 1906–11. An Analysis of its Economic and Social Character*, pp. 18, 25, 31, 35, 44–5, 49).

[4] e.g. Bryce MSS., P. 6, Sir John Gorst to James Bryce, 7 Feb. 1906. Moreover, the very fact that Balfour and Lansdowne arranged in advance to frustrate Liberal measures through the House of Lords suggests that the Government was expected to introduce radical reforms (R. C. K. Ensor, op. cit., pp. 386–7).

democratic politics, the penetration of sociology into urban problems, and the technological shortcomings of British industry had persuaded many Liberals to advance beyond the conventional radicalism of the Newcastle programme; and it has been shown that in the discussion of unemployment the ideas of the new radicalism were largely independent of, and to a certain extent in competition with, the old. But the new radicalism of the 1900s was by no means a homogeneous movement. It covered a wide spectrum of persons, opinions, and problems; it included 'imperialists' like Haldane and the 'little Englanders' of the *Daily News*, technocrats like Sir John Brunner, and sentimental proponents of 'back to the land'. At one extreme was Sir Christopher Furness, a bitter enemy of Keir Hardie-ite socialism, who nevertheless supported Brunner's proposals for public investment in industry and scientific education.[1] At the other extreme was Percy Alden, newly elected for Tottenham, whose views on public ownership and under-consumption were virtually indistinguishable from those of the I.L.P.[2] The largest radical group in the new Parliament were those who wanted the imposition of a tax on the site-value of land; and even they were divided into those who were mainly interested in the reform of central and the reform of local taxation.[3] 'Social reform' was in fact a politically divisive rather than cohesive influence in the Liberal party; and even after 1906 Liberal 'social reformers' were a federation of cliques rather than a uniform and clearly identifiable wing of the parliamentary party.

The strength of 'new liberalism' was, moreover, mainly a back-bench strength, and the aims of the party leaders were almost whiggish in their moderation. Campbell-Bannerman was anxious to avoid giving the Conservatives any excuse for accusing the Government of socialism;[4] and John Morley was still disposed to see all political, social, and

[1] Add. MS. 41225, ff. 142–4, Lord Ripon to Campbell-Bannerman, 23 Sept. 1906.

[2] *Hansard*, 4th series, vol. 161, col. 444.

[3] A. S. King, 'Some Aspects of the History of the Liberal Party, 1906–1914', Oxford D.Phil. thesis, 1962, pp. 75–8.

[4] Asquith MSS., vol. 10, f. 200, Campbell-Bannerman to H. H. Asquith, 21 Jan. 1906.

economic problems, including unemployment, as either the inevitable price of technical progress or as the aftermath of an unjust war.[1] Of the senior ministers Richard Haldane was identified with the 'national efficiency' school of administrative reformers; but he was in no position at the War Office to initiate major social reforms. In 1907 he appointed a representative committee to advise on the stabilization of employment in the ordnance factories;[2] but throughout this period the munitions industry was notorious for its irregularity and deliberate cultivation of a supply of surplus labour.[3] Asquith had acquired a 'progressive' reputation, mainly through association with Haldane and more remotely with the Webbs;[4] but his first year at the Treasury was spent in restoring the sinking-fund, reducing the national debt, and de-protectionizing the fiscal system; and there was no outward evidence to suggest that he was about to preside over a revolution in public finance. Of the ministers most directly concerned with social policy, Herbert Gladstone had a private interest in but no public commitment to social reform; and he was too overwhelmed by the responsibility of making sixty decisions a day at the Home Office to make serious policy innovations.[5] Lloyd George, the President of the Board of Trade, had not yet revealed that his active radicalism extended beyond Celtic nationalism and hostility to landlords, imperialists, and the established Church.[6] John Burns, the President of the Local Government Board, a hero of the unemployed demonstrations of twenty years before, was himself regarded as a symbol of social change.[7] Yet his attitude to social reform and to unemployment in particular was extremely ambiguous. He saw

[1] *Liberal Magazine*, Dec. 1905, pp. 670–1.

[2] *Report of the Bath T.U.C.*, 1907, p. 87.

[3] R. C. Trebilcock, 'A "Special Relationship"—Government, Rearmament and the Cordite Firms', *Economic History Review*, N.S. 19, no. 2 (Aug. 1966), 374–5.

[4] On Beatrice Webb's hopes of Asquith as an advanced radical and her subsequent disillusionment see *Our Partnership*, pp. 104, 112, 225, 227.

[5] Add. MS. 46118, H. Gladstone's MS. and typescript autobiography, f. 25; Bryce MSS., P. 6, H. Gladstone to James Bryce, 19 Mar. 1907.

[6] T. Jones, *Lloyd George*, chapters 1 and 2.

[7] Lucy Masterman, op. cit., p. 68: '. . . you expect to meet Queen Elizabeth, and lo! it is John Burns . . .'.

unemployment as part of a complex of urban problems which could be indirectly removed by the nation-wide adoption of self-help, combination, and technical education;[1] and he also ascribed it to the shortage of capital in the domestic economy, particularly in the building trade, caused by excessive military expenditure overseas.[2] At the same time Burns, by 1905, was convinced of the intemperance and personal inferiority of many of the unemployed:[3] and he regarded himself as the guardian of the morals of the poor against the 'virus of pampered dependency' which was being engendered by unconditional outdoor relief and the Unemployed Workmen Act.[4]

There was, moreover, within the Liberal party, even among the radicals, a widespread feeling that social improvement could best be achieved not by expensive social legislation but by an all-round reduction in public expenditure. Reform in this sense was primarily equated with peace and retrenchment. Francis Hirst, the future editor of the *Economist*, thought that depression and distress could be simply explained by over-taxation and wasteful expenditure on armaments; of which the outcome was 'dear money, lowered credit, less enterprise in business and manufacturers, reduced home demand and therefore reduced output to meet it, reductions in wages, increase of pauperism and unemployment'. He urged Campbell-Bannerman that 'to restore credit and to lower taxes is the first great remedy for unemployment and the first great mission of the Liberal government'.[5] This view was far more common than, for instance, that of Sir Leo Chiozza Money, who thought that unemployment could only be cured by nationalization, the equalization of resources and 'the organisation of services under public control'.[6]

---

[1] *Hansard*, 4th series, vol. 169, cols. 952–62.

[2] Add. MS. 46299, ff. 25–34, Burns's draft of a letter to the Editor of *The Times*, 24 Jan. 1906.

[3] John Burns, *The Straight Tip to Workmen. Brains Better than Bets or Beer* (Clarion Pamphlet No. 36), 1902; *Labour and Drink*, Lecture in the Free Trade Hall, Manchester, 31 Oct. 1904.

[4] Add. MS. 46324, Burns Diary, 12 May 1906.

[5] Add. MS. 41238, ff. 251–2, Francis Hirst to Campbell-Bannerman, 29 Dec. 1905.

[6] L. C. Money, *Riches and Poverty* (1906 ed., first publ. 1905), pp. 255–6;

It is not therefore surprising that in the policies of the Liberal government old-fashioned radicalism and financial retrenchment initially prevailed. In the election of 1906, unemployment had figured mainly as a side-issue in the debates on protection and land-revival; and although the King's speech promised an amendment of the Unemployed Workmen Act, the existence of distress committees and the revival of trade reduced the immediate urgency of the problem.[1] A Cabinet committee under John Burns, which was appointed in December 1905 to deal with unemployment, therefore decided in July 1906 to postpone remedial legislation until the report of the Royal Commission on the Poor Laws had been received.[2] In the spring the creation of a Royal Commission on Canals and Inland Navigations under Lord Shuttleworth removed these issues out of practical politics for two and a half years. In March Professor Oliver Lodge of Birmingham University tried to persuade Burns to give part of the parliamentary grant to the Midlands Reafforesting Association to employ the unemployed in the Black Country.[3] But Burns replied that the Association was not eligible for such a subsidy under the terms of the Unemployed Workmen Act.[4] In March 1907 Burns convened a conference of 'large employers of labour' on problems of decasualization,[5] but with no practical results. In Asquith's first Budget speech the problems of rating inequalities and the relation of central to local taxation were sympathetically shelved, as they were year by year throughout this period.[6] The only minister who suggested a substantial new departure in social administration was Lord Carrington, the

---

Leo Chiozza Money (1870–1944), financial journalist and contributor to *Daily News*; Liberal M.P. for North Paddington 1906–10 and East Northants. 1910–18; unsuccessful Labour candidate 1918.

[1] *Hansard*, 4th series, vol. 152, col. 24. The only M.P. who protested against the absence of immediate action on behalf of the unemployed was the ex-Secretary of the Primrose League and radical Conservative M.P. for Hoxton, the Hon. Claude Hay (ibid., cols. 876–81).

[2] CAB 41/30/35, Campbell-Bannerman to the King, 14 Dec. 1905; CAB 41/30/69, Campbell-Bannerman to the King, 13 July 1906.

[3] Add. MS. 46299, ff. 285–7, Oliver Lodge to John Burns, 20 Mar. 1907.

[4] Ibid., f. 288, W. Jerred to Oliver Lodge, 23 Mar. 1907.

[5] Beveridge MSS., L. ii. 218 a, John Burns to W. H. Beveridge, 21 Mar. 1905.

[6] *Hansard*, 4th series, vol. 156, cols. 283–4. See Appendix A, pp. 369–70.

President of the Board of Agriculture.[1] Carrington had for
the previous twenty years been experimenting with the
creation of smallholdings on his Lincolnshire estates and
had urged other landowners to follow his successful example
in stemming the flood of migration to the towns.[2] He now
proposed to convert his experience into public policy by
creating smallholdings on Crown lands and by turning the
State into a 'model landlord', represented by an amalgamation of his own department with the Office of Woods and
Forests. 'We should show . . . that land may be so managed
as to provide a greater amount of employment, much more
favourable to the occupiers than at present, without serious
detriment to the interests of the owners.'[3] This policy was
incorporated into the Smallholdings and Allotments Act of
1908, which set up a Small Holdings' Commission to
investigate the regional demand for smallholdings and endowed County Councils with powers of compulsory purchase
in order to provide land for these purposes.[4] In the next
seven years, 155,000 acres were acquired under this Act,
and the Land Inquiry of 1914 found that smallholdings provided considerably more employment per acre than larger
farms.[5] The Act was, however, primarily designed to establish
an agricultural 'ladder' for enterprising farm labourers; it
did not provide employment for the urban unemployed.

Otherwise, the Liberal government avoided the problem
of unemployment during its first two years of office. When
pressed by their followers to take action, the Liberal leaders
reasonably pleaded that it was necessary to wait for the
results of the experiments under the Unemployed Workmen
Act and the reports of the Royal Commissions on the Poor
Laws and on Canals and Afforestation.[6] Four factors

1 Charles Wynn-Carrington (1843–1928), President of the Board of Agriculture
1905–11, Lord Privy Seal 1911; first Marquess of Carrington 1912. On Carrington's
private agricultural experiments see *Charity Organisation Review*, N.S. 5 (Mar.
1899), 155.
2 Add. MS. 41212, f. 288, Lord Carrington to Campbell-Bannerman, 5 Nov. 1905.
3 Ibid., ff. 304–5, Memorandum by Lord Carrington on 'Agriculture Legislation', Feb. 1906.
4 8 Edw. VII, c. 36: an Act to consolidate the enactments with respect to Small
Holdings and Allotments in England and Wales.
5 *The Land. The Report of the Land Inquiry Committee*, i. 193–4.
6 *Hansard*, 4th series, vol. 161, cols. 426–8.

operated, however, to prevent the problem of unemployment from falling out of political discussion altogether. These were the criticisms of the Labour party in the House of Commons; the constant glare of publicity that surrounded the investigations of the Royal Commission on the Poor Laws; the growing conviction of a majority of the Central Unemployed Body that the problem called for national rather than local administration; and finally from the middle of 1907 onwards the onset of depression and the revival of the familiar protectionist equation between unemployment and Free Trade.

THE REVIVAL OF THE 'RIGHT TO WORK'

The views of Labour members on unemployment were far from unanimous; their slant on the problem varied according to personal experience, political doctrine, and the situation of the people they represented. The socialists among them were committed to the view that unemployment was a necessary part of the capitalist system; but nevertheless the I.L.P. and certain sections of the T.U.C. had been advancing the theory that unemployment might be reduced by a redistribution of consuming power within a capitalist society.[1] Many trade unionists on the other hand still hoped that the problem might be solved by a limitation of hours and overtime;[2] while labour representatives who were also involved in local politics—like Will Crooks the M.P. for Woolwich and chairman of the Poplar guardians—were in the short term primarily concerned with extending the power of the Poor Law and of local distress committees to provide work for the unemployed.

[1] Recent historians have pointed out that the Labour movement's unemployment policies of the 1920s anticipated the Keynesian analysis of the problem (D. Mackay, D. Forsyth, and D. Kelly, 'The Discussion of Public Works Programmes, 1917–1935; Some Remarks on the Labour Movement's Contribution', *International Review of Social History*, 11 (1966), 8–17). But the idea that unemployment was caused by shortage of demand, which could be remedied by counter-depressive public works, was central to the unemployment policy of the I.L.P. from 1895 onwards (Tom Mann, *The Programme of the I.L.P. and the Unemployed* (Clarion Tract, No. 6), June 1895; and *Report of the 13th Annual Conference of the I.L.P.*, April 1905, pp. 32–5).

[2] *Report of the Liverpool T.U.C.*, 1906, pp. 139–40, 162.

In spite of these widely differing approaches to the problem, labour leaders managed to achieve a high degree of outward coherence on unemployment policy.[1] At the end of December 1904 a joint conference of Labour M.P.s, the Parliamentary Committee of the T.U.C., and the General Federation of Trade Unions called for the reissue of the Chamberlain circular and the regulation of public employment.[2] In January 1905 a further conference was summoned at Liverpool before the annual meeting of the Labour Representation Committee, to discuss the allied problems of unemployment and public provision for school feeding.[3] At this conference Keir Hardie resumed the prophetic mantle of the 'member for the unemployed'. In a long speech which defined unemployment as 'the root cause from which most of the troubles in the labour world sprang', he denied that it arose from the personal faults of workmen, the unnatural expansion induced by the South African war, the exhaustion of domestic resources, or the competition of immigrant labour. Instead he focused attention on the decline in the purchasing power of the working class, whose money wages since 1900 had been reduced by nearly £60,000,000 a year; and 'every reduction in wages rendered the unemployment difficulty more acute by lessening the demand for labour'.[4]

The conference passed three resolutions, which prescribed a local and national policy for labour. The first resolution called for the abolition of monopoly and 'such an organisation of industry as will prevent alternate periods of overwork and unemployment'. The second urged labour representatives in local politics to persuade local authorities to adopt the standard wage, the eight-hours day, and the

[1] Frank Bealey, 'Keir Hardie and the Labour Group', *Parliamentary Affairs*, 10 (1956–7), 225.

[2] *22nd Quarterly Report of the General Federation of Trade Unions*, pp. 10–11. Co-operation between the three groups was soon to be formalized by the creation of a Joint Board on policy-making issues (H. Clegg, A. Fox, and A. Thompson, *A History of British Trade Unions since 1889*, pp. 382–3).

[3] See *Charity Organisation Review*, 19 (Feb. 1906), 59; and 20 (Sept. 1906), 121–2, for a critical account of the overlapping of the movements for school feeding and public provision for the unemployed.

[4] *Labour Representation Committee. Report of the Fifth Annual Conference*, Appendix I, p. 63.

regulation of public works, and to petition Parliament for the abolition of disfranchisement for unemployed paupers. The third resolution, drawn up by Will Crooks, prescribed an unemployment policy for Labour members of Parliament: the extension of the powers of local authorities to acquire land, public investment in afforestation and land reclamation, and the transfer of responsibility for the unemployed from the Poor Law to a Ministry of Labour.[1]

These resolutions formed the core of subsequent Labour policy on unemployment, which was outlined to the Prime Minister by a deputation of Labour M.P.s and trade unionists on 7 February 1905.[2] Balfour's response was entirely negative, but when Walter Long's Unemployed Workmen Bill was first published in April 1905 Labour members hoped that some at least of their proposals were about to be implemented.[3] Crooks was particulary enthusiastic about the support that the Bill gave to labour colonies, and Hardie chose to interpret it as a recognition of the right to work, although Ramsay Macdonald, the secretary of the L.R.C., was more sceptical.[4] Labour hopes were dashed, however, by the withdrawal of the clause authorizing the payment of wages out of the rates to men employed on relief works and in labour colonies; and by the refusal of the Government to pay standard daily wages or to make the Act compulsory outside London.[5] Widespread demonstrations in support of the L.R.C.'s criticisms were promoted, but to no avail. After August 1905, therefore, the policy of Labour leaders was specifically directed towards an amendment of the Unemployed Workmen Act. In November George Lansbury protested to Gerald Balfour against the disqualification of workmen who had been employed on relief works in two preceding years, or who had been relieved by the guardians in the previous twelve months.[6] In December Campbell-

---

[1] Ibid., pp. 64–8.
[2] *Report of a deputation of the Parliamentary Committee of the T.U.C., the General Federation of Trade Unions and Labour M.P.s to A. J. Balfour on Unemployment and Balfour's Reply*, 7 Feb. 1905 (printed and bound with the *G.F.T.U. Sixth Annual Report*).
[3] Above, pp. 162–3.
[4] *Hansard*, 4th series, vol. 161, col. 432.   [5] Above, pp. 162–3.
[6] Lansbury MSS., vol. 2, f. 146, Gerald Balfour to George Lansbury, 29 Nov. 1905.

Bannerman and John Burns received a deputation—headed by Ramsay Macdonald and by Burns's ex-colleague, the veteran S.D.F. leader Harry Quelch—which protested against the inquisitorial investigation into character and industrial record that was to be imposed on applicants for employment relief.[1] The new Prime Minister was non-committal, but he implied that the amendment of the Act was part of the Liberal agenda; and Burns agreed to modify the offensive inquiry papers by omitting reference to fore-men and 'responsible persons'.[2]

Labour strength in Parliament was increased to twenty-nine by the January election; but Labour members were temporarily dissuaded by the promises of the new Government from pressing for an immediate amendment of the Unemployed Workmen Act or for the implementation of their own policy. This was partly because the electoral alliance between Liberals and Labour tended to obscure the fundamental difference in the aims and policies of the two parties. These differences gradually became more apparent; but now and for some years to come Labour members tended to overestimate the extent to which they could influence Liberal policy, while Liberals tended to underestimate the true independence of the new party.[3] Thus it was only by degrees that the Labour party publicly emancipated itself from the Liberal approach to unemployment. When the new Parliament assembled a committee of Labour M.P.s and trade unionists was set up to inquire into causes of and remedies for unemployment;[4] and at the annual conference in February 1906 Labour members reaffirmed their dislike of the existing structure of the Unemployed Workmen Act.[5] But not until the debate on the Easter adjournment was the non-appearance of the amending Bill brought up by Hardie in the House of Commons.[6] Burns refused to name a day

---

[1] Add. MS. 46323, Burns Diary, 13 Dec. 1905.
[2] *Annual Register*, Dec. 1905, pp. 238–9; *Report of the Liverpool T.U.C.*, 1906, p. 65. Reference was henceforth to be made only to previous employers.
[3] e.g. John Morley, *Recollections*, ii. 269; G. Riddell, *More Pages From My Diary*, pp. 38–9.
[4] *Report of the Liverpool T.U.C.*, 1906, p. 66.
[5] *Labour Party. Report of the Sixth Annual Conference*, p. 55.
[6] *Hansard*, 4th series, vol. 155, cols. 1364–5.

for its introduction,[1] and thereafter Labour members grew increasingly restive. In May they drew up suggested amendments, which were expounded in a series of deputations to Burns, and on three separate occasions he was asked to name a date for the Act's amendment.[2] It was unfortunate that Burns interpreted this campaign as a personal attack—a revival of old feuds between himself and Keir Hardie, and his fellow-engineer, George Barnes.[3] Coinciding, as it did, with Davy's revelations in Poplar, it hardened his hostility to any substantial extension of the Unemployed Workmen Act. He adopted the argument that better administration rather than new legislation was necessary to extend the benefits of the Act to distressed areas.[4] In July Labour members endorsed his decision not to extend the existing Act and welcomed the parliamentary grant of £200,000 as a recognition of the principle that unemployment relief should be subsidized from national funds.[5]

Even so, the Labour party was slow to formulate alternative legislative proposals on unemployment. During the summer of 1906 the attention of the Joint Board of the Labour party, the Parliamentary Committee of the T.U.C., and the General Federation of Trade Unions was mainly focused on the problem of persuading the Central (Unemployed) Body that its new labour exchanges should only register vacancies that conformed to standard wages and conditions.[6] The debates at the Liverpool Trades Union Congress in September suggested that many trade unionists were still inclined to see the restriction of hours and the banning of systematic overtime as the most effective measures for reducing the number of unemployed.[7] Throughout the winter of 1906–7 Labour members were surprisingly quiescent about the Unemployed Workmen Act. It played little part in the borough elections of November 1906, although it is possible that resentment against the extra rates levied under the Act helped to swing London to the Moderates

[1] Ibid., col. 1369.   [2] Ibid., vol. 157, cols. 191, 911, 1271.
[3] Add. MS. 46324, Burns Diary, 30 May 1906.
[4] Hansard, 4th series, vol. 161, cols. 427–9.
[5] Ibid., cols. 429–33.
[6] C.U.B. Minutes, i. 209, 20 July 1906.
[7] Report of the Liverpool T.U.C., 1906, pp. 139–40, 162–3.

in the County Council elections of March 1907.[1] Criticism came rather from borough councils in poor areas, like Poplar, Woolwich, Bermondsey, and St. Pancras: or from fringe groups like the Cambridge Fabians.[2] Of the Labour members of Parliament only Keir Hardie kept up a grumbling attack.

In February 1907, however, Labour members were disturbed by the absence of any reference to unemployment in the King's speech, and Thorne introduced an amendment regretting the Government's failure to introduce remedial legislation.[3] Will Crooks, re-emerging from the shadow of the Poplar inquiry, seized the opportunity to attack the inequalities of the rating system,[4] and J. O'Grady, the member for Leeds, called for a national programme of afforestation.[5] Percy Alden suggested that England needed Congested Districts Boards, modelled on those of Ireland, to redevelop depressed areas and to direct labour back to the land.[6] He also urged the Government to subsidize trade-union insurance for the unemployed.[3] John Burns replied that the problem was being dealt with by organic change—by home colonization, afforestation, and the improvement of hours and wages.[7] He thought that direct payments to the unemployed were a disguised subsidy to low wages, and suggested that if the working classes refrained from spending over £75,000,000 a year on drink they would be able to support themselves whilst unemployed.[8] Burns's view was endorsed by his Conservative predecessor, Walter Long;[9] but Keir Hardie replied that if all the unemployed became

---

[1] On the L.C.C.'s resentment against the levying of extra rates under the Act see T. 171/10, Report of the Finance Committee of the L.C.C. on 'London and the Imperial Exchequer', 22 June 1910, p. 15.

[2] *C.U.B. Minutes*, i. 327, 31 Oct. 1906; ii. 32, 21 Dec. 1906; E. T. (ed.), *Keeling Letters and Reminiscences* (1918), p. 13.

[3] *Hansard*, 4th series, vol. 169, cols. 923–8.

[4] Ibid., col. 950.   [5] Ibid., col. 932.

[6] Ibid., cols. 935–6. A similar proposal was put forward by the Conservative Free Trader, Ernest Hatch, who suggested that a statutory Board for the Unemployed, modelled on the Irish Congested Districts Boards, should be created out of the machinery of the Unemployed Workmen Act (E. Hatch, *A Reproach to Civilisation: A Treatise on the Unemployed and some suggestions for a possible solution*, 1906).

[7] *Hansard*, 4th series, vol. 169, cols. 952–9.

[8] Ibid., col. 962.

[9] Ibid., cols. 962–4.

sober they would, under existing industrial conditions, merely displace those who were employed.[1]

At the beginning of March the Joint Board set up two sub-committees to consider the political and economic aspects of unemployment and to draft proposals for a Bill.[2] The return of the unexpended balance of the parliamentary grant to the Treasury at the end of the financial year completed Labour's disillusionment with Liberal policy towards the unemployed. The significance of Burns's action was primarily symbolic, since between 1906 and 1908 the L.G.B. authorized loans for relief works amounting to several times the value of the parliamentary grant.[3] It was, however, interpreted as evidence of his total assimilation to a 'departmental' point of view.[4] In the Commons, Keir Hardie led a bitter attack on the President of the Local Government Board, who defended himself on the ground that much of the money had been badly spent, and that distress committees were refusing to pull their weight by using local resources to supplement the grant.[5] A significant minority of Burns's own party, however, were evidently unhappy with the situation, notably Percy Alden and Charles Masterman;[6] and Burns both in public and in private, adopted an increasingly paranoid tone of self-defence.[7]

The reports of the sub-committees on unemployment were submitted to the Joint Board in May, and Ramsay Macdonald and Isaac Mitchell were instructed to convert them into a Bill.[8] The Labour party's Unemployed Workmen Bill was introduced into the Commons on 9 July 1907.[9]

[1] Ibid., col. 965. The amendment was negatived by 247 votes to 207 (ibid., col. 972).

[2] Report of the Bath T.U.C., 1907, p. 116.

[3] Loans authorized by the L.G.B. for public works between 1906 and 1908 were as follows: June–Oct. 1906, £3,530,000; June–Oct. 1907, £3,589,000; June–Oct. 1908, £4,388,000 (Hansard, 4th series, vol. 194, col. 1161).

[4] Report of the Bath T.U.C., 1907, p. 140.

[5] Hansard, 4th series, vol. 171, cols. 1853–4, 1876–9.

[6] Ibid., cols. 933–7; 1865–7. Alden was in favour of appointing a Select Committee on Unemployment to prepare legislation.

[7] e.g. Add. MS. 46325, Burns Diary, 27 Mar. 1907.

[8] Report of the Bath T.U.C., 1907, p. 120.

[9] Hansard, 4th series, vol. 177, cols. 1446–8. The Bill was introduced by Ramsay Macdonald, who claimed that it had been given a mandate by the return of Peter Curran in the recent by-election at Jarrow.

This Bill proposed to set up by Order in Council a central unemployment committee—consisting of representatives of the trade unions, the L.G.B., and the Boards of Agriculture, Trade, and Education—which would be assisted by a secretary and a professional staff of unemployment commissioners appointed by the L.G.B. Each county council was to appoint a local unemployment committee, partly from its own members, partly from experts in agriculture and industry. Local committees were to conduct censuses of the unemployed, organize labour exchanges, regulate local public works, assist emigration and migration, promote industrial education and re-training, and report cases of refusal to work to courts of summary jurisdiction. The central unemployment authority was to co-ordinate the work of local unemployment authorities and to plan schemes of public employment. Funds were to be provided partly by Parliament and partly out of the rates, in order to prevent local extravagance. But if a local unemployment authority declared that an area was suffering from exceptional unemployment, then the central government would be obliged to sanction emergency schemes entirely at the cost of the Exchequer. The most crucial item in the Bill was its third clause, which provided that when a man had been registered as unemployed, it was the duty of the local authority to supply him with work or maintenance. 'This clause recognizes the right of the unemployed workman to demand an opportunity to work' wrote Ramsay Macdonald in an I.L.P. pamphlet, expounding the principles of the Bill. 'If the local authority has been so lax in its duty so as to be unable to offer him relief work, it ought to be compelled to keep his body and soul together.'[1]

The most striking features of the Bill were the 'right to work' clause, the provision for the employment of 'experts', and the emphasis upon the responsibility of existing local authorities. 'It was a great mistake to try and create a separate organisation for dealing with the unemployed' wrote Ramsay Macdonald. 'Such an organisation looks important, but in practice it becomes insignificant.

[1] J. Ramsay Macdonald, *The New Unemployed Bill of the Labour Party*, I.L.P. pamphlet 1907, p. 6. The account of the Bill is here summarized from this pamphlet.

Unemployed schemes must become part and parcel of the ordinary duties of the ordinary local administrative bodies'.[1] This view foreshadowed the role prescribed for county councils by the Minority Report of the Royal Commission on the Poor Laws. The relationship envisaged between local unemployment committees and elected local authorities, designed to combine responsibility to the electorate with freedom of action for the expert, was strikingly similar to that laid down by the Webbs as desirable in all branches of social administration.[2] It was in marked contrast to the social policy of the Liberals, who throughout this period were inclined to mistrust and to circumvent locally elected councils and administrative authorities.

Apart from the 'right to work' clause, the Labour party's Bill contained nothing that was conspicuously 'socialist' or inconsistent with the mainstream of Liberal thinking about unemployment. But the 'right to work' clause was sufficient to alienate most Liberal support, and the Bill did not reach the stage of a second reading in 1907. 'UNDER THIS BILL, STRIKE PAY BECOMES A CHARGE UPON THE RATES', an anonymous contributor warned readers of the Liberal magazine.[3] Throughout the autumn and winter of 1907, however, local 'right to work' committees were created to promote the principles of the Bill. The Bill was reintroduced in March 1908 by the Liberal M.P., Philip Whitwell Wilson, as no Labour member had been successful in the ballot to bring in a private member's bill.[4] John Burns warned his colleagues that

the essential features of the measure are at once so dangerous and so far-reaching in their consequences that it is impossible to accept it. . . . The Bill appears to contemplate that, side by side with independent industry relying upon free contract between capital and labour, there is to grow up an artificial system of industry in which labour is to claim as its right that work is to be executed at the public cost, not because it is wanted or will be remunerative, but as an excuse for paying wages, and the rate-payer or taxpayer is to be bound to supply the capital.

[1] Ibid., p. 9.
[2] Below, p. 299.
[3] *Liberal Magazine*, Apr. 1908, p. 190.
[4] *Hansard*, 4th series, vol. 183, col. 541.

He thought that such a system must ultimately result in the State becoming the sole employer of labour: a conclusion so distasteful that 'I do not consider that the Government can assent to its second reading'.[1] Asquith, who was on the brink of becoming Prime Minister, agreed that 'the obligation to provide work for all applicants at the public expense and at the standard rate of wage' was an 'obviously inadmissable proposal'.[2] Nevertheless, the Bill was introduced for a second time on 13 March. Whitwell Wilson criticized the conventional horror of national workshops which was evoked whenever state intervention in industrial organization was proposed. 'I do think that "Paris in 1848" has done duty enough' he told the House of Commons. 'You take a city in a state of revolution, with barricades in the street, and public opinion in an absolutely electrified condition and you say that that is a fair parallel to a country which has enjoyed 60 years of unmistakeable municipal progress and municipal development. . . .' He proposed that the Bill should be put into operation with a new programme of public building and naval construction.[3] Asquith protested, however, that the Bill implicitly proposed not merely the right to work, but the right to work at trade-union wages; and he echoed Burns's argument that it would lead to 'the complete control by the State of the full machinery of production'.[4] Two 'Lib–Lab' M.P.s, Maddison and Vivian, put forward an amendment stating that the Bill would undermine trade unionism and increase unemployment, and suggesting that the Government should delay legislation until it had received the report of the Poor Law Commission.[5] A significant minority of Liberals and Nationalists supported the Bill but it was defeated by 267 votes to 118.[6]

[1] CAB 37/91/33, 'The Unemployed Workmen Bill', by John Burns, 9 Mar. 1908. In the Cabinet discussion of the Bill Burns recorded that he 'stuck to [his] guns of resisting the bedlam Bill establishing the right to shirk' (Add. MS. 46326, Burns Diary, 9 Mar. 1908).

[2] CAB 41/31/49, H. H. Asquith to the King, 11 Mar. 1908.

[3] Hansard, 4th series, vol. 186, cols. 15–16.

[4] Ibid., cols. 85–6.

[5] Ibid., cols. 28–49.

[6] Ibid., cols. 91–4. 65 Liberals, 2 Conservatives, 1 Independent Socialist, and 16 Irish Nationalists supported the Bill. 60 per cent of the Liberals and 62 per cent of all M.P.s who voted for the Bill had been returned to Parliament during or after

THE ROYAL COMMISSION ON THE POOR LAWS

The contribution of the Royal Commission on the Poor Laws to the sociological and economic analysis of unemployment has already been discussed. But the published reports of the Commission were of less immediate influence on social policy than the political manœuvring of members of the Commission while it was still in session, and the public anticipation of the reforms that it would recommend.

Since the 1880s, when for the first time the penalty of disfranchisement became a potentially meaningful deterrent to the receipt of poor relief, conservative and radical social reformers had been suggesting that Poor Law administration should be publicly investigated and the principles of 1834 either revised or reaffirmed.[1] The need for such an inquiry was clearly more urgent after 1894, when guardians were increasingly faced with the often conflicting claims of democracy and deterrence. During the 1890s witnesses before the Old-Age-pension inquiries called in question not merely the moral and political aspects of the deterrent principle, but the validity of the economic argument that gratuitous payments to the poor necessarily depressed the level of wages.[2] These problems came to a head in the early 1900s, when a series of private members' bills proposed to stimulate thrift by a general enforcement of the guardians' discretionary power to discount friendly society benefits when granting poor relief.[3] These bills received the support of

the general election of 1906 (Calculated from *Dod's Parliamentary Companion*, 1908). Liberal supporters of the Bill included Charles Masterman, John Burns's Parliamentary Secretary, who voted for it against the advice of Sidney and Beatrice Webb (W. H. Beveridge, *Power and Influence*, p. 67).

[1] Sidney Webb, 'The Reform of the Poor Law', *Contemporary Review*, 58 (July 1890), 95; Sir William Chance, *The Better Administration of the Poor Law* (1895), p. 223.

[2] Above, p. 212. See also *Report of Old Age Pensions Conference held at Birmingham, 25 March 1899*, p. 7, speech of Charles Booth.

[3] Guardians had been given a discretionary power to discount friendly society benefits by the Outdoor Relief (Friendly Societies) Act, 1894. Bills to make the exercise of this power obligatory in the case of benefits of up to 5s. a week were introduced annually between 1900 and 1904. They were supported by the National Conference of Friendly Societies, the Association of Poor Law Unions, and by private insurance interests; they were opposed by the COS, by Sir Edward Brabrook,

Walter Long, against the advice of a majority of his inspectors and permanent officials.[1] They argued that the Bill was not an encouragement to thrift but 'an endowment of inadequate and unsuccessful thrift', which would almost certainly be followed by a demand for the abolition of disfranchisement;[2] that it would demoralize Friendly Societies without elevating the Poor Law;[3] and that it pandered to the 'strong disposition on the part of some guardians to give relief in aid of poverty rather than of destitution'.[4] A Lancashire guardian, F. R. Bentham, who was later to become a member of the Royal Commission, pointed out that the principle of the Bill would be capable of wide and dangerous extension.

Why should not a trades union member be eligible for the same right as the friendly society member? He has contributed to his society and received out-of-work pay which has saved the rates, he is self-respecting, his provision has prevented him from coming under the degrading influence of the Poor Law etc., etc. . . .[5]

In spite of the support of the Government the Bill was annually rejected by the House of Lords until 1904. In 1904, however, it was supported by a majority of the peers—possibly because Lord Halsbury, the Lord Chancellor, pointed out that it was essentially a 'financial bill', any interference with which would constitute a breach of Commons' privilege.[6] The strongest critic of the Bill was Lord Wemyss, Chairman of the Liberty and Property Defence League, who reminded his colleagues of a time when 'labourers used to say "Damn work, blast work, why should

the Chief Registrar of Friendly Societies, and by the Society of Poor Law Workers (H.L.G. 29/84, vol. 76, ff. 75–98).

1 H.L.G. 29/84, vol. 76, ff. 99–100, Memorandum by W. H. L., 28 Apr. 1902; ibid., ff. 114–16, Walter Long to the House of Lords, May 1903. Long argued that the Bill would promote thrift and remove the existing inequity that arose from the pursuit of different policies in different parishes. He thought that in any case the increase of relief would be slight since Friendly Societies contained such a small proportion of 'labouring men'.

2 Ibid., ff. 141–50, J. S. Davy to Sir Samuel Provis, 10 June 1903.

3 Ibid., ff. 158–71, Baldwyn Fleming to Sir Samuel Provis, 18 June 1903.

4 Ibid., ff. 199–200, Philip Bagenall to Sir Samuel Provis, 17 June 1903.

5 Ibid., f. 201, Copy of a letter from F. Bentham to the Editor of *The Pilot*, 21 May 1903.

6 Ibid., f. 237.

I work when I can get 12/- off the rates" '.[1] On the third
reading of the Bill, he moved that no such change in the
Poor Law should be made without 'a full, independent
public inquiry into the working of the present law'. The
permanent officials were, however, opposed to such an
inquiry,[2] and the Bill passed into law in August 1904.[3]

Meanwhile progressive administrators were increasingly
conscious of the fact that, whatever its moral and political
value, the principle of deterrence might be detrimental to
health and economic efficiency;[4] and in the spring of 1905
Mr. Long conceded the need for a statutory form of unemployment relief outside the sphere of the Poor Law.
The Unemployed Workmen Act, however, was the source
of much misunderstanding among both the supporters and
the opponents of the Conservative government, and aroused
expectations of social reform to which the Conservative
leaders were not prepared to commit themselves.[5] It was
to avoid this commitment that the Royal Commission on
the Poor Laws and the Relief of Distress[6] was set up under

[1] *The Times*, 13 Aug. 1904.
[2] H.L.G. 29/84, vol. 76, ff. 234–5, Local Government Board note on Lord
Wemyss's amendment moved on 3rd reading of Gretton's Bill, May 1904.
[3] 4 Edw. VII, c. 22.
[4] Above, pp. 147–8.
[5] CAB 41/30/5, Arthur Balfour to the King, 1 Mar. 1905.
[6] The precise reasons for the appointment of the Commission are conjectural.
The Webbs stated that there was no crisis in Poor Law administration and no public
demand for an inquiry. They therefore ascribed it to the conjunction at the L.G.B.
of a reactionary head of the Poor Law division, J. S. Davy, and a President, Gerald
Balfour, who was 'a philosopher and recognised the public advantage of precise
discrimination between opposing principles'. But this explanation is unconvincing
since (*a*) there *was* an administrative crisis in many poverty-stricken unions and had
been since the 1890s (above, pp. 145–50); and (*b*) there had been much public comment on the breakdown of the traditional functions of the Poor Law, and demand
for the revival or abandonment of the principles of 1834 (e.g. *Report of the Special
Committee of the COS on the Relief of Distress due to Want of Employment*, 1904,
pp. 12–18; G. Lansbury, *The Principles of the English Poor Law*, address to the
Central Poor Law Conference, printed as a pamphlet, 1897). Sidney Webb himself
had remarked upon 'the political instability of the existing system of Poor Relief'
in a democratic context (S. Webb, 'The Reform of the Poor Law', *Contemporary
Review*, 58 (July 1890), 95–120); and there had been a demand for a public inquiry
in the Lords in the previous year (above, p. 247); (*c*) J. Brown, 'Ideas concerning
Social Policy and their influence on Legislation in Britain 1902-11' (London Ph.D.
thesis 1964), pp. 190–1, has shown that the crucial pressure on Balfour for the
appointment of a Royal Commission came not from Davy and Gerald Balfour but
from Herbert Samuel and from Walter Long.

the chairmanship of Lord George Hamilton in December
1905.[1]

The association of Poor Law and unemployment problems in the same inquiry was an indication of the extent to
which unemployment was still primarily regarded as a
problem of distress rather than of economic policy or industrial organization. It has been shown that both the Conservative and the Liberal governments used the existence of
the commission as a plausible excuse for postponing unemployment legislation. But the controversial nature of the
problem under review and the polemical character of some
of the commissioners meant that no subject that they touched
faded entirely from public discussion. The Commission
contained some of the foremost social theorists of the day,
including Charles Booth and the matriarch of organized
charity, Octavia Hill, both of whom were rather out of
touch with new trends in social policy by 1905.[2] Among the
other commissioners were Mrs. Beatrice Webb, the Fabian
socialist and co-author of a new science of social institutions;
Russell Wakefield, the Christian Socialist chairman of the
C.U.B.; George Lansbury, the advocate of labour colonies
and co-operative production; and C. S. Loch, together with
four other members of the Charity Organisation Society.
Representing the 'official mind' were Sir Samuel Provis,
Sir Henry Robinson, and J. Patten Macdougall, the permanent heads of the English, Irish, and Scottish Local
Government Boards;[3] while boards of guardians were represented by F. R. Bentham and by the eccentric Oxford
economist, the Reverend Lancelot Phelps.[4]

[1] Lord George Hamilton (1845–1927), Conservative M.P. 1868–1906; First
Lord of the Admiralty 1885–6, 1886–92; Chairman of London School Board 1894–
5. His *Parliamentary Reminiscences and Reflections*, 2 vols., contain no reference to
the Poor Law Commission.

[2] Booth throughout took a very conservative view of Poor Law Reform (B.
Webb, *Our Partnership*, p. 357). He resigned early in 1908 because he felt unable to
agree with his colleagues and too weak to disagree with them (Passfield MSS.,
ii, 4, d, item 2, B. Webb to M. Playne, 2 Feb. 1908). For Octavia Hill's views on
Poor Law Reform see C. Edmund Maurice, *Life of Octavia Hill*, ch. XI.

[3] Sir Henry Robinson, *Memories: Wise and Otherwise*, pp. 212–14.

[4] E. P. Donaldson, 'Provost Phelps and the Poor Law Commission, 1906–9',
Oriel Record, 1959, 15–25.

The other members were the O'Conor Don, who died and was replaced by the
Bishop of Ross in 1906; Dr. Arthur Downes, Senior Medical Inspector of the Poor

The history of this Royal Commission has been both illuminated and distorted by the Webbs' account of the ideas that were put forward and the events that occurred. It was a characteristic of all the Webbs' writing that, in order to extrapolate the 'curve of History' from the dark mass of historical facts they tended to impose an artificial simplicity and coherence upon the actual state of affairs.[1] Nowhere is this more apparent than in their recounting of situations such as the Royal Commission on the Poor Laws, where they were protagonists of conflict and the agents as well as the chroniclers of historical events.[2]

From the start Beatrice Webb was convinced that the official and charitable factions on the Commission were seeking for a return to the principles of 1834,[3] and in order to forestall such a reaction she set out to secure at all costs the renunciation of the deterrent principle and the abandonment of the category of 'destitution'.[4] For the policy of deterrence the Webbs wanted to substitute a policy of 'prevention', 'compulsion', and 'universal provision';[5] they proposed to abolish the distinction between pauper and non-pauper services, to break up 'the present unnatural aggregation of the Poor Law into its compound parts', and to consolidate each branch of social administration under specialist committees of the county councils which would deal uniformly with the whole community.[6]

The Webbs saw these two principles—the establishment of the principle of 'prevention' and the abolition of a separate

---

Law Division of the L.G.B.; William Smart, Professor of Political Economy at Glasgow University; and Mrs. Helen Bosanquet, wife of the philosopher and a member of the COS. Francis Chandler, Secretary of the Amalgamated Society of Carpenters and Joiners, was added after protests against the absence of a trade union representative in Dec. 1905.

[1] On the Webbs as historians see E. Hobsbawm, 'The Fabian Society 1884–1913', Cambridge Ph.D. thesis, p. 63. On their distortion of Poor Law History see Asquith MSS., vol. 78, ff. 147–8.

[2] Mrs. Webb significantly 'seized upon the historical survey' of pauperism as her first task on the Royal Commission (B. Webb's Diary, 15 June 1906, quoted in Our Partnership, p. 343).

[3] B. Webb's Diary, 2 Dec. 1905, quoted in Our Partnership, p. 322.

[4] S. and B. Webb, English Poor Law Policy, pp. 296–7.

[5] Ibid., pp. 297–9.

[6] Asquith MSS., vol. 76, ff. 102–37, 'Notes on the Proposed Transfer of the Poor Law to the County and County Borough', n.d., Sidney Webb's MS.

destitution authority—as the two main questions around
which controversies within the Royal Commission revolved.[1]
But the polarization of opinion within the Commission was
far less extreme than the Webbs liked to suggest. Moderate
reformers like Hamilton and Phelps were just as anxious as
Beatrice Webb that the Commission should not be throttled
by conservative elements and that the report should not be
drafted by members of the COS;[2] and the definition of social
problems ultimately adopted by a majority of the Commission was in fact remarkably similar to that put forward by
the Webbs. 'We found that the Poor Law . . . failed to
satisfy modern needs, as being designed with a view solely
to deterrence' wrote Lancelot Phelps in an explanatory letter
to James Bryce in March 1909.

In practice, this had largely given way to other principles, but . . .
the attempt to graft these on to the old stem led to confusion and
maladministration . . . we had to combine in one the old principles of
deterrence and the new demand for curative and restorative treatment.[3]

The majority of the Commission also wanted to abandon
the old Poor Law structure and to turn the county into the
basic unit of social administration—'mainly because, altho'
not ideal, the county was there already'.[4] But, instead of
amalgamating the old Poor Law services with other statutory social services, they wished to retain an all-purpose
destitution authority, composed of councillors, paid officials,
and organized voluntary workers, to deal with different
kinds of distress among persons and families relieved at the
public expense. This was to be a two-tier institution consisting, firstly, of a Public Assistance Authority which
would be a statutory committee of the County or County
Borough Council; and, secondly, of Public Assistance Com-

---

[1] S. and B. Webb, *English Poor Law Policy*, pp. 278–81, 297–304.

[2] Phelps MSS., unsorted, Lord George Hamilton to Lancelot Phelps, 12 Oct.
1908.

[3] Bryce MSS., Box E.28, Lancelot Phelps to James Bryce, 26 Mar. 1909.

[4] Ibid., Lancelot Phelps to James Bryce, 26 Mar. 1909. It was decided that 'a
change was necessary in the whole spirit and impulse of local administration', and
that this could not be brought about without the abolition of the Guardians
(Phelps MSS., unsorted, Lord George Hamilton to Lancelot Phelps, 26 May
1909).

mittees, based geographically on the old Poor Law unions, which would conduct inquiries into the condition of applicants for relief.[1] The Webbs ascribed this proposal to the desire of organized charity to maintain its stranglehold on the relief of the poor,[2] and to a residual belief among the majority that 'destitution' was an autonomous social problem, existing separately and requiring different treatment from other kinds of social, economic, industrial, and physical distress.[3] They condemned it as inefficient and expensive and tending to perpetuate the wasteful duplication of services for the pauper and non-pauper working class.[4] But this indictment was only partly justified, since the arguments put forward in favour of a destitution authority were not entirely negative. Some at least of the commissioners believed that such an authority was desirable in order to make full use of existing resources and because different social problems in the same family ought not to be to treated in isolation.[5] Moreover, the Webbs themselves believed in the 1900s that the character of public dependants should be subject to moral constraints; and one of their reasons for objecting to the retention of a voluntary element in the social services was their conviction that only a public authority could exercise the necessary 'element of compulsion and disciplinary supervision' over the recipients of public relief. 'It is no use', stated Beatrice Webb, 'letting the poor come

[1] Cd. 4499/1909, *RC on the Poor Laws, Majority Report*, pp. 604–7, 623–5.

[2] Asquith MSS., vol. 76, ff. 109–10, S. Webb, 'Notes on the Proposed Transfer of the Poor Law to the County and County Borough Councils', n.d.

[3] S. and B. Webb, *English Poor Law Policy*, pp. 280–1.

[4] Ibid., pp. 313–14. The majority meanwhile condemned the Webbs' proposals as 'neither politically or administratively practicable' because they would involve the creation of a large staff of professional officials at great public expense (Asquith MSS., vol. 78, ff. 5–7, Memorandum by Lord George Hamilton on 'The Royal Commission on the Poor Laws and Relief of Distress', with a covering note to H. H. Asquith, 5 Mar. 1908). See also Austen Chamberlain, *Politics from Inside*, p. 238.

[5] Cd. 4499/1909, *Majority Report*, Part IX, para. 13. B. Webb, *Our Partnership*, p. 407. It is interesting to note that a decade later the Webbs had partially come round to the majority point of view. 'Mrs Sidney Webb came to see me some six months back' wrote Hamilton to Phelps early in 1918. 'She was in a very repentant and attractive mood. She admitted that she had made a mistake in trying to break up the family and in persistently objecting to our proposal to make for the future the Home Assistance Committee the foundation of Poor Law Relief and Poor Law assistance . . .' (Phelps MSS., unsorted, Lord George Hamilton to Lancelot Phelps, 17 Jan. 1918).

and go as they think, to be helped or not as the charitable choose.'[1]

Nevertheless, it was true that the majority were not prepared in the last resort to abandon the principle of deterrence; and although they accepted the principle of prevention it was not given first priority in their plans for reform. 'We take our stand', wrote Lancelot Phelps, 'on the old principle that destitution is the sole claim for relief' and on 'the hard truth that self-caused poverty is a crime.'[2] Moreover, the variations of opinion among the majority were necessarily much greater than among the small handful of commissioners who followed Mrs. Webb; and since their proposals were based on compromise they were inevitably less consistent and less clear-cut.

It was within this context of controversy over the aims and organization of Poor Law reform that the commissioners developed their views on unemployment. During its first eighteen months the Commission concentrated mainly on Poor Law problems,[3] and unemployment was considered merely as part of the spectrum of pauperism. Mrs. Webb was appalled by the view of the Poor Law inspectors that all forms of assistance to the unemployed were tantamount to 'relief in aid of wages' and that the able-bodied could only be dealt with by a literal enforcement of the principle of 'less eligibility'.[4] It was she who insisted that the Commission should inquire into poverty as well as pauperism, and into 'the sum total of the legal obligation of the Guardians and their officers' as well as local variations in Poor Law administration.[5]

When the commissioners turned to unemployment, they found that they 'had not nearly so much available as [they] had on Poor Law topics, and those who had knowledge had not wisdom'.[6] In July 1906, the Commission therefore appointed two assistant commissioners, Cyril Jackson and

---

[1] Passfield MSS., ii, 4, e, item 11, B. Webb to Georgina Meinertzhagen, Mar. 1911.
[2] Bryce MSS., Box E. 28, Lancelot Phelps to James Bryce, 26 Mar. 1909.
[3] B. Webb, *Our Partnership*, p. 386.
[4] S. and B. Webb, *English Poor Law History*, II. ii. 477–80, 500.
[5] Lansbury MSS., vol. 29, ff. 16–22, 'Suggestions', by B. Webb, 6 Jan. 1906.
[6] Bryce MSS., Box E. 28, Lancelot Phelps to James Bryce, 29 Mar. 1909.

the Reverend J. C. Pringle, to inquire into methods of relieving the unemployed outside the Poor Law.[1] In six months these two investigators produced an exhaustive factual account of relief experiments during the previous twenty years;[2] but most of their material was completely undigested, and the authors were rather narrowly preoccupied with the artificial dilemma of whether unemployment was caused by faults of character or faults in the organization of industry.[3] Beatrice Webb disliked this Report, not because she dissented from its criticism of previous relief experiments, but because she feared it would be used as a justification for restoring the unemployed to the exclusive control of the Poor Law.[4] She herself was convinced that 'the Workhouse Test has broken down, and must . . . be dismissed as no longer practicable or even desirable'.[5]

During the early stages of the Commission, however, the Webbs had no definite alternative remedy for unemployment to propose.[6] It was significant that their plan for a 'National Minimum', first advanced in 1897, had prescribed treatment for the unemployable but not for the unemployed.[7] For a while they hoped that John Burns might be persuaded to tackle the problem, if his permanent officials could be circumvented[8] and if his 'strong vigorous and audacious character' could be tempered by Fabian efficiency.[9] In July

[1] Sir Cyril Jackson (1863–1924); Chief Inspector to the Board of Education 1903–6; Progressive member of the L.C.C. 1907–13. John Christian Pringle (1872–1938); member and historian of the Metropolitan Visiting and Relief Association; Secretary of the COS 1914–19.

[2] Cd. 4795/1909, *The Effects of Employment or Assistance Given to the 'Unemployed' since 1886 as a Means of Relieving Distress outside the Poor Law*, 30 Jan. 1907.

[3] Ensor MSS., J. C. Pringle to R. C. K. Ensor, 23 July 1906.

[4] Ibid., B. Webb to R. C. K. Ensor, 4 May 1907.

[5] Lansbury MSS., vol. 29, ff. 37–8, B. Webb to Lord George Hamilton, 10 Oct. 1906.

[6] An imaginary letter from Campbell-Bannerman to his Cabinet colleagues, written by Sidney Webb in Jan. 1905, contained detailed proposals for other branches of social policy; but on the problem of unemployment, Burns was advised to await the reports of the Poor Law Commission. ('The Liberal Cabinet— An Intercepted Letter', *National Review*, Jan. 1906, 789–802. On Sidney Webb's authorship of the letter see Fabian Society MSS., Box 4, S. Webb to J. Pease, 16 Dec. 1905).

[7] S. and B. Webb, *Industrial Democracy* (1897 ed.), pp. 784–7.

[8] Add. MS. 46287, ff. 293–4, B. Webb to John Burns, ? Mar. 1906.

[9] B. Webb's Diary, 9 Feb. 1 06, quoted in *Our Partnership*, p. 330.

1906 Burns informed the Commission that he could only wait until the autumn of 1908 before introducing unemployment legislation;[1] and he and Mrs. Webb agreed that the crux of the problem was 'the question of the wayward man'.[2] But each of them thought the other excessively 'doctrinaire',[3] and co-operation between them was made impossible by Burns's resistance to permeation and his immersion in trivial administrative routine.[4]

Above all, however, the Webbs were anxious that the intellectual vacuum that surrounded unemployment should not be allowed to lead to the revival of a purely deterrent policy, simply for want of viable alternatives.[5] In October 1906 Mrs. Webb confessed to Lord George Hamilton that she was baffled by the difficulty of devising curative treatment for the able-bodied; but she was determined to 'get the best brains to work on the problem without being in any way shackled by the old formulas', and to conduct a comprehensive inquiry into all aspects of the problem of the unemployed. She admitted that 'the solution will probably be different for different classes, different persons, or perhaps different for different persons on different days'. But she was inclined to think that 'compulsory technical training or military or other training . . . absorbing the whole time of the man from 6 a.m. to 10 p.m., without taking him away from his home at night, might conceivably be the best thing for some of this class.'[6]

Nevertheless, it proved difficult to impose this line of approach to the problem on politicians or on other members of the Royal Commission. In May 1907 Mrs. Webb tried to persuade Reginald Mackenna,[7] the President of the

1 B. Webb's Diary, 9 Feb. 1906, quoted in *Our Partnership*, p. 348.

2 Add. MS. 46324, Burns Diary, 27 Oct. 1906.

3 B. Webb's Diary, 2 Dec. 1905, quoted in *Our Partnership*, p. 325. Add. MS. 46324, Burns Diary, 27 Oct. 1906.

4 B. Webb's Diary, 30 Oct. 1907, quoted in *Our Partnership*, pp. 393–4. Burns's diaries suggest that an excessive amount of his time as President of the L.G.B. was spent on the kind of 'fieldwork' that should have been left to officials and inspectors.

5 Ensor MSS., B. Webb to R. C. K. Ensor, 4 May 1907.

6 Lansbury MSS., vol. 29, ff. 37–8, B. Webb to Lord George Hamilton, 10 Oct. 1906.

7 Reginald Mackenna (1863–1943), Liberal M.P. for North Monmouthshire 1895–1918; Financial Secretary to the Treasury 1905–7; President of the Board

Board of Education, that responsibility for the able-bodied should ultimately rest with his department, and that it was desirable to create 'something in the nature of an "Industrial School" absorbing both the working hours and the leisure of the Out-of-Work in educational discipline and recreative treatment'.[1] But Mackenna was apparently dismayed by the Webbs' efforts to thrust all kinds of expensive and intransigent social problems into the domain of education.[2]

Moreover, in April 1907 Lord Hamilton rejected the Webbs' scheme for an inquiry into all aspects of the problem of the able-bodied, and substituted instead his own plan, which—according to Beatrice Webb—rambled indiscriminately over all aspects of the problem but excluded any further investigation of the treatment of the unemployed under the Poor Law.[3] The Webbs therefore decided to conduct their own inquiry and convened an informal committee, consisting of Barnett, Beveridge, Maynard, and Robert Ensor, to discuss remedies and to consider 'all ways of dealing with the Able-bodied or persons assumed to be Able-bodied, including even the casual ward, and therefore vagrancy. . . . Only in that way shall we get a statesmanlike grip of the question.'[4]

All the members of this committee had prior experience in the investigation and relief of unemployment—Barnett in the Mansion House schemes of the 1880s and 1890s, Beveridge and Maynard on the Central (Unemployed) Body and Ensor from settlement work in the slums of Manchester.[5] Beveridge already had a carefully prepared

of Education 1907–8; First Lord of the Admiralty 1908–11; Home Secretary 1911–15; Chancellor of the Exchequer 1915–16; subsequently Chairman of the Midland Bank.

[1] Passfield MSS., ii, 4, c, item 81, B. Webb to Reginald Mackenna, 30 May 1907.

[2] B. Webb, *Our Partnership*, p. 379, describes his reaction to the Webbs' suggestion that all problems concerning children should be transferred to the Education Authority.

[3] B. Webb's Diary, 23 Apr. 1907, quoted in *Our Partnership*, p. 378.

[4] Ensor MSS., B. Webb to R. C. K. Ensor, 4 May 1907; W. H. Beveridge, *Power and Influence*, p. 62.

[5] Robert Ensor (1877–1958), lawyer, journalist, historian, and research assistant to Seebohm Rowntree; previously a writer for the *Manchester Guardian*, and had worked at the Ancoats settlement in Manchester; legal advisor to Will Crooks in the Poplar case. In 1904 Ensor had toured the casual wards of Scotland disguised

plan for the national treatment of unemployment through labour exchanges, which was not entirely compatible with the aims of the Webbs.[1] Indeed, the Webbs were looking for an institutional method of reforming the unemployed just at a time when Beveridge was coming to the conclusion that such a method was unnecessary, and that insurance rather than labour reformatories should be the administrative counterpart of a system of labour exchanges.[2] The Webbs, on the other hand, were basically sceptical about the utility of labour organization without a general collectivist framework for the prevention of unemployment and the retraining of the unemployed; and this view was reinforced by the discovery of the Commission's special investigators, Rose Squire and Arthur Steele-Maitland, that 'Casual Labour is the main cause of Pauperism, and remedy a legal minimum term of employment'.[3]

Nevertheless, the Webbs were intrigued by Beveridge's scheme for labour exchanges, and in August 1907 they invited him to spend a week-end with them at Bernard Shaw's house in Ayot St. Lawrence. He there persuaded them that labour exchanges were an essential administrative feature of any remedial system that they might propose and were equally compatible with a collectivist or *laissez-faire* system of government.[4] Beatrice Webb later told Beveridge that this conversion played a crucial part in crystallizing the ideas not merely of the Webbs but of the whole Royal Commission.[5] Henceforth the Webbs campaigned for a compulsory version of Beveridge's plan for a voluntary system of labour exchanges,[6] although they continued to be

as a tramp and was surprised at the high proportion of 'genuine unemployed workmen' among so-called vagrants. At this time a leader-writer for the *Daily Chronicle* and member of the I.L.P.

[1] Above, pp. 205–7. [2] W. H. Beveridge, *Power and Influence*, pp. 59–60.
[3] Fabian Society MSS., Box 5, B. Webb to E. Pease, n.d.; ibid., Box 5 (a), B. Webb to E. Pease, 1 May 1907. Both Rose Squire and Steele-Maitland were nominees and confidantes of Mrs. Webb (Rose Squire, *Thirty Years in the Public Service*, p. 116). Their report was published as Cd. 4653/1909, *Report on the Relation of Industrial and Sanitary Conditions to Pauperism*.
[4] W. H. Beveridge, *Power and Influence*, pp. 62–3. [5] Ibid., pp. 63–4.
[6] Ibid., p. 65: '. . . from October onwards, Sidney was putting labour exchanges into his own form, attributing the original idea always to me, but freeing me from responsibility for the "Utopian" plan which he was about to boom . . .'

concerned with the reformation as well as the organization
of the casual labourer and the unemployed.[1]

With this end in view the Webbs spent the following
winter 'immersed in Farm Colonies, Labour yards and
Distress Committees', searching for an institution that might
be used to improve the character and situation of workmen
who became unemployed.[2] They extracted information from
John Burns about the treatment of the able-bodied under the
Poor Law;[3] and they visited the labour colonies at Hadleigh
and Hollesley Bay, taking with them 'one or two practical
agriculturalists' to advise them in their criticism of these
institutions.[4] They commissioned R. H. Tawney and Nettie
Adler to investigate the employment problems of school-
leavers and child labour;[5] and with the assistance of secret
funds from the Fabian Society they carried out a private
inquiry into all aspects of unemployment among able-
bodied workmen.[6] 'The difficulty of solving the question
oppresses me,' wrote Mrs. Webb. 'I dream of it at night,
I pray for light in the early morning. I grind, grind, grind
all the hours of the working-day to try to get a solution.'[7]

On the basis of this research, the Webbs gradually worked
out a comprehensive plan for the prevention of unemployment and the relief of the unemployed. But at the same time
Mrs. Webb—partly through overstrain and partly through
her ill-disguised contempt for the sociological inefficiency
of her colleagues—became almost completely estranged
from the majority of the Commission.[8] The whole debate on

[1] Beveridge MSS., L. ii. 218b, B. Webb to W. H. Beveridge, n.d. (notepaper
headed Ayot St. Lawrence).

[2] Passfield MSS., ii, 4, c, item 114. B. Webb to Graham Wallas, 29 Nov. ? 1907.

[3] B. Webb's Diary, 22 Oct. 1907, quoted in *Our Partnership*, p. 392.

[4] Passfield MSS., ii, 4, c, item 115, B. Webb to M. Playne, 9 Dec. 1907. For the
Webbs' reactions to Hadleigh and Hollesley Bay see above, pp. 134–5, 197–8.

[5] Their research was published by the Women's Industrial Council as N. Adler
and R. Tawney, *Boy and Girl Labour* (1909).

[6] Fabian Society MSS., Box 5, B. Webb to E. Pease, n.d.; and Box 5 (a),
B. Webb to E. Pease, 1 May 1907; and Box 5 (a), B. Webb to E. Pease, 12 July
1907; Ensor MSS., 'Investigation into the Cause and Treatment of Ablebodied
Destitution', 1907.

[7] B. Webb's Diary, 15 Nov. 1908, quoted in *Our Partnership*, p. 419.

[8] Ibid., p. 402. Phelps MSS., unsorted, Lord George Hamilton to Lancelot Phelps,
22 Aug. 1909. 'From first to last she has declined whilst in the Commission to merge
her individuality in it, but claims the right of unrestricted free action outside in connection with matters under the consideration of the Commission. She is . . . hopeless.'

unemployment within the Commission took place under exaggerated conditions of tension and conflict, largely generated by Mrs. Webb's determination 'to prepare an atmosphere for our able-bodied proposals'.[1]

This situation was, however, strictly unnecessary, since in the absence of practical alternatives the commissioners were more prepared to listen to the Webbs on unemployment than on any other topic. In October 1907, when Russell Wakefield introduced the Webbs' proposal for a county unemployment authority, passing it off as his own, the rest of the Commission accepted it without a qualm.[2] Mrs. Webb anticipated that her 'revolutionary scheme of dealing with unemployment' would complete her alienation from the rest of the Commission;[3] but when they accepted a large part of this scheme, she proved to be a 'determined Minority Report-writer' and decided to present a separate report in any case.[4] 'We are preparing a very elaborate, very scathing and fully reconstructive report' wrote Sidney Webb to Ensor in September 1908. 'We intend to make it the basis of a prolonged campaign with the object of re-settling, on the lines suited to a collectivist state, the whole provision for the necessitous, including a complete provision for the unemployed.'[5] And to Edward Pease he confided that he was 'mapping out a complete revolution', which involved 'much Local Government reconstruction' and 'the Final Solution' of the problem of the unemployed.[6]

This report was composed by Sidney Webb under great pressure during the autumn and winter of 1908, and signed in February 1909 by Mrs. Webb, Russell Wakefield, Francis Chandler, and George Lansbury. The report proposed to eliminate under-employment by measures of de-casualization and by the compulsory[7] registration of all kinds of unemployed workmen at the local branches of a national

---

1 B. Webb's Diary, 26 Nov. 1907, quoted in *Our Partnership*, p. 396.
2 Ibid., pp. 390–2.   3 Ibid., p. 396.
4 Phelps MSS., letter 214, Lancelot Phelps to H. R. Boyce, 12 Feb. 1908.
5 Ensor MSS., S. Webb to R. C. K. Ensor, 17 Sept. 1908.
6 Fabian Society MSS., Box 4, S. Webb to E. Pease, 12 Aug. 1908.
7 i.e. compulsory only for workmen unable to find work and unable to support themselves whilst unemployed; the Webbs did not rule out the possibility of agencies other than the National Labour Exchange helping to find employment.

labour exchange.[1] Cyclical unemployment was to be reduced to a minimum by decennial programmes of public works designed to flatten out the variations of the trade cycle.[2] Mothers of young children would be withdrawn from the labour market; and juvenile labour would be halved by the provision through Local Education Authorities of compulsory part-time technical education.[3] Maintenance and industrial retraining would be provided for surplus workmen and penal labour colonies would be established for those who refused to work.[4] To win the co-operation of trade unionists, the report proposed that state subsidies should be given to all unions that paid an out-of-work donation, subject to checks on malingering and guarantees of the good behaviour of those relieved.[5] The supervision of all these functions, together with most of the existing industrial responsibilities of the Board of Trade, Home Office, and Local Government Board would be transferred to a new central government department under a separate 'Minister of Labour'.[6] The minority's policy for the unemployed turned out in fact to be less 'collectivist' than coercive, centring on the compulsory organization of the labour market, education and discipline for the unemployed, and penal repression for the wilfully idle. Intellectually it owed more to utilitarianism than to socialism; and as a statement of political belief it belonged almost entirely to the Webbs, since Wakefield and Lansbury signed mainly in personal

[1] Cd. 4499/1909, *RC on the Poor Laws, Minority Report*, pp. 1183–8.
[2] Ibid., pp. 1195–8.   [3] Ibid., pp. 1190–5.
[4] Ibid., pp. 1205–6.
[5] Ibid., pp. 1199–1201, 1211–12. The rather grudging support of the minority for subsidized trade-union insurance should be compared with Sidney Webb's views on the subject nineteen years earlier: 'The free and independent elector will never submit to the "regimentation", identification and restrictions on locomotion which any scheme of general insurance must necessarily involve. No Government is at all likely to attempt to collect compulsory insurance premiums from men already supporting their trade unions and friendly societies, their benefit clubs and their building societies and paying, moreover, a not inconsiderable poor rate . . .' (S. Webb, 'The Reform of the Poor Law', loc. cit., p. 105).
[6] Ibid., pp. 1208–14. Unemployment was the one social problem which the Webbs thought could better be dealt with directly by a national rather than local authority. They proposed, however, that if a national scheme was thought 'premature', unemployment could be treated by a committee of the County Council, modelled on those recommended for other branches of administration (ibid., pp. 1179–80).

protest against the majority's disparagement of the Central (Unemployed) Body and the Poplar Guardians.[1]

The Majority Report of the Royal Commission, drafted by a committee consisting of Phelps, Professor Smart, and the Bishop of Ross,[2] included virtually every proposal on unemployment which the Webbs had made, with the exception of the central planning of public works and the creation of a Ministry of Labour. Like the minority they recommended the limitation of juvenile labour and the improvement of technical education, training schemes for the unemployed, and domestic maintenance for their families.[3] Both reports condemned the principle of the 'right to work' at standard wages as 'absolutely subversive of self-respect, self-exertion, and independence, and . . . detrimental to the industrial efficiency of the community'.[4] The majority were more positively in favour of unemployment insurance, and they recommended that a committee should be appointed to draft an insurance scheme which would include unskilled and casual as well as organized workmen.[5] They preferred voluntary to compulsory labour exchanges;[6] and although they proposed that training schemes should be backed up by voluntary labour colonies and compulsory 'detention colonies' they were more sceptical than the Webbs about the political feasibility of reforming the habitually and wilfully unemployed.[7] The outstanding difference between the two reports was that the majority proposed to

---

[1] Phelps MSS., letter 233, Lancelot Phelps to H. R. Boyce, 16 Mar. 1909.

[2] Phelps MSS., letter 229, Lancelot Phelps to H. R. Boyce, 10 Dec. 1908.

[3] Cd. 4499/1909, *Majority Report*, Part IX, paras. 127–8; Part VI, paras. 624, 627.

[4] Ibid., Part IX, para. 142; Passfield MSS., ii, 4, d, item 88, B. Webb to Lady Betty Balfour, Nov. 1910.

[5] Cd. 4499/1909, *Majority Report*, Part IX, paras. 133–4.

[6] Ibid., Part VI, paras. 473–528. The majority hoped that the lack of success of earlier non-compulsory exchanges could be overcome by rigorously separating them from relief mechanisms, preserving impartiality in trade disputes involving both workers and employers in their management, and making use of telephones and cheap rail fares (ibid., paras. 506–7).

[7] Ibid., Part VI, paras. 636–50, 653–8. Phelps, moreover, was basically sceptical about the existence of an 'unemployable' class. 'What I felt very strongly was . . . that there is no surplus labour. Everyone who has come into the world hitherto has produced more than he consumed . . .' (Bryce MSS., Box E. 28, Lancelot Phelps to James Bryce, 29 Mar. 1909).

resort to public works only in times of crisis, whereas the
minority believed that public works could be used not merely
to relieve but to forestall depressions.[1] Otherwise, as
Beveridge later commented, there was no real reason why
the unemployment proposals of the Royal Commission
should not have been contained in one report rather than
two.[2] Indeed, Thomas Hancock Nunn, the Chairman of
the Hampstead COS, went further than the Webbs in
seeing unemployment as a problem of 'education' and 'prevention', and attached to the Majority Report a private
memorandum on the need for public works, decasualization,
the raising of the school-leaving age, and the adaptation
of popular education to the requirements of industry.[3]

After her failure to convert the Commission to a total
'break-up of the Poor Law', however, Mrs. Webb lost
interest in forging any kind of consensus that fell short of a
total acceptance of her own ideas. On the question of the
Poor Law, this attitude was perhaps justified because two
great principles—prevention and universal provision—were
at stake.[4] Even so, it is probable—as Mrs. Webb admitted
to Hamilton nine years later[5]—that her views would have
been more politically effective if she had concentrated on
permeating her fellow commissioners, instead of taking such
an uncompromisingly independent line. But no major principle was involved in the discussion of unemployment, and
in adopting an aggressive minority position the Webbs

[1] Ibid., Part IX, para. 40. The only member of the Commission who positively
dissented from a public-works policy was Octavia Hill (Cd. 4499/1909, p. 678).

[2] Beveridge, *Unemployment* (1930 ed.), p. 261.

[3] Cd. 4499/1909, pp. 712–18, 'Memorandum by Mr. T. Hancock Nunn in
Regard to Unemployment'.

[4] Harold Baker, the lawyer engaged by Asquith to digest the two reports on
behalf of the Cabinet, commented that 'the underlying question which is really
at issue [is] whether relief should be made available for the poor generally instead
of for the destitute and necessitous only. This question will not be found to be
explicitly raised by either Report, but it is evident that it is the basis of their discussion. Their silence may be intentional or unintentional: in either case the issue
necessarily arises out of the two views presented in the Reports' (Asquith MSS.,
vol. 78, f. 140, ? early 1909. This memorandum is unsigned, but written in Baker's
MS.: cf. Asquith MSS., vol. 35, f. 188). It should be noted that when Baker referred
to 'universal provision' he did not mean what is understood by the term in the 1970s,
but 'gratuitous state services for the working classes and not for the destitute alone'
with 'charge and recovery from all able to pay' (Asquith MSS., vol. 78, f. 169).

[5] Phelps MSS., unsorted, Lord George Hamilton to Lancelot Phelps, 17 Jan. 1918.

were tilting at a windmill. Both sides agreed that the Poor Law was no longer relevant to the problem;[1] and both sides were groping to find a pragmatic solution. The majority were concerned with the 'moral factor' in unemployment; but so were the Webbs.[2] The political economists among the majority paid conventional tribute to the 'play of natural forces' and tended to be fatalistic about the inevitability of cyclical depressions;[3] but this had little influence on their practical recommendations for dealing with the unemployed.

Instead of trying to consolidate this wide area of agreement, however, Mrs. Webb after the autumn of 1907 virtually ignored the views of her fellow commissioners and concentrated on the conversion of the leaders of political life. She had by this time lost faith in John Burns;[4] but Asquith, Haldane, Churchill, and the Balfours were courted with the break-up of the Poor Law and the substitution of a 'national minimum', and early in 1908 several members of the Labour party gave 'almost a promise of active support'.[5]

For a time this policy promised to pay dividends. Asquith agreed to forecast the break-up of the Poor Law when introducing Old Age Pensions in his Budget speech of 1908.[6] Members of the parliamentary Labour party seemed 'inclined . . . to take up our views, despairing of the practicability of their own . . .'.[7] Early in February Mrs. Webb renewed her acquaintance with Winston Churchill, who was 'very anxious to be friends' and accepted an invitation to dine with the Webbs in order 'to discuss our scheme for dealing with unemployment'.[8] A few weeks later Churchill

---

[1] Cd. 4499/1909, *RC on the Poor Laws, Majority Report*, Part VI, paras. 325–37.
[2] Above, pp. 43, 251.
[3] Bryce MSS., Box. E. 28, Lancelot Phelps to James Bryce, 29 Mar. 1909.
[4] B. Webb's Diary, 19 May 1908, quoted in *Our Partnership*, p. 411.
[5] Ibid., p. 399.
[6] B. Webb's Diary, 10 Feb. 1908, quoted in *Our Partnership*, p. 402. This promise was only half kept. Asquith referred to the 're-classification' of paupers but not to the actual break-up of the Poor Law. He did, however, anticipate that the Commission would advocate the removal of the care of the aged from the Poor Law (*Hansard*, 4th series, vol. 188, col. 466).
[7] Passfield MSS., ii, 4, d, item 2, B. Webb to M. Playne, 2 Feb. 1908.
[8] Passfield MSS., ii, 4, d, item 4, B. Webb to M. Playne, 22 Feb. 1908. According to C. F. G. Masterman, Churchill at this time was 'full of the poor whom he has just discovered. He thinks he is called by providence—to do something for them' (Lucy Masterman, op. cit., p. 97). But see below, p. 264.

published in the *Nation* his proposals for a 'minimum standard', the organization of the labour market, the training of juveniles, and 'the development of certain national industries' as the 'means of counter-balancing the natural fluctuations of world trade'.[1] 'Altogether we are feeling rather happy about our [policy?] of "Permeation",' wrote Mrs. Webb to one of her sisters, shortly after a select political dinner at the London School of Economics, '. . . everyone seeming convinced that they have "to move on" and . . . really grateful to anyone whose knowledge they trust, telling them which way to go.'[2]

These events encouraged the Webbs to believe that when social reform was eventually introduced they would be its secret arbiters; that the Poor Law would be replaced by 'prevention', specialist services and the national minimum, and that the unemployed would be dealt with by the organization of the labour market and the regulation of public employment.[3]

As events turned out the Webbs' ideas on unemployment were far more immediately influential than their ideas on the Poor Law. The political and financial objections to the reform of the Poor Law on the lines suggested by either the Majority or the Minority Report were manifold. Into the vacuum in unemployment policy created by the rejection of the Right to Work Bill, however, it was the Webbs who introduced a new fourfold programme, based on labour organization, reformatory training, subsidized insurance, and public works. But in promoting these ideas the Webbs made a serious miscalculation; they underestimated the intellectual independence of the politicians whose policies they were trying to dictate; and they were misled by the caution of the Liberal cabinet during its first two years of office into doubting whether the existing Government would

[1] *Nation*, 7 Mar. 1908, Winston Churchill to the Editor, pp. 812–13.
[2] Passfield MSS., ii, 4, d, item 2, B. Webb to M. Playne, 2 Feb. 1908.
[3] 'The net impression left on our mind is the scramble for new constructive ideas' recorded Beatrice Webb, after a series of encounters with politicians in Feb. 1908. 'We happen just now to have a good many to give away, hence the eagerness for our company. Every politician one meets wants to be coached—it is really quite comic—it seems to be quite irrelevant whether they are Conservatives, Liberals or Labour Party men—all alike have become mendicants for practicable proposals' (quoted in *Our Partnership*, p. 402).

ever introduce social legislation.[1] Hence they were taken aback when legislation was introduced in a form very different from that they had prescribed. Moreover, the presentation of a Minority Report was both a strength and a weakness. It enabled the Webbs to make a clear statement of principle; but it also enabled politicians to play off one report against the other, and to borrow policies from both reports without acknowledging their debt.

LIBERAL VOLTE-FACE

Politicians in fact proved willing to accept ideas but not dictation from the Webbs. This was particularly true of Winston Churchill, who was singularly ill fitted to play Trilby to a Svengali in the guise of Sidney and Beatrice Webb. Churchill had been spasmodically interested in social reform since before he left the Conservative party, seeing it as a prerequisite of military success and imperial expansion.[2] He was considering the possibility of introducing labour organization and social security on the German model some time before he renewed his connection with the Webbs in February 1908.[3] His ideas on unemployment were spelt out in his letter to the *Nation* on 7 March and in a letter to Asquith a week later, where 'dimly across the gulfs of ignorance', he described 'the outline of a policy which I call the Minimum Standard'. This policy would apply to housing and unemployment, electoral and rating reform, and the administration of Old Age pensions. Provision for the unemployed would include the training and discipline of juvenile labour; labour exchanges, decasualization, and the 'curative treatment' of redundant labourers 'as if they were hospital patients'; the provision of state employment through military service; the regulation of the hours of labour; and state intervention in industry to increase the demand for labour during periods of depression. Finally, Churchill proposed that the great mass of voluntary thrift institutions

---

[1] B. Webb's Diary, 24 Mar. 1908, quoted in *Our Partnership*, p. 406.
[2] He had written an unpublished review to this effect of Seebohm Rowntree's *Poverty: A Study of Town Life* (R. Churchill, *Winston Churchill: Young Statesman 1901–14*, pp. 30–2).
[3] Wilson Harris, *J. A. Spender*, p. 81; R. Churchill, op. cit., pp. 300–1.

which had grown up in England should be buttressed by 'a sort of Germanized network of state intervention and regulation'. Such a programme would be 'national rather than departmental', although the Local Government Board would necessarily be the 'fountain' of such social innovation.[1]

Churchill's scheme clearly owed much to the Webbs; it also contained echoes of salvationism, German state socialism, and the strain of 'regimentation' which has already been noted as a characteristic of Edwardian social ideas. When writing this letter Churchill disclaimed any personal desire or capacity to introduce such a programme; but when he was promoted to the Presidency of the Board of Trade in April 1908 he instantly started work on the kind of programme that he had prescribed for the Local Government Board.[2]

This programme formed part of the remarkable volte-face that occurred in Liberal domestic policy during the course of 1908 when, having for two years postponed the consideration of 'social reform' until it had received the reports of the Royal Commission, the Government turned its attention to social problems over a year before these reports appeared. The new policy, although clearly influenced by the Royal Commission, was in many ways quite independent of the Majority and Minority Reports; it had, moreover, very little in common with the kind of policy that the most advanced radicals, such as Masterman, Alden, and Herbert Samuel had been putting forward since 1902. The first outward evidence of a new trend was Asquith's public promise of Old Age Pensions in February 1907;[3] but the full extent of the change was not fully apparent until the Cabinet re-shuffle after Campbell-Bannerman's retirement in the spring of 1908. The reasons for this change were complex, and can be explained partly by external political pressures, partly by a shift in the internal distribution of power

[1] Asquith MSS., vol. 11, ff. 10–15, W. S. Churchill to H. H. Asquith, 14 Mar. 1908.

[2] On the conflicting evidence as to whether or not Churchill wanted to be President of the L.G.B. see Violet Bonham Carter, *Winston Churchill as I knew Him*, p. 159; B. G. Gilbert, *The Evolution of National Insurance in Great Britain*, p. 249. The concluding paragraph of Churchill's letter suggests that he may have been exaggerating the arduousness of local-government affairs in order to persuade Asquith to offer him the L.G.B. as a Secretaryship of State.

[3] *Hansard*, 4th series, vol. 169, cols. 222–7.

within the Government and partly by a gradual redefinition of the social questions at stake.

Firstly, it must be remembered that the Royal Commission on the Poor Laws had been called into existence to deal with a crisis in the administration of the Poor Law, and that this crisis intensified after 1905. The erosion of old norms of Poor Law administration continued in boroughs like West Ham and Poplar even after the Local Government Board inquiries of 1906. When unemployment increased in the winter of 1907–8 the Poplar guardians again came in conflict with the ratepayers by allowing outdoor relief for an indefinite period to the families of unemployed workmen who entered the workhouse under the Modified Workhouse Test. The Poplar Municipal Alliance claimed that this policy was 'encouraging men to abstain from work . . . and attracting the undesirable and unemployable to Poplar';[1] and early in 1909, in the face of protests from Lansbury and Buxton, John Burns ordered the Poplar guardians to limit relief to eight weeks in any one year to families assisted under the Modified Workhouse Test.[2]

There is no reason to suppose that Liberal ministers did not genuinely hope that the Royal Commission would provide a solution to this kind of politically embarrassing situation, whether by a reconstruction of the Poor Law or by a reversion to the principles of 1834. By the middle of 1907, however, the Government was well aware 'through the Sidney Webb manœuvres',[3] that both sides of the Royal Commission would recommend a dismantling of the old Poor Law structure; and while they were glad to make use of the ideas of the Webbs, it was clearly desirable to give any measures of social reform a distinctly 'Liberal'

[1] Ensor MSS., Poplar Borough Municipal Alliance to the Poplar Guardians, 30 July 1908.
[2] Buxton MSS., unsorted, G. Lansbury to H. H. Asquith, 10 Jan. 1909; Sydney Buxton to John Burns, 30 Jan. 1909. In Aug. 1908 Burns had come to the conclusion that Crooks and Lansbury were trying to use the Poplar crisis 'as a means of converting L.G.B. to deal in their silly way with Unemployed problem. Burning a house down to get roast pig is trivial compared with this reckless destruction of morals in an individual district . . .' (Add. MS. 46326, Burns Diary, 28 Aug. 1908).
[3] Phelps MSS., unsorted, Lord George Hamilton to Lancelot Phelps, 16 Aug. 1908.

identity. Even so, the Government was taunted with being unduly under the influence of Sidney Webb, 'who despises the electorate and who thinks that only trained government officials are fit to carry on the affairs of the country'.[1]

Secondly, although Liberal leaders were publicly committed to the idea of Poor Law reform, it became increasingly clear that the kind of reform envisaged by both the majority and the minority of the Royal Commission was within the foreseeable future technically impossible. During the Cabinet discussions of Old Age Pensions in 1907 Haldane had sketched out an agenda of social reforms closely akin to those that were being devised by the Webbs;[2] and in August 1908 he wrote to Asquith that 'as to the [Poor Law] the objective is clear enough. We know what we want and the question is how to guide Burns into doing it.'[3]

Nevertheless, the reform of the Poor Law along the lines suggested by the Royal Commission was doomed to be fatally waterlogged among the unsettled questions of local taxation and local finance. These problems centred upon the distribution of financial responsibility for local services between the State and the local authority, between rich and poor localities and between different types of property in a local-government area.[4] Since 1901 successive Chancellors had promised a reform of the system of local taxation, and many local authorities were consequently reluctant to accept new responsibilities until a greater share of local burdens had been transferred to the national exchequer.[5] In 1906 Campbell-Bannerman was hoping to introduce a Valuation Bill, which was seen as a prerequisite of rating reform,

---

[1] *Hansard*, 5th series, vol. 10, col. 924. See also Add. MS. 46301, f. 121, John Burns to H. G. Wells, 16 May 1910: 'The new helotry in the servile state run by the archivists of the School of Economics means a race of paupers in a grovelling community ruled by uniformed prigs. Rely upon me saving you from this plague.'

[2] Asquith MSS., vol. 74, ff. 176–8, Memorandum on 'Old Age Pensions' by R. B. Haldane, 6 Sept. 1907.

[3] Asquith MSS., vol. 11, ff. 162–5, R. B. Haldane to H. H. Asquith, 9 Aug. 1908.

[4] See Appendix A, pp. 369–70.

[5] T. 171/10, Report of the Finance Committee of the L.C.C. on 'London and the Imperial Exchequer', 22 June 1910.

during the parliamentary session of 1907.[1] However, the reform of local taxation was one of the most difficult of all problems of domestic policy, far exceeding Poor Law reform in legal and financial complexity and in the strength of the vested interests involved.[2] By the autumn of 1908 John Burns was ready to introduce an extremely cautious measure, which would have unified the national and local systems of valuation but made no provision for the separate assessment of the site value of land.[3] This Bill was not, however, introduced into Parliament;[4] and the problem was taken over by Lloyd George, who included land valuation and a tax on undeveloped site values in the controversial Budget of 1909.[5] The 1909 Finance Act was not passed until the autumn of 1910; and in 1911 the questions of the relationship of imperial to local taxation and the relative burdens of rich and poor local authorities were referred to a Treasury departmental committee, which did not report until 1914.[6] Between 1909 and 1914, therefore, the possibility of creating new social services on the lines suggested by both Reports of the Royal Commission on the Poor Laws, by transferring the work of the guardians to committees of the county councils, was rendered virtually impossible by the unsettled question of the local rates.[7] Throughout this period it proved

[1] Add. MS. 41239, f. 100, Note by Arthur Ponsonby, 26 July 1906; seen by H. H. Asquith, John Burns, and Sir John Sinclair. Valuation bills had been introduced annually by Liberal backbenchers in 1902–5 (*Liberal Magazine*, Aug. 1907, p. 440).

[2] The House of Lords rejected a Land Values (Scotland) Bill in 1907; and in 1908 a similar Bill was so severely mutilated in the Upper House that it had to be withdrawn (*Liberal Magazine*, Sept. 1907, pp. 524–5; Jan. 1909, p. 753).

[3] CAB 37/95/122, 'Valuation Bill' by John Burns, 10 Oct. 1908.

[4] Presumably because of pressure on Asquith from 244 Liberal M.P.s to introduce a much more drastic measure of land taxation in Nov. 1908 (A. King, 'Some Aspects of the History of the Liberal Party in Great Britain, 1906–14', Oxford D.Phil. 1962, p. 78).

[5] CAB 37/97/16, Memorandum by Lloyd George on 'Taxation of Land Values', 29 Jan. 1909.

[6] Cd. 7315/1914, *Departmental Committee on Local Taxation, Final Report*, pp. 100–4. This committee recommended that local inequalities in the yield of rates should be levelled out by a considerable increase of state subventions for 'semi-national' services.

[7] T. 171/10, Unsigned notes on Mr. Hayes Fisher's Amendment to the Address, no date, ?Mar.–Apr. 1912. On the interdependence of Poor Law reform and local taxation reform see T. 171/87, 'Budget 1914. Extracts from speeches promising local taxation reform since 1908'.

easier to create new public authorities and to introduce entirely new forms of social policy, however complicated and controversial, than to add to the existing burdens of county councils or to make other reforming adjustments in the existing system of local administration.

On the other hand, social policies of an entirely new kind had been made possible by the innovations in the system of direct taxation, which had been declared possible by a Select Committee of 1906[1] and were introduced into the revenue system by the Budgets of 1907–9. The differentiation between 'earned' and 'unearned' incomes under £2,000 and the addition of a 'supertax' to the graduated estate duty by the Budget of 1907 greatly increased the potential yield of direct taxation—although this was not fully apparent until the Budget of 1909.[2] The arguments in favour of this measure were threefold. Firstly, and probably least important, was the argument that graduation and differentiation were measures of direct social justice which would enable the Government to alleviate the 'consumption taxes' that pressed most heavily upon the incomes of the poor.[3] Secondly, progressive taxation was seen as a prerequisite of social, naval, and military reforms.[4] It was a measure, observed Haldane to Asquith, that the spirit of the age had called for as unmistakably as it had once called for the extension of the franchise; and he urged his colleague that

we should boldly take our stand on the facts and proclaim the policy of taking, mainly by direct taxation, such toll from the increase and growth of ... wealth as will enable us to provide for (1) the increasing cost of social reform (2) National defence and also (3) to have a margin in aid of the Sinking fund. The more boldly such a proposition is put

1 H. of C. 365/1906, *SC on the Income Tax, Report*, para. 30. This committee under the chairmanship of Sir Charles Dilke, concluded that both graduation of the income-tax and differentiation between 'earned' and 'unearned' incomes was administratively feasible. Members of the committee were inclined to favour a 'supertax' rather than a graduation of the standard rate.

2 The main immediate effect of the 1907 Budget was to relieve taxation on professional incomes by 25 per cent and it was strongly criticized by Labour leaders for pandering to 'the city clerk and the small gentry' (*Report of the Bath T.U.C.*, 1907, p. 86). But it was definitely framed with a view to financing social reform (T. 171/3, Conspectus of the Budget, 1907–8).

3 T. 171/3, 1907–8 Budget, Notes in Sir Edward Hamilton's MS.

4 CAB 37/87/24, Memorandum on Supertax by W. B., 26 Feb. 1907.

the more attractive I think it will prove. It will commend itself to many timid people as a bulwark against Nationalisation of Wealth. . . .[1]

Thirdly, and as a political argument probably decisive, graduation was seen as a means of countering and check-mating the Conservative demand for a system of protection. 'While the present position continues, it puts a powerful weapon in the hands of the advocates of Tariff Reform' wrote a partisan Civil Service official in February 1907.

They are able to point to the admitted desire of the Government to effect social changes, a desire which is thwarted by the want of elasticity in our present system of revenue. No better answer to the most specious argument on their side could be found than a fiscal change which would enable additional revenue to be raised at need without an increase—indeed with a diminution—of the proportion of burden falling upon the most numerous classes.[2]

This argument was underlined when, contrary to precedent, the yield of the income-tax continued to rise through the commercial depression of 1908–9.[3] Asquith himself told St. Loe Strachey in May 1908 that he had 'realised from the first that if it could not be proved that social reform (not Socialism) can be financed on Free Trade lines a return to Protection is a moral certainty. This has been one of the mainsprings of my policy at the Exchequer.'[4]

Finally, by the winter of 1906–7 Liberal ministers were already realizing that they might not enjoy a full six-year term of office, and that a General Election might be provoked by a conflict with the House of Lords, comparable in constitutional significance to the reform crisis of 1832.[5] It was therefore necessary to oppose Conservative forces by making the appeal of radicalism as broad as possible, and a small but influential minority of Liberals assumed that

---

[1] Asquith MSS., vol. 11, ff. 162–5, R. B. Haldane to H. H. Asquith, 9 Aug. 1908.
[2] CAB 37/87/24, Memorandum on 'Supertax' by W. B., 26 Feb. 1907.
[3] Josiah Stamp, *British Incomes and Property* (1927 ed., first published 1916), pp. 316–19.
[4] A. Gollin, *Proconsul in Politics*, p. 152; H. H. Asquith to St. Loe Strachey 9 May 1908.
[5] Campbell-Bannerman considered, and rejected, the idea of a dissolution of Parliament after the first defeat of the Education Bill in the Lords at the end of 1906 (R. Jenkins, *Mr. Balfour's Poodle*, pp. 25–6).

this could be done by accelerating the Government's progress towards social reform. This view was strengthened at the end of 1907 by the worsening trade situation and the apparent need to outbid Conservative promises of full employment and social improvement under a system of protection.[1] In reality this line of reasoning was almost certainly a miscalculation, since there is no reason to suppose that constructive social legislation was any more attractive to a majority of the electorate in 1907–8 than at any previous time. J. A. Spender thought that the by-election swing against the Liberals early in 1908 was actually caused by the anticipation of expensive social reforms;[2] and Francis Hirst again warned Campbell-Bannerman that the Government could regain its popularity only by a traditionalist policy of financial retrenchment.[3] Winston Churchill assured Asquith that social reform was a democratic imperative, which would 'not only benefit the State, but fortify the party';[4] but there was probably more truth in the misgivings of John Morley, who feared that nothing could be done for the working class without alienating the lower-middle-class backbone of Liberal support.[5]

Nevertheless, by the beginning of 1908 social reform had become both financially feasible and in the judgement of many Liberals politically advantageous. In March the realignment of the Cabinet after the retirement of Campbell-Bannerman, who had always been a brake on social reform,[6] released the Disraelian ambition of Churchill and Lloyd

---

[1] Asquith MSS., vol. 13, ff. 110–11, D. Lloyd George to H. H. Asquith, 28 Dec. 1912. B. B. Gilbert, *The Evolution of National Insurance in Great Britain*, pp. 209–21, states that it was Liberal by-election losses to Labour at Durham and Colne Valley in July 1907 that frightened the Government into social reform; but Liberal leaders were much more concerned at the prospect of a swing to the Conservatives than to Labour (Bryce MSS., Box E. 28, R. B. Haldane to James Bryce, 12 Mar. 1907; *Annual Register*, 1907, p. 251).

[2] Bryce MSS., Box E. 28, J. A. Spender to James Bryce, 9 Mar. 1908.

[3] Add. MS. 41240, ff. 153–4, Francis Hirst to Campbell-Bannerman, 9 Nov. 1907.

[4] Asquith MSS., vol. 11, f. 253, W. S. Churchill to H. H. Asquith, 29 Dec. 1908.

[5] Bryce MSS., Box P. 6, John Morley to James Bryce, 6 Jan. 1908.

[6] Asquith MSS., vol. 10, f. 200, Campbell-Bannerman to H. H. Asquith, 21 Jan. 1906; Add. MS. 45988, f. 213, Campbell-Bannerman to H. Gladstone, 23 Jan. 1906.

George to transform the Liberals into the 'party of the nation'.[1] It was in this context of new ideas and new policy-makers that the movement for a national policy on unemployment came at last to fruition.

[1] Bryce MSS., Miscellaneous English Correspondence C, W. S. Churchill to James Bryce, 18 Jan. 1906. On Churchill's and Lloyd George's fondness for national parties and coalitions see Violet Bonham Carter, op. cit., pp. 205–6.

VI

A SCIENTIFIC POLICY FOR THE UNEMPLOYED

THE emergence of a new unemployment policy in the Liberal party coincided with the most acute commercial depression since 1879.[1] It was particularly severe among skilled artisans in the heavy industrial areas of the north-east, to whom the Liberals traditionally looked for support.[2] There was a heavy drain on trade-union funds, and although the skilled unions were financially stronger than in the 1870s and 1880s, many union leaders were no longer willing to assume that the cost of unemployment should be borne entirely by the workman.[3] On 15 October Victor Grayson, the Socialist M.P. for Colne Valley, created a scene in the House of Commons by declaring that he had an 'unemployed mandate' to demand instant legislation and refusing to sit down when the Speaker ruled that discussion of the question should be postponed.[4] In November the metropolitan Poor Law inspector reported that there was 'every indication that pauperism and unemployment will give serious trouble . . . there appears to be throughout London a general organisation which, for political ends, is exerting pressure on Boards of Guardians to give undue relief'.[5]

[1] Adjustments in the construction of the Board of Trade's index figure concealed the fact that in the shipbuilding and engineering trades unemployment was almost certainly more severe in 1908–9 than in 1886 (N. B. Dearle, 'English Statistics of Unemployment', *International Conference on Unemployment*, Paris 1910, Tome 3, Report No. 25, pp. 2–5).

[2] 4,719 workmen from the engineering, shipbuilders, and metal trades were registered under the Unemployed Workmen Act in 1907–8, and 17,028 in 1908–9 (Beveridge, *Unemployment* (1930 ed.), pp. 168 and 449).

[3] In Sept. 1908 the T.U.C. voted by 921,000 votes to 559,000 in favour of government 'grants in aid' to trade unions who paid out of work benefit (*Report of the Nottingham T.U.C.*, 1908, pp. 166–9).

[4] *Hansard*, 4th series, vol. 194, cols. 495–7. A similar demonstration was staged in the L.C.C. (Lucy Masterman, *C. F. G. Masterman*, p. 110).

[5] CAB 37/96/143, 'Pauperism and Unemployment', by Mr. Oxley, 3 Nov. 1908, pp. 2–3.

The situation was under discussion in the Cabinet from the late summer of 1908. In August Churchill warned his colleagues that the conjunction of unemployment, rising food prices, and falling wages had inflicted 'a period of unusual severity' upon the working class, and that conditions were likely to get worse rather than better; although he derived 'sinister consolation' from the fact that the situation was just as bad in France, Germany, and the United States.[1] In September Asquith published a declaration of intent by sending an open letter to the Liberal candidate at the by-election in Newcastle—an area particularly affected by the shipbuilding slump—pledging the Government to 'the early presentation of practical legislative proposals' on behalf of the unemployed.[2] This letter seems to have been the signal for a spate of proposals and political promises. Haldane at the end of September publicly advocated unemployment insurance and military recruitment for the unemployed.[3] Under pressure from Churchill and Lloyd George, the First Lord of the Admiralty, Reginald McKenna, agreed to bring forward shipbuilding contracts that were scheduled for the following year; his aim was 'not merely to give employment when work is scarce, but to get cheap rates for my contracts in the present condition of trade'.[4] Churchill, addressing his new constituents at Dundee, condemned the 'gross and . . . increasing evil of casual labour' and proposed compulsory technical training for juveniles up to the age of eighteen. He deplored 'the lack of any central organisation of industry or any central and concerted control either of ordinary Government work, or of extraordinary relief works'.[5] And John Burns, who throughout the summer had

---

[1] *Board of Trade Reports on Employment and Trade, 1905–9*, vol. D, W. S. Churchill to the Cabinet, 8 Aug. 1908.

[2] *Liberal Magazine*, Oct. 1908, p. 554. [3] Ibid., pp. 554–5.

[4] Asquith MSS., vol. 20, f. 83, D. Lloyd George to R. McKenna, 11/12 Sept. 1908; ibid., ff. 85–90, R. McKenna to D. Lloyd George, 12 Sept. 1908; R. Churchill, *Winston Churchill, The Young Statesman 1901–14*, p. 287. These contracts amounted to £2,000,000, of which £200,000 would be spent in wages before 31 Mar. 1909 (CAB 41/31/68, H. H. Asquith to the King, 20 Oct. 1908). McKenna's biographer was apparently unaware that he accepted the proposal (S. McKenna, *Reginald McKenna 1863–1943. A Memoir*, p. 79). McKenna was also keen to start work on the building of minesweepers and an 'experimental airship', but this latter proposal was vetoed by the Treasury for the financial year 1909–10.

[5] *Liberal Magazine*, Oct. 1908, p. 588.

been arranging loans to local authorities with which to ward off winter unemployment, proposed that the whole of the nation's public expenditure, amounting to £300,000,000 a year, should be arranged so as to counterpoise the fluctuations of private commerce.[1]

Nevertheless, Burns was reluctant to concede that any short-term measures could be taken against unemployment that were not already being executed by his department. This attitude provoked a minor Cabinet crisis in the middle of October 1908, and 'J. B. was outvoted and practically superseded by the appointment of a Cabinet Committee on unemployment. J. B. said nothing was the matter and nothing could be done. W. S. C. arose and said something must be done, it was a burning question'.[2] The controversy centred chiefly on the extension of the use of the rates and the relaxation of the conditions for relief under the Unemployed Workmen Act. Arthur Henderson, the Chairman of the Labour party, urged Burns's Parliamentary Secretary, Charles Masterman, to introduce a special supplementary estimate and 'permission for this year for the payment of wages by a special rate levied locally'.[3] This proposal was endorsed by Churchill and Sidney Buxton; but it was overruled by Burns, who insisted that it would be an openended concession and an intolerable burden on 'poorer districts with a low rateable value'.[4] He was prepared to concede, however, that regulations for the receipt of benefit should be relaxed and that the parliamentary grant to Distress Committees under the Unemployed Workmen Act should be increased to £300,000 in 1908–9; and this compromise was accepted by the Cabinet on 19 October 1908.[5]

Nevertheless, other Cabinet ministers were clearly dissatisfied with Burns's treatment of the problem; and for several months previously alternative policies for dealing

[1] Ibid., p. 582; *Hansard*, 4th series, vol. 194, cols. 1667–8.

[2] Lucy Masterman, op. cit., pp. 110–11. This Cabinet meeting took place on 14 Oct. 1908.

[3] Ibid., p. 111.

[4] CAB 37/95/125, J. Burns to H. H. Asquith on 'The Unemployed', 16 Oct. 1908.

[5] CAB 41/31/68, H. H. Asquith to the King, 20 Oct. 1908.

with unemployment, which circumvented Burns and his over-cautious officials, had been under consideration by Churchill and Lloyd George. In August 1908 Lloyd George had paid his much publicized visit to Germany and had come back enthused not only with the idea of sickness insurance but—according to contemporary authorities— with the untried possibilities of compulsory insurance against unemployment.[1] In October he told George Riddell, the proprietor of the *News of the World* that

his idea (was) to form a board in each trade which will make a levy in prosperous times upon employers and workmen, and apply the sums contributed to alleviate distress in times of depression. His suggestion (was) that the Board should be formed of employers and workmen with an independent chairman.[2]

But it was Churchill who pressed home the doctrine of 'social organization' and persuaded Asquith that he had found a 'scientific' solution for the problem of unemployment,[3] which would unite the nation behind the Liberals and which the 'House of Lords will not dare oppose'.[4] In his first Cabinet post Churchill was anxious to make a name for himself. As he had told Sir Charles Dilke two years earlier, he was not 'among those who having gained power do not wish to use it';[5] and he brought to the study of social problems the same grasp of strategic principle, the same contempt for tactical difficulties that later characterized his approach to military, naval, and diplomatic affairs. It was Churchill who brushed aside the stolid conviction of John Burns that unemployment could be adequately relieved by self-help and by the advance of municipal socialism. Moreover, as he confided to Beveridge, he had 'a reason for getting

[1] Both Lloyd George and Churchill accused the other of having stolen the idea of unemployment insurance from himself (Arthur C. Murray, *Master and Brother. Murray of Elibank*, pp. 88–9; R. Churchill, *Winston Churchill. Young Statesman 1901–1914*, p. 306). The question had, however, been discussed at length in the Royal Commission on the Poor Laws for nearly a year before Lloyd George's German visit (below, pp. 300–1).

[2] G. Riddell, *More Pages from My Diary*, p. 3.

[3] R. Churchill, op. cit., pp. 278, 304.

[4] Asquith MSS., vol. 11, f. 252, W. S. Churchill to H. H. Asquith, 29 Dec. 1908.

[5] Add. MS. 43877, f. 53, W. S. Churchill to Sir Charles Dilke, 24 Jan. 1906.

on quickly with Labour Exchanges . . . he had not himself many years to live; he expected to die young, like his father Randolph'.[1]

On 20 October Churchill, in reply to a question in the House of Commons, announced that the Board of Trade was considering remedial measures to deal with 'the general questions of unemployment, under-employment and casual labour'.[2] In November he and Lloyd George consulted Sidney Webb and the Parliamentary Committee of the T.U.C., and a trade union deputation was sent to report on the treatment of unemployment in the German empire.[3] The English trade unionists were greatly impressed by the visible prosperity of the German working class, and reported that the Berlin Labour Exchange was 'admirable in every respect'.[4] On state unemployment insurance, which was still at the discussion stage, they were non-committal, pointing out the problems that would arise over trade disputes, the verification of unemployment, and the reluctance of efficient and regular workmen to subsidize the inefficient and irregularly employed.[5]

The outline of Churchill's plan for an interdependent scheme of labour exchanges and compulsory insurance against unemployment in certain trades—'to organise the mobilities and stabilities of labour'[6] was submitted to the Cabinet early in December 1908. On 29 December he described in a letter to Asquith some of the details of his Bismarckian vision of social organization, which included labour

1 Beveridge MSS., D. 047, Notes for a speech on the fiftieth anniversary of the Employment Exchange service, 1959; Lucy Masterman, op. cit., p. 67.

2 *Hansard*, 4th series, vol. 195, col. 492.

3 CAB 37/96/159, 'Unemployment Insurance: Labour Exchanges', by W. S. Churchill, p. 6.

4 Braithwaite MSS., II, (i), 'Workmen's Insurance Systems in Germany. Report of Deputation', pp. 3–5. The deputation reported that the German unemployed 'seemed to lack that dejection and absolute misery that . . . is so frequently met within the streets of English towns. A spirit of sturdy self-reliance seemed to manifest itself, even in the demeanour of those out of work and was ascribed by one informant to the fact that most of the men had undergone military service . . . and had thereby gained in physique and moral strength . . .'.

5 Braithwaite MSS., II, (i), 'Workmen's Insurance Systems in Germany, Report of Deputation', pp. 6–7.

6 CAB 37/96/159, 'Unemployment Insurance: Labour Exchanges', by W. S. Churchill, p. 1.

exchanges and insurance, 'special expansive state industries', the development of roads and forests, a classified
Poor Law system, and compulsory education up to the age
of seventeen.[1] Draft bills were submitted to the Cabinet on
labour exchanges in January and on unemployment insurance in April 1909;[2] and on 29 April Lloyd George in
his Budget speech announced the creation of two special
funds which would finance 'healthy and productive' employment and promote schemes 'which have for their purpose
the development of the resources of the country'.[3] These
three measures, for the organization of the labour market,
state-subsidized national insurance, and the development of
national resources, were the three main items of the new
unemployment policy adopted by the Liberals in 1909.
It is clear that they were originally conceived as part of a
single programme; but the introduction of the three measures will here be considered separately, since each of them
posed a different set of administrative questions and each
dealt with a different aspect of the problem of the unemployed.

THE ORGANIZATION OF THE LABOUR MARKET

In the establishment of a system of labour organization
three main issues were involved: whether such organization
should be local or national, whether it should be voluntary
or compulsory, and which government department should
exercise central administrative control. The resolution of
these questions was central to the planning and implementation of a uniform system of labour exchanges in 1908–10.

It has sometimes been suggested that the introduction of
the Labour Exchanges Bill was an exclusively departmental

---

[1] Asquith MSS., vol. 11, f. 252, W. S. Churchill to H. H. Asquith, 26 Dec.
1908.
[2] CAB 37/97/17, 'Labour Exchanges (together with Appendix on Labour Exchanges in foreign countries and rough drafts of heads of a Bill for their establishment in the United Kingdom)', Jan. 1909; CAB 37/99/69, 'Scheme for Unemployment Insurance and Draft Heads of a Bill for the Establishment of Unemployment
Insurance', Apr. 1909.
[3] *Hansard*, 5th series, vol. 4, cols. 489–98.

measure; and that, contrary to public expectation, it was taken over by the enterprising ministers and permanent officials of the Board of Trade at the expense of the reactionary and inefficient Local Government Board.[1] But neither of these views is entirely correct, since there was a long history of extra-governmental pressure for public exchanges, which since the 1880s had pointed towards the Board of Trade as the most appropriate government department to direct a system of labour organization.

Throughout the nineteenth century there were many philanthropic and commercial agencies which helped the unemployed to look for work;[2] but most of them were confined to a particular class of client, such as servants, governesses, and discharged soldiers and prisoners,[3] and their impact on the labour market appears to have been very slight.[4] However, a new type of labour exchange, which was neither charitable nor profit-making, was opened by Nathaniel Cohen at Egham, Surrey, in 1885.[5] The Egham Free Registry was designed to put employers in contact with local workmen of 'authenticated good character' and to diminish 'the waste of time and energy involved in the search for work'.[6] The main criterion employed in sending an applicant to a situation was not the extent of his distress but his suita-

---

[1] B. B. Gilbert, *The Evolution of National Insurance in Great Britain*, p. 258; J. A. M. Caldwell, 'The Genesis of the Ministry of Labour', *Public Administration*, 37 (1959), 371–82.

[2] Some of the earliest 'labour exchanges' were opened by the Co-operative Movement during the 1830s; they were workshops and markets for produce as well as for labour (W. H. Oliver, 'The Labour Exchange Phase of the Co-operative Movement', *Oxford Economic Papers*, N.S. 10 (1958), 355–67). The Corporation of Coventry had founded a labour exchange in the sixteenth century (Beveridge MSS., D. 047, Some Historical Notes on Early Attempts to set up Labour Exchanges). Beveridge also discovered a seventeenth-century plan for a national labour exchange which would also act as a marriage bureau, stock market, and land registry (W. H. Beveridge, 'A Seventeenth Century Labour Exchange', *Economic Journal*, 24 (Sept. and Dec. 1914), 371–6, 635–6).

[3] C. 7182/1873, *Agencies and Methods for Dealing with the Unemployed*, pp. 122–30, 141–3.

[4] H. of C. 86/1905, *Labour Bureaux*, Report to the President of the L.G.B. by Arthur Lowry, Nov. 1905, p. 19.

[5] Beveridge MSS., D. 047, W. H. Beveridge to Sir Bernard Waley-Cohen, 8 Feb. 1960. Cohen opened the Egham Free Registry to deal with a temporary crisis in the local building industry.

[6] *8th Annual Report of the Egham Free Registry for the Unemployed*, 1892, unpaginated.

bility for the post;[1] and during the nine years of its existence, the registry's success rate in placing workmen in situations was over 80 per cent.[2] In the same year a disillusioned Poor Law guardian, the Reverend Wickham Tozer, founded a labour bureau in Ipswich and by 1890 was receiving 400 applications a year from all parts of the country.[3] The first municipal labour bureau was opened by the Chelsea vestry in 1891.[4] By September 1892 there were seventeen private provincial bureaux, modelled on the registry at Egham,[5] and no less than thirty-one permanent or temporary labour registries had been opened by local authorities in England and Wales.[6]

Some of these registries had a permanent staff and offices, but most of them were associated with emergency relief works and conspicuously failed to find commercial situations for the unemployed.[7] The most successful were those that recommended clients on purely commercial grounds, in the belief that 'it would be fatal to the success of an Employment Agency if it were thought to be a mere register of inefficient men'.[8] The leaders of the labour-registry movement, however, pursued a common policy which was more significant than the success or failure of individual registries. They urged the Royal Commission on Labour that a network of labour bureaux should be established on a national scale;[9] and E. T. Scammell, the secretary of the Exeter Chamber of Commerce, put forward a plan for county labour exchanges, supervised by the Board of Trade and financed out of the technical education fund. These exchanges would register vacancies, publish statistics, and

---

[1] It is possible that Cohen was influenced in this policy by the experience of the employment register of the Jewish Board of Guardians, which failed in 1886 owing to its connection with charity (V. Lipman, *A Century of Social Service 1859–1959*, p. 48).

[2] *10th Annual Report of the Egham Free Registry for the Unemployed*, 1894.

[3] C. 7063–I/1893, *RC on Labour (Sitting as a Whole), Minutes of Evidence*, QQ. 6181–5.

[4] Ibid., Q. 2032.   [5] Ibid., Q. 1995.

[6] C. 7182/1893, *Agencies and Methods for Dealing with the Unemployed*, pp. 188–208.

[7] Ibid., p. 112.

[8] *9th Annual Report of the Egham Free Registry*, 1893.

[9] C. 7063–I/1893, *RC on Labour (Sitting as a Whole), Minutes of Evidence*, QQ. 6208–29.

organize a 'comprehensive national system' of emigration on the lines suggested by the 'Darkest England' scheme of General William Booth.[1]

The Royal Commission rejected the proposal that these responsibilities should be taken over by the State; but it proposed that the Board of Trade should conduct inquiries into the state of the labour market, publish more frequent employment statistics, and assist in the establishment of local and voluntary labour registries.[2] Llewellyn Smith, however, thought that the establishment of labour exchanges would merely increase the migration of workmen from the countryside;[3] and after investigating the labour-exchange system in Luxembourg he came to the conclusion 'that the system is not suitable to this country'.[4] In 1895 the question of labour exchanges was again raised before the Select Committee on Distress from Want of Employment by the secretary of the social department of the Church Army, who pressed for the creation of a national labour bureau under a separate minister answerable to Parliament; but the Chairman of the Committee, Campbell-Bannerman, replied rather enigmatically that this service was already supplied by the Board of Trade.[5]

Although their operations were numerically insignificant, the importance of the labour registries of the early 1890s should not be underestimated. Many of the problems which arose in the management of exchanges in the 1900s were already apparent in 1895, particularly those of keeping them free from the suspicion of charity and cultivating the good-will of trade unions and employers. Moreover, the pressure

---

[1] C. 7063-I/1893, QQ. 6387-6427; also C. 7083-III.A/1894, *RC on Labour (Sitting as a Whole), Minutes of Evidence*, Appendix CXVII, 'A National Labour Bureau, with affiliated Labour Registries', by E. T. Scammell, pp. 224-5.

[2] C. 7421/1894, *RC on Labour, Fifth and Final Report*, paras. 309-12.

[3] B.T. 13/23/E. 11499, Memorandum on Certain Recommendations included in the Final Report of the Royal Commission on Labour (Majority Report), by H. Llewellyn Smith, 29 June 1894. Llewellyn Smith was at this date Chief Commissioner for Labour in the Commercial, Labour and Statistical Department of the Board of Trade.

[4] B.T. 13/23/E. 11499, Recommendations of Majority (of the Royal Commission) which affect the Board of Trade, indicating those which immediately involve legislation, 12 Jan. 1895.

[5] H. of C. 365/1895, *SC on Distress from Want of Employment, Minutes of Evidence*, QQ. 10185-6.

for a national system of exchanges clearly showed that they were seen as complementary to the publication of labour statistics, and therefore as a logical extension of the functions of the Board of Trade. There was absolutely no suggestion at this stage of the movement that any other government department—except a hypothetical 'Ministry of Labour'—should be asked to intervene.

In the absence of any positive legislation to this effect, however, the registries established by local authorities were indirectly responsible to the Local Government Board. Their legal status was, moreover, called in question by some of the London vestries and district boards which denied that the formation of such bureaux was within their statutory powers.[1] In 1899 the London Government Act, which replaced vestries and district boards by borough councils, also introduced a rigorous auditing system whereby councillors were personally chargeable for unauthorized expenditure. The few metropolitan labour bureaux that had survived the trade boom of the late 1890s were declared *ultra vires* by the district auditors, and all except the Battersea bureau were closed.[2] In 1902 a Bill was therefore introduced into Parliament by the London Liberal M.P.s, legitimizing expenditure on labour bureaux out of the rates. The Bill was passed with the approval of the Local Government Board, which nevertheless made it clear that 'the principle of the measure is one for the Board of Trade rather than for the Local Government Board, who are only more particularly concerned with the proposal for the payment of the expenses out of the rates'.[3]

It has already been shown that the main purpose of the Unemployed Workmen Act of 1905 was to rationalize existing forms of relief; and the establishment of labour exchanges was therefore included among the powers con-

[1] H. of C. 365/1895, Q. 5015.
[2] Beveridge MSS., D. 047, Some historical notes on early attempts to set up labour exchanges.
[3] H.L.G. 29/69, vol. 63, f. 60, Unsigned memorandum, Feb. 1902. The Bill was passed as the Labour Bureaux (London) Act, 1902 (2 Edw. VII, c. 13). At the suggestion of the Board of Trade a clause was inserted stipulating that no bureau should 'give any unfair preference to any person on the ground that he is or is not a member of a Trade Union'.

ferred by the Act.[1] Thus it was almost entirely accidental that the first central department empowered to establish a national system of labour exchanges should have been the Local Government Board. The machinery created by the Unemployed Workmen Act was quite unsuited for such a task, and the Board had no kind of contact with the employers and workmen on whose co-operation the exchanges depended. In the eyes of the public the Local Government Board was 'tainted with Poor Law traditions and . . . managed by the Charity Organisation Society';[2] and, whether or not this was true, it was enough to deter employers and independent workmen from using exchanges which were under L.G.B. control.

The Board of Trade, on the other hand, had well-established channels of communication with both sides of industry. In 1907 it acquired a new Permanent Secretary, Sir Hubert Llewellyn Smith, who was well known as an expert on labour questions and in particular on the problem of the unemployed. It was popular with the business community, which since the 1890s had pressed for it to be converted into a 'Ministry of Commerce and Industry', headed by a minister with the status of a Secretary of State.[3] Confidential information about the condition of trade was received from all parts of the country and supplied free to business-men and manufacturers by the Commercial Intelligence Branch of the Board of Trade.[4] The staff of the Commercial, Labour

[1] Above, pp. 160–1.
[2] Hansard, 5th series, vol. 1, col. 1441.
[3] The Times, 4 June 1897, 6 May 1899, 3 Apr. 1903. The status of the Board of Trade and Local Government Board was examined in 1904 by a committee under Lord Jersey, which recommended that the President of the Board of Trade should be turned into a 'Minister of Commerce and Industry', with the rank and salary of a Secretary of State (Cd. 2121/1904, Report of the Committee appointed to consider the position and duties of the Board of Trade and the Local Government Board, p. iv). The salaries of the President and permanent staff of the Board of Trade were raised to those of a Secretary of State's office by the Board of Trade Act 1909, although its title was not changed. The status and salaries of the L.G.B. remained those of an inferior department.
[4] The Commercial Intelligence branch was established in 1899 to edit consular reports, maintain contact with Chambers of Commerce, and to collect and supply British manufacturers with confidential information about trade openings at home and abroad. By 1907 it was answering 9,000 inquiries a year (B.T. 13/46/E. 20079, Information re Work done by the Board of Trade during 1907–8, for Mr. J. A. Pease, 6 Aug. 1908).

and Statistical Department increased by over 40 per cent between 1902 and 1909;[1] and during this period officials of the department carried out pioneering inquiries into wages, living conditions, productivity, and methods of dealing with unemployment in foreign countries.[2]

It was therefore quite consistent with the previous history of the labour-exchange movement that Beveridge in June 1907 should have proposed that a national system be established by the Board of Trade in preference to any other department;[3] and it was equally consistent with the existing capacities and functions of the Board. On 24 June 1907 Percy Alden suggested in the Commons that information about German exchanges should be circulated by the Board;[4] and during the winter of 1907–8 Beveridge prepared several memoranda on aspects of the unemployment problem, which were submitted to the Poor Law Commission on behalf of the Board of Trade.[5] Finally, at a meeting between Sir Samuel Provis and senior Board of Trade Officials on 29 July 1908 it was agreed that, 'as regard the readjustment of work' between the two departments, 'the subject of Labour Exchanges ought to be considered as a question of Employment and not of Relief, and consequently should be dealt with by the Board of Trade'.[6]

The cause of labour organization was taken up by Winston Churchill in the spring of 1908; and the Webbs persuaded him to recruit the assistance of Beveridge, who

[1] The C., L. and S. department employed 98 officials of all grades in June 1902 and 140 in Apr. 1909 (Establishment lists, Board of Trade Library).

[2] Cd. 6320/1912, *Final Report on the First Census of Production of the United Kingdom* (1907); Cd. 1761/1903 and Cd. 2337/1905, *British and Foreign Industrial Trades Conditions*; Cd. 3864/1908, *Cost of living of the Working-Classes . . . in the Principal Towns of the United Kingdom.* Above, p. 205.

[3] Beveridge MSS., Coll. B, vol. xiv, item 20, 'Memorandum as to the Future of Labour Exchanges', by W. H. Beveridge, submitted to the L.G.B., July 1907. Above, p. 205.

[4] *Hansard*, 4th series, vol. 176, col. 1899.

[5] Cd. 5068/1910, *RC on the Poor Laws, Minutes of Evidence*, Appendix XXI. Sections A, B, and K of this Appendix were written by Beveridge (above, p. 23).

[6] T. 1/11093/6763/19536, Note of an interview with Sir Samuel Provis with regard to Status, unsigned, 29 July 1908. At this meeting the general question of the allocation of duties between the two departments was discussed, and the Board of Trade agreed to hand over to the L.G.B. its control over the sanctioning of loans to local tramway authorities.

entered the Board of Trade as a temporary civil servant in July 1908.[1] Beveridge while still an undergraduate had decided that singlemindedness was the key to success as a social reformer,[2] and since the beginning of 1906 his career had revolved around the fixed idea of the organization of the labour market. His concern for reform was inspired less by philanthropic emotion than by a passion for efficiency and by an almost obsessive dislike of social and individual waste. He was confident that labour exchanges would in-augurate a new era of industrial efficiency, and he later claimed that his sole ambition in entering the Board of Trade was to convert his vision of an organized labour market into legislation.[3]

Beveridge produced his first brief on the theory and practice of labour exchanges before the end of July 1908.[4] He laid down as basic principles that they should be nation-wide and industrial rather than local and charitable, and that central control should be exercised by the 'depart-ment of industrial intelligence'. Exchanges could be used to manage an insurance system, to dovetail seasonal occupa-tions, to promote technical education, and to retrain work-men who had become redundant through changes in in-dustrial structure. Moreover, in spite of his emphasis on the 'business' aspect of exchanges, Beveridge invested them with functions that were also implicitly moral. By ex-hausting all possibilities of employment they would impose on the unemployed workman an industrial version of the principle of deterrence. They would eliminate the casual labourer who 'remains financially and morally beyond the possibility of thrift and organisation', and they would 'enable the idle vagrant to be discovered unmistakeably and sent to an institution for disciplinary detention'.

The cost of such a scheme Beveridge estimated at

---

[1] W. H. Beveridge, *Power and Influence*, pp. 68–9.
[2] Beveridge MSS., L. i. 209, W. H. Beveridge to Jeanette Beveridge, 6 July 1898.
[3] W. H. Beveridge, *Power and Influence*, p. 69.
[4] Beveridge MSS., Coll. B, vol. xiv, item 23, c. 20 July 1908. A précis of this memorandum was circulated by Churchill to some of his Cabinet colleagues before the end of July (R. Churchill, *Winston S. Churchill*, ii, Companion, Part 2 (1907–11), pp. 827–31).

£130,000 a year; and he recommended that a special branch of the Board of Trade should be established to put it into operation. At a local level, control could be exercised directly, through a 'purely national system' like the Post Office; or indirectly, through the issue of grants-in-aid to semi-autonomous local bodies. Beveridge thought that the former alternative would secure greater prestige and national uniformity, but at this stage he was inclined to favour the latter, as likely to command greater local interest and co-operation. He recalled, however, that 'Winston Churchill would have none of this second alternative. He decided for a national scheme directly under the Board.'[1]

The plan that was eventually submitted to the Cabinet, however, was neither localized nor purely centralized, but regional.[2] It was proposed to create nine or ten regional divisions and two or three lesser divisions in 'minor and outlying' areas. In the principal town of each division there would be a 'clearing house', controlled by a Divisional Chief who would be responsible for the co-ordination and development of local labour exchanges in that region. Local offices would be graded into first- and second-class exchanges and sub-offices, according to the population of the district they were serving; and in remote areas labour-exchange facilities would be organized through post offices and other existing agencies. The whole system would be directed by a 'National Clearing House', which would receive regular returns from divisional clearing houses, and 'become the centre of general information as to industrial tendencies'. It was envisaged that the 'brains and driving power . . . of the Labour Exchange movement' would be the Divisional Chiefs, who would not only control the regional clearing house, but would be constantly in touch with local exchanges, with the central authority and with each other. 'They must be thoroughly responsible Officers combining initiative with business or administrative experience, and able to hold their own with employers and work people.'

[1] Beveridge MSS., D. 047, Notes for an address at the fiftieth anniversary of employment exchanges, 1 Feb. 1960.

[2] CAB 37/97/17, Memorandum on 'Labour Exchanges', Jan. 1909. The following account of the Board of Trade's scheme is summarized from this paper.

Each first-class exchange would provide separate facilities for skilled and unskilled, female and juvenile applicants, and accommodation would be spacious enough to allow every registered applicant to call at the labour exchange every day. Registration would be entirely voluntary and entirely free, although trade unions would be encouraged to hold their meetings on exchange premises for a small fee. Second-class exchanges would provide similar though less spacious accommodation; and sub-offices would be merely branches of larger exchanges, often with nothing more than a clerk–caretaker and a single waiting-room 'to serve as places of call in a scheme of decasualisation'. Workmen who registered at exchanges would be invited but not obliged to give details of their past industrial histories; and applicants would be recommended for situations who appeared to be 'industrially best qualified to perform the work'.

Labour exchanges would not lay down conditions about the level of remuneration, but trade unions would be allowed to use the exchanges to advertise the standard rate; and every workman sent to a situation would be informed of the wages and conditions that it entailed, so that the onus of blacklegging would fall on the individual rather than on the exchange. Notice of trade disputes would be placarded in all the public rooms of an exchange; and at each principal exchange an advisory committee of representative workmen and employers would be appointed by the Board of Trade 'to settle occasional difficulties as to disputes and the like'. Finally, it was recommended that special juvenile employment committees should be set up, including members of local education authorities, to advise on choice of occupation, to recommend legislative improvements in juvenile employment, and to cultivate 'a public opinion amongst employers, parents and children against premature earning in blind-alley occupations'.

Churchill's announcement to Parliament on 17 February 1909 that he intended to introduce a national system of voluntary exchanges was welcomed by all parties.[1] Its only critics were the advocates of the Minority Report of the Poor Law Commission, who argued that such a system would be

[1] *Hansard*, 5th series, vol. 1, cols. 193–4; vol. 5, cols. 514–19.

unworkable without compulsion.[1] Churchill, however, claimed that there was an overriding argument against compulsion in the existing state of affairs.

To establish a system of compulsory labour exchanges, to eliminate casual labour, to divide among a certain proportion of workers all available employment before you have made preparation for dealing with that surplus, would be to cause administrative breakdown, and could not fail to be attended with the gravest possible disaster. There-fore until poor law reform, which falls to the department of my right hon. friend Mr. Burns . . . has made further progress, to establish a compulsory system of Labour Exchanges, would naturally increase and not diminish the miseries with which we are seeking to cope. We have, therefore, decided that our system . . . shall be voluntary in its character.[2]

In other words, Churchill shrank from the social con-sequences of the evolutionary logic that Booth and Beveridge had prescribed for the labour market during the previous seventeen years.[3] He might have added that the political difficulties involved in imposing compulsory registration on the labour market would be almost insuperable, parti-cularly among workmen in skilled and highly organized trades.

Although the Labour Exchanges Bill was an overtly un-controversial measure, its passage involved delicate negotia-tions with the Treasury and with employers and workmen. The Treasury queried Beveridge's estimate that the scheme would cost £130,000 a year; and on 21 January 1909 Llewellyn Smith appointed a small committee, consisting of Beveridge, Charles Rey, F. H. Macleod, and the depart-mental accountant, George Fry, to inquire into the 'organisa-

---

[1] On the differences between the Webbs' plan and the Board of Trade's plan see B. B. Gilbert, 'Winston Churchill versus the Webbs: the Origins of British Unemployment Insurance', *American Historical Review*, 71, no. 3 (Apr. 1966), 857–9. Professor Gilbert rightly points out that labour exchanges were being planned by the Board of Trade well before the Commission reported and that the Webbs' influence has been much exaggerated. The Board of Trade's plan for non-compulsory labour exchanges was, however, virtually identical with that put forward by the *Majority Report* (Cd. 4466/1909, paras. 507–28). This was not surprising, since the recommendations of the majority were directly based on evidence submitted by Beveridge and the Board of Trade in 1907–8.

[2] *Hansard*, 5th series, vol. 5, col. 506.

[3] Above, pp. 20, 22.

tion and finance' of the labour-exchanges scheme.[1] This committee estimated that 227 exchanges of various kinds would be required in ten divisions, employing a staff of 863 officials, plus inspectors, temporary relief staff, and ten civil servants for the National Clearing House. Estimates of both capital and annual expenditure depended, however, on whether labour-exchange premises were to be purpose-built or merely hired and converted. Board of Trade officials desired the former,[2] the Treasury and the Office of Works the latter, at least until the scheme had proved itself with some measure of success.[3] Eventually it was decided that at least the ten divisional offices and the first-class exchanges should have specially designed premises; and the estimated cost of the system was raised to £200,000 a year for the first ten years and £180,000 after 1919.[4]

The Labour Exchanges Bill was introduced into Parliament on 20 May and passed its second reading without a division on 16 June 1909. Its terms were broad and brief. The Board of Trade was empowered to establish, take over, and maintain exchanges, to collect and publish information about employment, and to frame regulations for management which were to be approved by Parliament. The cost was to be borne by Parliament with the approval of the Treasury; and persons who made false statements to exchange officials were to be liable to fines of up to £10.[5]

The Bill aroused an immediate protest from the trade unions, who were disturbed by the absence of protection against blacklegs and strikebreakers. On 19 March 1909 a special conference of trade-union delegates, convened by the Parliamentary Committee at Caxton Hall, had passed a resolution in favour of national labour exchanges 'provided that the management boards contain an equal proportion of

[1] LAB 2/211/LE. 16918, Labour Exchanges Committee Report, 27 Mar. 1909.

[2] Ibid., Diagram of Suggested Labour Exchange.

[3] W. H. Beveridge, *Power and Influence*, p. 78. On the suspicion with which the Treasury regarded the labour-exchanges venture, see S. Tallents, *Man and Boy*, pp. 180, 184–5.

[4] Beveridge MSS., Coll. B, vol. xiv, item 24, Labour Exchanges Bill, Financial Statement.

[5] LAB 2/211/LE. 733, Draft of a Bill for the Establishment of Labour Exchanges and other purposes incidental thereto, clause 5, n.d.

employers and representatives from Trade Unions'.[1] A
pamphlet issued at the end of May by the General Federation of Engineering and Shipbuilding Trades—one of the
largest and richest groups of unions—showed that many
union secretaries thought exchanges would undermine the
position of organized labour.[2] Churchill was anxious to
placate the unions, realizing that the business reputation of
the exchanges would depend on attracting the custom of
highly skilled workmen; and he therefore summoned a
series of conferences with the Parliamentary Committee of
the T.U.C. and the Shipbuilding and Engineering unions
in June and July 1909.

These conferences were very skilfully managed, since
Churchill appeared to compromise with the unions but in
fact conceded nothing that had not already been accepted
by the Cabinet earlier in the year. He convinced the Parliamentary Committee that organized labour would be adequately protected if exchanges gave full publicity to trade
disputes and allowed the unions to put up notices which
stated the prevailing standard rate. He refused to accept the
principle that all workmen's representatives on advisory
committees should be trade unionists, since groups like the
National Free Labour Association would claim that exchanges were being used for political purposes; but he
pointed out that in practice trade unionists would nearly
always be appointed, since they formed the vast majority
of workmen who were active in public affairs. And he promised that labour-exchange appointments would not be
monopolized by 'university men', but would be given to men
from different classes and occupations who were 'in touch
with the practical issues of life'.[3]

The Federation of Engineering and Shipbuilding Trades,
confident in their numbers and financial self-sufficiency,
were more aggressive than the Parliamentary Committee.[4]

[1] *Report of the Ipswich T.U.C.*, 1909, p. 54.

[2] Pamphlet No. 58 of the Federation of Engineering and Shipbuilding Trades,
issued by the general secretary, William Mosses, to Executive Councils, 27 May
1909 (copy in LAB. 2/211/LE. 500).

[3] LAB 2/211/LE. 500, Typescript report of the first conference with the Parliamentary Committee of the T.U.C., 17 June 1909.

[4] LAB 2/211/LE. 500, Typescript report of the conference with the Federation
of Engineering and Shipbuilding Trades, 18 June 1909.

The delegates who met Churchill on 18 June made it clear that they resented the public provision of advantages that the unions had acquired by their own efforts; and they objected strongly to the proposal that advisory committees should contain impartial persons and local dignitaries as well as representatives of capital and labour. 'We do not want these benevolent persons at all,' said J. M. Jack, the spokesman of the deputation '. . . what we want is men who have come through the thick of battle.'[1] Churchill insisted, however, that such committees had to be as catholic as possible in order to cultivate goodwill; but he agreed that problems arising from trade disputes should be settled only by direct representatives of employers and organized labour.

The result of these negotiations was that on 22 June when the Bill came before a Standing Committee, Churchill introduced a new clause, stipulating that no workman who registered at any exchange should 'suffer any disqualification or be otherwise prejudiced' for refusing a vacancy caused by a trade dispute, or for refusing wages 'lower than those current in the trade in the district where the employment is found'.[2] This amendment was substantially accepted by the Parliamentary Committee of the T.U.C. on 8 July 1909.[3]

In August Churchill held a similar conference with representatives of the Engineering Employers' Association and the Shipbuilding Employers' Federation.[4] The employers objected that the workmen themselves had no desire for a labour-exchange scheme and that its cost would be out of all proportion to its utility. Moreover, they feared that exchanges would become centres of industrial discontent and that workmen would be demoralized by no longer having to search for work. 'Loss of incentive to personal effort . . . is detrimental to character,' said

---

[1] J. M. Jack, Secretary of the Scottish Ironmoulders, a representative of Old Unionism and consistent opponent of state intervention.

[2] LAB 2/211/LE. 733, Report of Standing Committee C, printed 8 July 1909.

[3] LAB 2/211/LE. 500, Typescript of the second conference with the Parliamentary Committee of the T.U.C., 8 July 1909.

[4] LAB 2/211/LE. 500, Typescript report of the conference with the Engineering Employers' Association and the Shipbuilding Employers' Federation, 18 Aug. 1909.

Alexander Siemens,[1] the spokesman of the deputation, 'and an employer would rather employ a man who has striven to help himself than one who has relied on the benevolence of the State . . .'. Frederic Henderson,[2] the President of the Shipbuilding Employers' Federation, argued that a surplus of labour was essential to industrial efficiency; otherwise 'the men become loose in their habits, and drink, and they become independent of their employers because they can get a job anywhere they like'. Churchill's answers to the employers were directly contrary to those he had given to the unions a month earlier. He denied that exchanges would enable unionists to bring pressure to bear on unorganized workmen, because strict and impartial discipline would be maintained among workmen waiting for jobs.[3] Moreover, he pointed out that 'if anybody had said a year ago that the trade unions would have agreed to a government labour exchange sending 500 or 1,000 men to an employer whose men are out on strike . . . [nobody] would have believed it at all'. Since the whole labour-exchange scheme had been designed from the start to allay the fears of employers, however, there were no substantial concessions that could be made to the employers' associations, short of abandoning the entire scheme. This Churchill had no intention of doing: and the Bill passed into law without further adjustment in September 1909.

During the autumn of 1909 Beveridge and Llewellyn Smith worked to get the scheme into operation by the be-

[1] Alexander Siemens (1847–1928), German-born electrical engineer (naturalized 1878) and advocate of 'self-help'. Siemens had fought as a private in the Franco-Prussian war and worked as an engineering apprentice, before becoming Manager of the Electric Light department of Siemens Bros. in 1879; President of the Institute of Civil Engineers 1910–11.

[2] Frederic Ness Henderson (1862–1944), Chairman of the Iron Trades Employers' Insurance Association Ltd., and of the North West Rivet Bolt and Nut factory, Airdrie, and director of several companies; knighted 1918.

[3] This was directly contrary to the private view of Churchill's officials. See Beveridge MSS. (first deposit), Parcel 2, Folder C (ii), item 16, 'The influence on Trade Unionism of Labour Exchanges and Unemployment', c. May 1909: '. . . the tendency of successful labour market organisation on neutral lines to prepare the ground for other organisation (trade unionism) is, I think, so certain that it would be impossible conscientiously to recommend the use of public labour exchanges to an employer who regarded collective bargaining as the curse of British industry.'

ginning of 1910. Pressure on the Treasury to increase the staff of the Commercial, Labour and Statistical Department[1] led to a major re-organization of departmental responsibilities within the Board of Trade. Labour exchanges, trade boards, industrial conciliation, and the Census of Production were separated from the Commercial department and placed under the control of a newly constituted Labour department.[2] Within the Labour department, a 'Labour Exchanges branch' was established at Caxton House, with Beveridge as 'director', Charles Rey and Lord Basil Blackwood as general manager and assistant general manager of labour exchanges, and three upper-division clerks recruited from other departments of the Board of Trade.[3]

Eleven regional divisions were established, with headquarters in London, Bristol, Nottingham, Birmingham, Newcastle, Leeds, Liverpool, Manchester, Cardiff, Glasgow, and Dublin.[4] Churchill was determined to appoint officials

---

[1] B.T. 11/2/C7761, Minute with regard to the necessity for an increase in staff in the Commercial, Labour and Statistical department, 4 Dec. 1909.

[2] B.T. 15/57/F6644, Estimates File, A Note on the Labour department, n.d., 1910. For a detailed discussion of the impact of labour exchanges and insurance on the administrative structure of the Board of Trade see J. A. M. Caldwell, 'Social Policy and Public Administration, 1909–11', Nottingham Ph.D. thesis, 1956, Ch. IV. It should be noted that Sir George Askwith, the Assistant Secretary at the head of the new Labour department, was subsequently a bitter critic of the labour-exchanges system (G. Askwith, *Industrial Problems and Disputes*, pp. 272–81).

[3] Charles Rey (1877–1968), private secretary to Llewellyn Smith 1908; General Manager of Labour Exchanges 1909 and of Unemployment Insurance 1912; transferred to Ministry of Munitions 1915–17; Director-General of National Labour Supply 1918; Assistant Secretary, Ministry of Labour 1918.

Basil Blackwood (1870–1917), third son of the 1st Marquess of Dufferin; colonial administrator 1900–9; subsequently Secretary of the Development Commission (below, pp. 345–6); illustrator of Hilaire Belloc's *Bad Child's Book of Beasts*.

The three subordinate officials were (a) Thomas Phillips (1883–1965); entered Board of Trade 1906; subsequently Permanent Secretary of the Ministries of Pensions and National Insurance. (b) Stephen Tallents (1884–1958); entered Board of Trade 1909; subsequently an official with the Ministries of Munitions and Food; administrator for B.B.C. 1935–41; Assistant Secretary, Ministry of Town and Country Planning 1943–6. His autobiography *Man and Boy* (1943), pp. 178–91, contains an interesting account of the work of the new department. (c) Umberto Wolff (1886–1940) (subsequently Humbert Wolfe), Deputy Secretary, Ministry of Labour, and minor poet. His volumes of autobiography, *Now a Stranger* (1933) and *The Upward Anguish* (1938) contain little reference to his work at the Board of Trade.

[4] C. F. Rey, 'The National System of Labour Exchanges', *Report of the Proceedings of the National Conference on the Prevention of Destitution*, 30 May 1911, p. 396. After the establishment of unemployment insurance, the number of divisions

with business experience and therefore avoided the usual method of examination by the Civil Service Commission. A small committee, consisting of Stanley Leathes the first Civil Service commissioner, D. H. Shackleton, the trade union M.P., and W. H. Mitchell, a Conservative businessman, was appointed to select officials on grounds of commercial or administrative experience rather than academic merit.[1] The divisional chiefs included Richard Bell, the veteran trade unionist, J. B. Adams, the polar explorer, and W. S. Cohen, who had been an administrator in the Orange River Colony.[2] Nearly all the labour-exchange staff were recruited from outside the civil service, and 'a motley crowd they were—trade unionists, Fabians, social reformers, men and women from factory and workshop, and a sprinkling from the public schools.'[3] The details of labour-exchange management and the maintenance of neutrality in trade disputes were laid down by Board of Trade regulations, which were designed to create 'a perfectly colourless soulless piece of commercial mechanism. It is not intended to twist one side or the other one inch in the economical course of events, only to remove friction.'[4] It was agreed with the Treasury and Home Office that loans for travelling expenses should be advanced to workmen who were sent to situations more than seven miles distant and that these should subsequently be recovered by deductions from the

was reduced to eight. Nottingham, Liverpool, and Newcastle lost their divisional offices, and the divisional office for Scotland was transferred from Glasgow to Edinburgh. The number was reduced to seven after the creation of Eire in 1922, and Northern Ireland was henceforth administered direct from Whitehall (Beveridge, *Unemployment* (1930 ed.), p. 297).

[1] T. 170/8, Comparison of expenditure in 1906 and 1911/12. A note on labour-exchange appointments.

[2] W. H. Beveridge, *Power and Influence*, pp. 76–7. The other divisional chiefs included men who had worked as policemen, soldiers, colonial administrators, and local-government officials. Miss Maude Marshall, a member of the Whitechapel COS, who had supported subsidized unemployment insurance before the Royal Commission on the Poor Laws (below, p. 301), was appointed as organizing officer for Women's Employment.

[3] J. B. Adams, 'Reminiscences of a Divisional Controller', *Minlabour*, 14, no. 1 (Jan. 1960), 5.

[4] LAB 2/211/LE. 500, Typescript report of the conference with the Engineering Employers' Association and the Shipbuilding Employers' Federation, 18 Aug. 1909.

workmen's wages.[1] Arrangements were made to take over private labour bureaux and the exchanges of the Central (Unemployed) Body.[2] Sixty-one exchanges were opened in February 1910,[3] a figure which by February 1911 had risen to 175; and 1,400,000 applications for work were registered during the first year. By February 1914 there were 423 exchanges in the United Kingdom, which registered over 2,000,000 workmen a year.[4]

UNEMPLOYMENT AND NATIONAL INSURANCE

The idea of national insurance against unemployment was by no means entirely original. Over a hundred years earlier Jeremy Bentham had suggested that artisans should be endowed with state-subsidized annuities which could be realized in periods of distress.[5] But throughout the nineteenth century the only institutions that made communal provision against unemployment were a small number of Friendly Societies and skilled trade unions. Friendly Societies had statutory power to provide unemployment insurance, but only five registered Friendly Societies had taken advantage of this power by 1894.[6] All of these

[1] LAB 2/177/LE. 312, G. R. Askwith to the Secretary of the Treasury, 13 Dec. 1909; T. Heath to the Permanent Secretary of the Board of Trade, 15 Jan. 1910. The Railway Companies were asked and refused to grant cheap tickets for workmen travelling in search of work.

[2] A special committee of the C.U.B. was established in October 1909 which arranged to transfer the London staff and exchanges to the Board of Trade (LAB 2/211/LE. 341, 'Report of the Special Committee appointed to meet representatives of the Board of Trade with regard to the existing system of Labour Exchanges' (registered 13 Dec. 1909)). This transfer was approved by the Treasury, provided that the C.U.B.'s premises were leased and not purchased and the staff given no guarantee of security of tenure (LAB 2/211/LE. 345, T. L. Heath to the Permanent Secretary of the Board of Trade, 10 Dec. 1909).

[3] The opening of exchanges was postponed from Jan. to Feb. 1910 so as not to coincide with the General Election.

[4] Beveridge MSS., D. 030, 'Labour Exchanges and Unemployment Insurance. Report of the Proceedings of the Board of Trade under the Labour Exchanges Act 1909 and under Part II of the National Insurance Act 1911, to July 1915' (not printed as a command paper), paras. 16 and 234, and table V.

[5] *Jeremy Bentham's Economic Writings* (ed. W. Starke), ii. 50.

[6] E. Brabrook, *Provident Societies and Industrial Welfare* (1898), p. 42; G. Drage, *The Unemployed* (1894), p. 22. By 1906 thirty-one Friendly Societies in England and Wales, with a membership of 28,029, were specifically registered for the purpose of giving out-of-work benefits. These societies paid £7,887 to unemployed members in 1906, nearly half of this sum being accounted for by the

were affiliated orders with national connections; and their purpose in giving benefit was not so much to relieve distress from unemployment as to encourage industrial liberty and to facilitate the transfer of members from branch to branch.[1]

Insurance against enforced unemployment was therefore almost entirely the province of trade unions, which were peculiarly fitted to cover such a risk, since their members could keep a check on malingering, help each other to find jobs, and—in the event of a deficit—subscribe to a general levy. Financial records of unemployed benefits paid by the Amalgamated Society of Engineers dated from 1851, by the Printers and Bookbinders from 1856, and by the Carpenters and Joiners from 1860;[2] but certain unions had probably maintained informal out-of-work funds from a much earlier date.[3] Benefits might be paid in cash or kind, and might assist a workman to travel in search of work or to maintain him in his normal place of residence till trade revived. Their payment was conditional on the signing of a vacant-book in the possession of a union branch secretary, who could verify the fact that a man was unemployed and also inform him of suitable opportunities for work.[4] The origins of the travel or 'tramp' benefit are obscure, and it may have originated in the assistance given to journeymen to travel around workshops, gaining the varied experience

Provident Association of Warehousemen, Travellers and Clerks of Cheapside (Beveridge MSS., Parcel 2, Folder C (ii), item 27. 'Friendly Societies Providing Unemployed Benefit', by W. H. Beveridge, 27 May 1909).

[1] C. 7063/1894, *RC on Labour (Sitting as a Whole)*, *Minutes of Evidence*, Appendix LIII, p. 76, para. 20. On Friendly Society provision for 'tramp' benefit see P. H. J. H. Gosden, *The Friendly Societies in England 1815–1875*, pp. 76–7, 221–3.

[2] Cd. 5068/1910, *RC on the Poor Laws*, *Minutes of Evidence*, Appendix XXI (B), pp. 607–8.

[3] A. Aspinall, *The Early English Trade Unions*, p. 93; Neil J. Smelser, *Social Change in the Industrial Revolution*, p. 335; Wladimir Woytinsky, *Three Sources of Unemployment*, I.L.O. Studies and Reports, Series C (Employment and Unemployment), No. 20, pp. 19–20.

[4] This system was by no means foolproof (e.g. J. B. Jefferys, *The Story of the Engineers*, pp. 60–1). The A.S.E. regularly published the names of members who had defrauded the society or refused to refund benefits 'improperly received' (*Amalgamated Engineers' Report and Monthly Record of Facts and Figures Relating to the Society*, 1908–11).

necessary to a skilled craftsman.[1] While depressions and
labour markets were localized it was an effective means of
increasing labour mobility and minimizing unemployment;
but when local depressions coalesced into nation-wide
fluctuations, the usefulness of tramp benefit declined.[2] The
'stationary' unemployed benefit became common in skilled
unions during the latter half of the nineteenth century.
Rules, rates of contribution, and qualification for benefit
varied from union to union and at different times in the
same union.[3] Both types of benefit were primarily designed
to deter unemployed workmen from undermining the level
of wages and only secondarily to relieve distress;[4] and in
many unions 'out-of-work' benefit was not clearly distinguished from payments during strikes and lockouts.[5] On
the other hand, certain unions paid out-of-work benefit to
workmen unemployed through old age, sickness, or accident
who presented little threat to the standard rate.[6]

However, trade-union unemployment insurance never
covered more than a small minority of workmen. By 1891,
682,025 organized workmen were eligible for some kind of
out-of-work benefit—less than half the trade-union movement and less than a twentieth of the labour force as a whole.[7]

[1] E. Hobsbawm, 'The Tramping Artisan', in *Labouring Men*, pp. 34-9.

[2] Ibid., pp. 47-8. Hobsbawm also ascribes the decline of tramp benefit to the
decline of a fatalistic acceptance of trade depression among trade unionists.

[3] Cd. 5703/1911, *Tables showing the Rules and Expenditure of Trade Unions
in Respect of Unemployed Benefits and also showing Earnings in the Insured Trades.*

[4] S. and B. Webb, *Industrial Democracy*, pp. 161-2. But the Webbs possibly
overemphasized this point. The rules of the A.S.E., born in the midst of the depression of 1851, suggested that the reduction of unemployment was the primary
aim of combination, prior even to the maintenance of the standard rate (J. B.
Jefferys, op. cit., pp. 32-3).

[5] Of the trade unions whose rules were analysed by Beveridge in 1909, ninety-
six gave out-of-work benefit for the specific purpose of relieving workmen unemployed through a strike or lockout in the same factory or in another branch of
the trade (Cd. 5703/1911, *Tables Showing the Rules and Expenditure of Trade
Unions etc.*, pp. 18-338).

[6] Ellic Howe and H. E. Waite, *The London Society of Compositors: A Centenary
History*, p. 260. Of the unions examined by Beveridge in 1909 eleven gave out-of-
work benefit to workmen unemployed through 'contagious disease' (Cd. 5093/
1911, pp. 18-228). John Burns in 1895 thought that some of the money paid by
the A.S.E. as sick benefit should really be classified as unemployed benefit, since
the sickness arose from privation through want of work (H. of C. 365/1895, *SC
on Distress from Want of Employment*, QQ. 4958-61).

[7] C. 7182/1893, *Report on Agencies and Methods for dealing with the Unemployed*,
p. 18; B. R. Mitchell and P. Deane, *British Historical Statistics*, p. 60.

This proportion rapidly increased over the next seventeen years; and by 1908, 1,059 unions with 2,357,381 members, paid unemployed or travel benefit or gave assistance with removal.[1] Even so, workmen insured against unemployment were a privileged élite, mainly composed of skilled and highly paid workmen; and outside the élite mutual provision against unemployment was virtually non-existent. Protection against unemployment had been left out of the comprehensive social security scheme introduced into Germany in the late nineteenth century;[2] and Canon W. Blackley, who was the first English writer to put forward systematic proposals for state insurance, had excluded provision for unemployment from his campaign against 'national improvidence' in 1879.[3] An experiment in compulsory mutual provision against unemployment had been conducted in the Swiss canton of St. Gall during the 'embroidery crisis' of 1894–6; but this had been wrecked by an excess of bad lives and maladministration,[4] and it was often inferred from this experience that compulsory unemployment insurance was impracticable and that distress from unemployment was an uninsurable risk.

Elsewhere on the Continent, however, unemployment insurance schemes on a voluntary basis had met with more success. Since 1896 workmen in Cologne had been invited to pay small premiums into an Unemployment Insurance Society, from which they were entitled to benefits subsidized by the municipality during the winter months.[5] Some Ger-

---

[1] Cd. 5703/1911, p. 13.

[2] Joseph L. Cohen, *Insurance Against Unemployment* (1921), p. 140. Proposals for 'compulsory saving' for unemployment, through weekly deductions from the wages of insured workmen were, however, under discussion by German social reformers in the mid 1890s (J. G. Brookes, 'Insurance of the Unemployed', *Quarterly Journal of Economics*, 10 (1895–6), 341–8).

[3] W. Blackley, 'Thrift and National Improvidence', a sermon preached in Westminster Abbey, 28 Sept. 1879, reprinted in M. J. J. Blackley (ed.), *Thrift and National Insurance* (1905), pp. 128–39. Blackley argued that no system of mutual providence could meet the 'vast mass of misery to the able-bodied caused by sudden loss of labour', but he claimed that the provision of insurance for old age and sickness would greatly increase the capacity of Poor Law, charity, and private saving to meet sudden disasters like loss of work (W. L. Blackley, 'Compulsory Providence as a Cure for Pauperism', ibid., pp. 112–13).

[4] E. Hofman, 'Insurance against Unemployment in Switzerland', *International Conference on Unemployment*, Paris 1910, Tome 3, Report 46, pp. 16–18.

[5] Cd. 2304/1905, *Report to the Board of Trade on Agencies and Methods for Deal-*

man manufacturers paid redundancy benefits to workmen
discharged against their will; and this was made compulsory
in certain branches of the German chemical industry in 1910.[1]
The most successful voluntary insurance scheme was that
founded in Ghent in 1901, where the municipality contributed to out-of-work benefits paid by local trade unions. This
scheme was managed by a joint committee of masters and
workmen; beneficiaries were required to register daily with
their union secretaries; and the unions were themselves
responsible for the suppression of 'malingering'.[2] In 1907 the
Belgian government decided to make a national contribution
to the scheme;[3] and the 'Ghent system' had been used as a
model for municipal and national insurance schemes in
Germany, France, Italy, Holland, Denmark, and Norway by
1909.[4]

In England, on the other hand, the question of public
insurance against unemployment scarcely entered into the
discussion of social problems before 1907. The Liverpool
corporation's commission on unemployment in 1894 inquired of local trade-union secretaries whether the working
classes would be favourable 'towards a compulsory system of
national or municipal insurance . . . for . . . out-of-work allowances'. The replies were inconclusive, some trade unionists
thinking that it would meet with 'general support', others
that it would be impracticable or that working-class incomes

---

ing with the Unemployed in Certain Foreign Countries, by D. F. Schloss, pp. 3–11.
Under this scheme skilled workmen paid weekly premiums of 4¾d., and unskilled
workmen 3½d.; in return they received benefits, after a two-day 'waiting-period',
of 2s. a day for twenty days and 1s. a day thereafter, up to a maximum period of
eight weeks. Whilst in receipt of benefit a workman was required to report to the
Insurance Office twice daily, and to accept any job in his own line of trade that was
not caused by a trade dispute and not paid at a lower rate than in his previous
situations. Benefit was not payable to workmen unemployed through sickness,
trade dispute, or misdemeanour. The scheme was supervised by an Executive
Committee containing representatives of workmen, employers, and municipal
officers. Many features of the Cologne scheme were subsequently incorporated
into the English system.

[1] Johannes Feig, 'The Struggle Against Unemployment in the German Empire
at the Present Day', *International Conference on Unemployment*, Paris 1910, Tome 2,
Report No. 1, pp. 43–4.

[2] Cd. 2304/1905, pp. 186–94.

[3] T. Théate, 'Unemployment in Belgium', *International Conference on Unemployment*, Paris 1910, Tome 2, Report No. 7, p. 22.

[4] Joseph L. Cohen, op. cit., pp. 92–6, 115–17, 121–2, 124–7, 144–8.

were too precarious for such a scheme.[1] The subject was ignored by the Royal Commission on Labour which reported in the same year and by the Select Committees of 1895 and 1896. In 1896 the British Consul in Düsseldorf sent to Lord Salisbury a detailed report on the functions and management of the newly established Cologne scheme; and this was laid before Parliament but not debated in June 1896.[2] David Schloss drew attention to the Ghent scheme in a Board of Trade blue-book in 1904.[3] But the question was not brought into the public discussion of unemployment until July 1907, when Beveridge argued in the *Morning Post* that the necessary machinery for such a scheme could be established through a labour-exchange system.[4] This idea was elaborated in his evidence to the Royal Commission on the Poor Laws in the autumn of 1907, when he suggested that 'in the extension of unemployment insurance you have one of the great general methods of dealing with this problem'.[5] Continental experience showed that voluntary insurance required either an external subsidy or an institutional framework such as a trade union or a labour exchange to administer benefits and check malingering;[6] and Beveridge suggested that a subsidized voluntary system, together with improved technical education, should be substituted for the prevailing chaos of 'tiding-over' and 'relief works'.[7] He compared the cost and effectiveness of the Central (Unemployed) Body with the one of the benefit-paying London unions, demonstrating that during a six-month period the latter had relieved over twice as many 'man-weeks' of unemployment at less than half the cost per head.[8] He proposed that 'such provident associations might be deliberately fostered by public grants';[9] but pointed out

[1] *Full Report of the Commission of Inquiry into the Subject of the Unemployed in the City of Liverpool, 1894,* Table following p. 119.

[2] C. 7920.–20/1896 *Foreign Office. Miscellaneous Series, No. 399. Reports on Subjects of General and Commercial Interest. Report on the Society for Insurance against Want of Employment in Winter and the General Labour Registry at Cologne* (presented to Parliament, June 1896).

[3] Cd. 2304/1905, *Report to the Board of Trade on Agencies and Methods for Dealing with the Unemployed in Certain Foreign Countries,* pp. 186–94.

[4] W. H. Beveridge, *Power and Influence,* p. 60.

[5] Cd. 5066/1910, *RC on the Poor Laws, Minutes of Evidence,* Q. 78024.

[6] Ibid., Q. 77833, para. 19.

[7] Ibid., QQ. 78228–31.   [8] Ibid., Q. 77832, para. 71

[9] Ibid., para. 70.

that this arrangement by itself would be inadequate, since it would not apply to unskilled or casual workmen. 'The disease of casual employment must be eradicated first if thrift and association are to become universal.'[1]

Beveridge's advocacy of some kind of subsidized insurance against unemployment was endorsed by several other witnesses before the Commission. Professor Sidney Chapman thought that unemployment was commercially uninsurable but that state subsidies should be given to benefit-paying voluntary associations.[2] Even some of the COS witnesses favoured compulsory trade-union insurance.[3] The most radical proposals on unemployment insurance, however, were put forward by Thomas Smith, the Mayor of Leicester and a local correspondent of the Board of Trade. Smith was himself a retired union official, but he thought that an insurance scheme which merely gave subsidies to unions would be politically impracticable and give an unfair advantage to one section of the community.[4] On the other hand, he thought that voluntary insurance was equally unfair, because the thrifty workman paid twice—through insurance for himself and through rates and taxes for other distressed workmen;[5] and he therefore proposed that the State should start a compulsory unemployment insurance scheme quite separate from the trade-union system. He thought that 'sometimes Governments can do what insurance companies cannot do', and denied that the problems of collecting data and calculating risk were insuperable difficulties.[6]

During the winter of 1907–8 Beveridge supplied the Commission on behalf of the Board of Trade with detailed accounts of trade-union unemployment insurance, showing that since 1892 the hundred leading trade unions in the United Kingdom had paid to their unemployed members an average of over £350,000 per annum.[7] More than half this

---

[1] Ibid., para. 72.   [2] Ibid., Q. 84791, para. 43.
[3] Ibid., QQ. 82272–4, Evidence of Miss Maude E. Marshall, secretary of the Whitechapel COS, and member of the Stepney Distress Committee and Central (Unemployed) Body.
[4] Ibid., Q. 86859.   [5] Ibid., QQ. 86807–8, 86836–9.
[6] Ibid., QQ. 86931–3.
[7] Cd. 5068/1910, *RC on the Poor Laws, Minutes of Evidence*, Appendix XXI (C), 'The Growth of Trade Unions with Particular Reference to the Payment of Unemployed Benefit', p. 624.

sum had been paid by building, shipbuilding, and engineering unions, although the printing unions had paid the largest amount per member. A majority of textile and mining unions had out-of-work funds; but these were very little used, since the normal method of meeting unemployment in these trades was by working 'short time'.[1]

Beveridge also surveyed continental unemployment-insurance schemes, which he classified into five main types. Firstly, independent trade-union insurance, as in the United Kingdom. Secondly, subsidized trade-union insurance modelled on the Ghent system. Thirdly, municipal subsidies to private saving, which had been introduced in Ghent, Bologna, and La Rochelle. Fourthly, voluntary insurance supported either by charitable subscription, as in Leipzig, or by a municipal subscription, as in Cologne. And, fifthly, direct compulsory insurance through a municipal fund, of which the solitary example was the ill-fated scheme in St. Gall.[2] Beveridge refrained from committing himself to any particular method of insurance, but he pointed out that none of the imitations of the Ghent system had insured semi-skilled or unskilled workmen, among whom 'the bulk of distress through unemployment is found in the United Kingdom'.[3] Moreover, he thought that the failure of the St. Gall scheme could not be regarded as a conclusive condemnation of compulsory insurance, and implied that such a scheme might well be successful if unemployment could be tested by a system of labour exchanges, if contributions were collected from employers rather than workmen, and if allowance was made for the varying frequency of unemployment in different trades.[4]

The Commission also received reports on compulsory insurance from three consultant actuaries, each of whom objected that it was impossible to calculate the risk of unemployment in the absence of reliable data about the age-distribution

---

[1] Cd. 5068/1910, *RC on the Poor Laws, Minutes of Evidence*, Appendix XXI (C), pp. 620, 625.

[2] Ibid., Appendix XXI (K), 'Insurance Against Unemployment in Foreign Countries', pp. 735–6.

[3] Cd. 5068/1910, *RC on the Poor Laws, Minutes of Evidence*, Appendix XXI (K), p. 737.

[4] Ibid., pp. 751–5.

of the unemployed.[1] The Majority Report proposed, however, that a committee should be set up to formulate a scheme on a 'trade basis', and that special attention should be paid to the problem of including not only trade unionists but unskilled and casual workmen.[2] The Webbs, on the other hand, were sceptical about the value of insurance, believing that it would be strictly unnecessary if proper measures were taken for prevention, and that 'unconditional benefit paid by the state *with no conditions*' would be 'under the present conditions of human will, sheer madness, whatever it may be in good times to come'.[3] They thought, however, that subsidies to trade-union insurance might be used as a bribe to persuade trade unionists to accept a policy of reformatory training for unorganized unemployed workmen;[4] and the Minority Report therefore recommended a direct adaptation of the Ghent system, arguing that this would also act as a great incentive to combination and mutual thrift.[5]

Churchill, however, was anxious that contributors to his insurance scheme should include not only workmen but employers. Beveridge's memorandum to the Poor Law Commission had suggested that insurance contributions should be collected and paid to the State by employers rather than workmen; and Churchill and Llewellyn Smith carried this principle a stage further by proposing that the employers' share of responsibility for unemployment insurance should be not only administrative but financial. 'Unemployment is primarily a question for employers' Churchill wrote to the

[1] Cd. 5077/1911, *RC on the Poor Laws, Statistics Relating to England and Wales*, Part XVI:

(1) Report by Mr. Thomas G. Ackland, F.I.A., Upon Insurance against Invalidity, Sickness Unemployment, etc., pp. 814–16, paras. 23–5.

(2) Report Upon Various Questions of Sickness and Life Assurance by Francis G. Neison, F.I.A., p. 857, para. 6.

(3) Report by Mr. George F. King, F.I.A., on Cost of Insurance against Sickness, Invalidity, etc., pp. 839–41, paras. 17–24.

[2] Cd. 4499/1909, *RC on the Poor Laws, Majority Report*, p. 421.

[3] B. Webb's Diary, 15 Nov. 1908, quoted in *Our Partnership*, p. 419.

[4] Ibid., p. 417: 'Insurance against unemployment had the great advantage that you could offer more freedom to the person who insured compared with the person whom you maintained and forced to accept training. Hence, insurance against unemployment *might* be subsidised by the state as a sort of "set off" to the trade unionists to get them to accept "maintenance with training for all the others".'

[5] Cd. 4499/1909, *RC on the Poor Laws, Minority Report*, pp. 1199–200.

Cabinet at the end of November 1908. 'Their responsibility is undoubted, their co-operation indispensable.'[1]

The levying of contributions from employers, however, precluded the adoption of even a modified version of the Ghent scheme since, as Llewellyn Smith pointed out, employers would take strong objection to subsidizing a scheme that was confined to the members of trade unions.[2] Such a scheme would, moreover, exclude the type of workman who stood most in need of assistance when unemployed. In the autumn of 1908 the Board of Trade therefore rejected the adaptation of the Ghent system which was being canvassed by the Webbs; and Llewellyn Smith drafted a scheme that proposed instead to introduce compulsory insurance for all workmen, skilled and unskilled, organized and unorganized, in three groups of trades—shipbuilding, engineering, and building and works of construction.[3] These were trades liable mainly to cyclical or seasonal fluctuations, where the states of 'employment' and 'unemployment' were quite distinct from each other, and where the average number of days lost by a workman in the course of a year was thought to lie between the chronically irregular trades on the one hand and the stable and 'short-time' industries on the other. They were, moreover, the industries that apart from casual labour had supplied most applicants to the distress committees under the Unemployed Workmen Act during the previous three years.[4]

Llewellyn Smith initially estimated that 3,000,000 workmen would be covered by the scheme.[5] A contribution of

[1] CAB 37/96/159, 'Unemployment Insurance: Labour Exchanges', by W. S. Churchill, 30 Nov. 1908 (circulated 11 Dec. 1908), p. 2.

[2] CAB 37/99/69, 'Memorandum on a Scheme for Unemployment Insurance', by H. Llewellyn Smith, p. 6. This memorandum was circulated to the Cabinet in April 1909; but it had been 'in fairly complete form sometime before that', probably as early as Oct. 1908 (Beveridge MSS. (first deposit), Parcel 2, Folder A, note by W. H. Beveridge).

[3] CAB 37/99/69, 'Memorandum on a Scheme for Unemployment Insurance', p. 3.

[4] Beveridge MSS. (first deposit), Parcel 2, Folder A, Questions and Answers (submitted to W. H. Beveridge by H. Llewellyn Smith), Oct. 1908, Answer 20.

[5] This was the figure suggested in Churchill's first Cabinet memorandum on the subject at the end of 1908 (CAB 37/96/159, 'Unemployment. Insurance: Labour Exchanges', by W. S. Churchill, 30 Nov. 1908, p. 6). By April 1909 this estimate had been reduced to 2¼ million by excluding workers under 20 years and self-employed persons (CAB 37/99/69, 'Memorandum on a Scheme for Unemployment Insurance', p. 3).

2*d*. a week would be levied from the insured workman, to which the State and the employer would each add 1*d*.[1] The employers' and workmen's contributions would be paid into a national 'Insurance Office' by the employer, who would deduct the workmen's share from their wages and record the transaction by stamping an 'insurance card' for each employee. When a workman was discharged he would receive his insurance card and transfer it to a labour exchange, which would ascertain that he was really unemployed, inquire into the causes of his dismissal, and help him to find alternative employment. After a 'waiting period' of one week the workman would become eligible for out-of-work benefits, graduated from 7*s*. 6*d*. to 5*s*. a week according to the duration of unemployment, for a maximum period of fifteen weeks in any one year.

Disputed claims would be referred to joint advisory committees, which would be established in each area for each trade group and consist of representatives of capital and labour with a barrister as impartial chairman. These committees could also act as labour-exchange advisory committees and recommend measures 'for the diminution of distress through unemployment (organised short time, extension of Insurance etc.)'. A quorum of each committee would meet twice a week and two members would be appointed as 'referees' to report on claims that were disallowed by local insurance officers.[2] Beveridge and Llewellyn Smith attached great importance to this 'judicial' aspect of the committee's functions, 'since the men are to be excluded from recourse to the ordinary law courts'.[3] In the event of an appeal, however,

---

[1] CAB 37/96/159, 'Unemployment Insurance: Labour Exchanges', p. 6. The criticisms of the scheme by Beveridge and others in the winter of 1908–9, were based on this calculation, which was altered in Apr. 1909 (below, p. 311).

[2] It was envisaged that members of these advisory bodies would also advise on methods of reducing claims, finding employment, arranging systematic short time, and promoting co-operation between capital and labour (CAB 37/99/69, 'Rough Draft of Heads of a Bill for the Establishment of Unemployment Insurance', section 18).

[3] Beveridge MSS. (first deposit), Parcel 2, Folder C (ii), item 17, 'Local Committees for Insurance and Labour Exchanges', by W. H. Beveridge, 4 Feb. 1909. Llewellyn Smith noted on this memorandum that advisory committees on insurance should as far as possible be ᵙᵇ-committees of the advisory committees to be attached to labour exchanges.

the final decision would lie with an Umpire appointed by the Board of Trade.[1]

In the event of a deficit the Insurance Office would apply to the Treasury for a loan; separate accounts would be kept for each trade-group, and the Board of Trade would have the power to adjust their benefits and contributions according to the state of the fund. The Board would be empowered to extend the scheme to new occupational groups by provisional orders; and in addition, to give rebates of up to one-third to trade unions that paid benefits to unemployed members, either inside or outside the insured trades. It was not envisaged that many casual workmen would be covered by the scheme; but, where they were, casual employment would be penalized by charging a whole week's contribution for each separate engagement. On the other hand, employers of casual labour would be allowed a rebate if they agreed to hire their workmen solely through an exchange.

This scheme was submitted to the Cabinet, together with the scheme for labour exchanges, early in December 1908. It had been designed, Churchill argued, to placate and encourage private agencies already working in the field of unemployment insurance and to promote rather than to replace private thrift. Benefits would be kept low so as to 'imply a sensible and even severe difference between being in work or out of work'. Exchanges and insurance were complementary to each other, since an exchange could test a workman's right to benefit, whilst insurance would provide an incentive to register at the exchange.[2] Later in December, however, Churchill informed Asquith that he had agreed with Lloyd George to postpone the introduction of unemployment insurance till the Treasury had prepared an invalidity scheme;[3] and Asquith rather reluctantly agreed that, in view of the complications surrounding the invalidity scheme, it would

[1] CAB 37/99/69, 'Rough Draft of Heads of a Bill for the Establishment of Unemployment Insurance', section 21 (2). In the final version of the Bill this was altered to an impartial Umpire appointed by the Crown (1 & 2 Geo. V, c. 55, section 89 (1)).

[2] CAB 37/96/159, 'Unemployment. Insurance: Labour Exchanges', by W. S. Churchill, 30 Nov. 1908 (circulated 11 Dec. 1908), pp. 3–6.

[3] Asquith MSS., vol. 11, ff. 239–44, W. S. Churchill to H. H. Asquith, 26 Dec. 1908.

probably be necessary to postpone national insurance until the following session.[1]

Llewellyn Smith's scheme was submitted to a series of critics, both official and unofficial, during the winter of 1908–9. Llewellyn Smith himself invented a list of fifty-two objections, criticizing particularly the tripartite system of contribution, the flat rate of benefit, the imposition of compulsion on certain trades, and the failure to discriminate between different grades of workmen. He suggested that insurance would positively increase unemployment by encouraging malingering and by discouraging employers from retaining workmen during a depression; and that it would compete with private thrift and reduce the attractions of trade-union organization.[2]

These objections were passed on to Beveridge to prepare a case for the defence. Beveridge justified the levy upon employers on the ground that they would profit from the prevention of physical deterioration amongst unemployed workmen and the improvement of discipline among the employed.[3] The contribution from the State was justified because the unemployed were supported by the community in any case, and a subsidy to insurance would merely be a substitute for less systematic forms of public relief.[4] Moreover, contributions from the State and employers were necessary for 'disarming trade union jealousy' and as a 'compensation for this invasion' of the unions' previous monopoly in the unemployment insurance field.[5] To regular and efficient workmen who complained that they were supporting the irregular and inefficient, it could be argued that this injustice was cancelled out by the contributions of employers and the State.[6] It was untrue, Beveridge contended, that the unskilled would be unable to afford insurance contributions,

---

[1] R. Churchill, *Winston S. Churchill*, ii, Companion Part 2 (1907–11), pp. 869–70, H. H. Asquith to W. S. Churchill, 11 Jan. 1909.

[2] Beveridge MSS. (first deposit), Parcel 2, Folder A, Questions and Answers (submitted to W. H. Beveridge by H. Llewellyn Smith), Oct. 1908.

[3] Ibid., Answer 5.

[4] Ibid., Answers 3 and 7.

[5] Beveridge MSS. (first deposit), Parcel 2, Folder A, 'Notes on Memorandum from H. Llewellyn Smith', by W. H. Beveridge, 28 Nov. 1908.

[6] Beveridge MSS. (first deposit), Parcel 2, Folder A, Questions and Answers (submitted to W. H. Beveridge by H. Llewellyn Smith), Answer 32.

since 'there are probably few workmen who do not spend at least as much as this on luxuries with which they could well dispense'.[1] He defended the flat rate of benefit on the ground that graduation according to need would involve an undesirable inquisition into the private affairs of workmen;[2] and graduation according to income was unnecessary, because highly paid workmen could obtain additional insurance through trade unions or other private institutions.[3] Compulsion was necessary because trade-union insurance covered so few workmen, and voluntary schemes attracted only bad risks.[4] Beveridge denied that insurance would actually promote unemployment, since all parties would have a vested interest in minimizing contributions to the fund;[5] nor would it displace trade-union insurance, since the State would merely provide a minimum of benefit as a basis for additional private thrift.[6] Finally, Beveridge repudiated the Fabian viewpoint that compulsory insurance would be unnecessary within a framework of efficient prevention. He concurred in the need for supplementary schemes of repression for the unemployable and retraining for the unemployed. But there were also 'men who are not in need of cure or change of any sort', whose main problem was the purely financial one of 'averaging wages over good times and bad'; and to assist men of this sort insurance was cheaper and more efficient than artificial employment, less personally demoralizing, and 'more honourable' than gratuitous relief.[7]

Other authorities to whom Llewellyn Smith submitted his scheme were, however, more critical. The Treasury insisted that parliamentary subventions could only be paid to trade unions that separated their insurance funds from their strike funds and submitted their accounts for inspection by the Board of Trade.[8] Sir Benjamin Browne, the Tyneside shipbuilder and philanthropist, objected that insurance was un-

---

[1] Beveridge MSS. (first deposit), Parcel 2, Folder A, Questions and Answers (submitted to W. H. Beveridge by H. Llewellyn Smith), Answer 6.
[2] Ibid., Answer 37.
[3] Ibid., Answer 36.　　　　　　　　　　　　　　[4] Ibid., Answers 11–12.
[5] Ibid., Answers 41–2.
[6] Ibid., Answer 13.　　　　　　　　　　　　　　[7] Ibid., Answer 15.
[8] Beveridge MSS. (first deposit), Parcel 2, Folder A, 'Unemployment Insurance. Criticisms', Comment by G. Barstow.

necessary for skilled workmen;[1] and Professor W. T. Ashley urged that insurance should be postponed until the machinery of labour exchanges had been established, and the true market situation of labour had been clarified by the process of decasualization.[2]

The points in the scheme that aroused most controversy, however, were compulsion and the system of contribution. C. J. Drummond, an old-fashioned trade unionist and labour correspondent of the Board of Trade, recognized that 'desperate remedies' were called for, but thought that 'a compulsory scheme scarcely appeals to our English ideas'.[3] Sidney Webb objected that there was insufficient statistical evidence to frame a sound scheme for unorganized workmen; and that, in the absence of systematic provision for the prevention of unemployment, the scheme would be flooded with bad risks, malingerers, and seasonally unemployed workmen. He suggested instead the series of measures that was shortly to be recommended by the Minority Report of the Royal Commission on Poor Laws: 'My wife and I . . . cannot help thinking that the compulsory labour exchange, plus subsidised voluntary insurance, and maintenance under disciplinary training for uninsured men in distress, solves more difficulties than compulsory insurance plus a voluntary labour exchange.'[4] But the Webbs' scheme had already been rejected by the Board of Trade several months earlier; and the only one of their detailed proposals ultimately adopted by Llewellyn Smith was the introduction of a rule that an individual's claim upon the fund should be limited to one benefit for every five contributions—thereby eliminating those who were either wilfully idle or chronically unemployed.[5]

---

[1] Ibid., Sir Benjamin Browne to H. Llewellyn Smith, 22 Feb. 1909. Browne ascribed unemployment to the emigration of capital in anticipation of redistributive legislation; his remedy was the provision of 'co-partnership' schemes which would enable workmen to save and invest in industry (Benjamin Chapman Browne, 'Unemployment' (14 Nov. 1908); and 'Co-partnership and Unemployment' (4 Dec. 1908), *Selected Papers on Social and Economic Questions*, pp. 162–4, 165–73).

[2] Ibid., Comment by Professor W. T. Ashley, 30 Dec. 1908. Ashley, who was Professor of Economic History at Birmingham University, was a leading academic exponent of tariff reform.

[3] Ibid., Comment by C. Drummond, 25 Jan. 1909.

[4] Ibid., Comment by Sidney Webb, 13 Dec. 1908.

[5] Below, p. 315.

More effective criticisms were levied against the financial structure of the scheme. George Barnes, the Labour M.P., proposed that the state contribution should be increased to offset the saving that would be effected in Poor Law expenditure;[1] and Percy Ashley, the Secretary of the Board of Trade's Commercial Intelligence Advisory Committee, thought that the contribution of employers should be raised, because Llewellyn Smith had exaggerated the extent to which the scheme gave them a financial interest in reducing unemployment.[2] On behalf of the employers, however, H. T. Holloway, the President of the Institute of Builders, protested that the levy on employers would be a tax on industry;[3] and the Treasury objected that the employers had no 'insurable interest' that justified the imposition of premiums of nearly £1,000,000 a year. 'The contribution is bound to affect either [the employer's] prices or his wages' commented George Barstow, a Treasury assistant secretary.

If as anticipated . . . it is taken out of the consumer, prices of his commodities must go up, this will reduce demand and therefore reduce employment: so that the scheme would create the very evil it is intended to mitigate. If, as is more likely, the result of the contribution is to reduce wages, then the workman really pays twice, once directly and once indirectly.[4]

Churchill himself thought that the whole burden of contribution would ultimately fall on the workmen and that 'the whole system will prove to be nothing more than wages-spreading'.[5] But Beveridge argued that the burden would fall equally on

[1] Beveridge MSS. (first deposit), Parcel 2, Folder A, 'Unemployment Insurance: Criticisms', Comment by G. N. Barnes, n.d. Barnes also objected that the 'waiting week' was unfair to workers in 'short-time' industries and to casual labourers: 'Take the casual worker in, say, the ship-repairing industry. He is in and out of work like a dog at a fair, often being out less than a week . . . he will be cut out by the scheme.'

[2] Ibid., Comment by Percy Ashley, 11 Jan. 1909.

[3] Ibid., Notes on Mr. Holloway's criticisms, by W. H. Beveridge, 25 Mar. 1909.

[4] Ibid., 'Unemployment Insurance: Criticisms', by George Barstow, n.d. Barstow, a Treasury official until 1936, subsequently became Chairman of the Prudential Assurance Company—in which position he led the opposition of the commercial insurance lobby to the Beveridge Report in 1942.

[5] Beveridge MSS. (first deposit), Parcel 2, Folder C (i), item 11, 'Note on Malingering', by W. S. Churchill, 6 June 1909.

both parties and give employers an interest in promoting labour exchanges, regularizing employment, and minimizing claims upon the fund.[1] He and Llewellyn Smith admitted that insurance was 'a new form of taxation', but insisted that this was 'perfectly reasonable';[2] it was, indeed, an argument in favour of the scheme that, instead of penalizing depressed areas, distressed workmen, and charitable individuals, the effects of unemployment would be shared by the whole community and 'each product [would] be made to bear in part at least the cost of the unemployment incidental to its production'.[3] Moreover, the Treasury could scarcely maintain its objection to a levy upon employers, since the Chancellor of the Exchequer was already contemplating a similar scheme of contributory insurance against invalidity and sickness.[4]

When Llewellyn Smith's scheme, together with the draft heads of an Unemployment Insurance Bill, was submitted to the Cabinet in April 1909 the employers' contribution had been raised to $2d$. and the State's to $1\frac{1}{3}d$. a week. Benefits, payable after the first week of unemployment, would be graduated downwards from $7s$. to $5s$. a week for fifteen weeks; the cost of such a scheme was equivalent to giving $6s$. for twenty weeks, but it was thought that by gradually reducing benefits 'an increasing pressure is put on the recipient of benefit to find work'.[5] The total cost to the State, including administrative expenses, was estimated at £1,000,000 a year for the first five years. In view of 'the admitted incompleteness and uncertainty of the data' Churchill asked for between £1,250,000 and £1,300,000 a year for the combined scheme of insurance and labour exchanges; but his advisers were 'confident that this is an extreme figure'.[6] These financial calculations were based on the expectation that unemployment would fluctuate between 4 per cent and 16 per cent, al-

---

[1] Beveridge MSS. (first deposit), Parcel 2, Folder A, Questions and Answers (submitted to W. H. Beveridge by H. Llewellyn Smith), Answer 5.

[2] Beveridge MSS. (first deposit), Parcel 2, Folder A, 'Unemployment Insurance. Confidential', by H. Llewellyn Smith, n.d.; and 'Notes on Mr. Holloway's Criticisms', by W. H. Beveridge, 25 Mar. 1909.

[3] Beveridge MSS. (first deposit), Parcel 2, Folder A, 'Unemployment Insurance: Confidential', by H. Llewellyn Smith, n.d.

[4] Below, p. 316.

[5] CAB 37/99/69, 'Memorandum on a Scheme for Unemployment Insurance', p. 4. [6] Ibid., pp. 1–2.

lowing each workman an average of 27 days of unemployment per year; but the average unemployment incurred by trade unionists in insured trades over the previous ten years had been only 5·6 per cent, or 17½ days per year, so that allowance was made for a wide margin of error.[1] These estimates had been approved by Thomas Ackland, a Vice-President of the Institute of Actuaries, who emphasized that ideally the risk of unemployment should have been computed on more detailed information about different age-groups and different occupations;[2] but he thought that the conjectural nature of the scheme was justified by the generous allowance for error, and by the provision for adjustment of rates of contribution to the fund.[3] In fact, Llewellyn Smith was quite undaunted by the actuarial problems of unemployment insurance; and he made it clear to Sidney Buxton a year later that it was anticipation of administrative difficulties rather than absence of statistical data that had persuaded him to limit the scheme in the first instance to a small and experimental section of the workforce.[4]

As a result of the circulation of this memorandum a further committee of the Cabinet was appointed on 26 April 1909 'to consider and report upon the question of insurance against unemployment'.[5] At the instigation of John Burns this committee asked Llewellyn Smith to prepare a paper explaining how the insurance fund was to be protected against 'malingering and other forms of imposition'.[6] Llewellyn Smith stated that there were five kinds of improper claim upon the fund.[7] Firstly, claims by those who were secretly employed; secondly, claims by those 'who have discharged themselves or been dismissed by personal fault'; thirdly, claims by those who had not tried to find work; fourthly, claims by men who were either unemployable or superfluous in their particular trades;

[1] CAB 37/99/69, 'Memorandum on a Scheme for Unemployment Insurance', p. 4.
[2] Ibid., pp. 17–20.
[3] Ibid., pp. 22–3.
[4] Buxton MSS., unsorted, H. Llewellyn Smith to Sydney Buxton, 24 Jan. 1911.
[5] CAB 41/32/11, H. H. Asquith to the King, 26 Apr. 1909.
[6] Add. MS. 46327, Burns Diary, 6 May 1909.
[7] Beveridge MSS. (first deposit), Parcel 2, Folder C (i), item 12, 'Protection of the Insurance Fund against Malingering and other Forms of Imposition', by H. Llewellyn Smith, 18 May 1909.

and fifthly, claims by workmen using someone else's insurance card. These forms of imposition would be prevented by daily registration at the exchange and inquiry into the causes of dismissal; by keeping the benefits deterrently low, and relying on the joint action of exchange officials, advisory committees, and public opinion to expose malingerers; by limiting the proportion of benefits to contributions paid in respect of each workman, and insisting upon a minimum period of employment between one sequence of benefits and the next; and by imposing severe penalties on those who broke the rules or who attempted to impersonate other insured workmen.

Llewellyn Smith claimed that these precautions were based on the rules of many benefit-paying unions, and in May 1909 Beveridge compiled an exhaustive collection of trade-union rules showing that in the vast majority of cases no benefit was given to workmen who voluntarily abandoned situations that conformed to the standard rate.[1] It was admitted that 'difficult cases would arise as to dismissal for minor offences—unpunctuality, not being kind to the horses (a common cause with carmen) quarrelsomeness, swearing, smoking, etc.'; and in such cases it was suggested that benefit should be suspended rather than entirely refused.

This might even apply to some of the graver offences. A man may, for instance, have lost a job just before a depression through drunkenness, yet after a certain time . . . it becomes increasingly difficult to regard the original offence as the true cause of unemployment. If trade were not so slack he would no doubt get work.[2]

Churchill, however, objected that the disqualification of men discharged for misconduct infringed the basic insurance principle that eligibility for benefit depended on the payment of a premium rather than the character of the insured. He thought that 'the qualifications for Insurance must be actuarial. You qualify, we pay. If you do not qualify it is no good coming to us. That is the only safe and simple plan upon which the administration of such a fund can be conducted.' Moreover,

[1] Cd. 5703/1911, *Tables Showing the Rules and Expenditure of Trade Unions.*
[2] Beveridge MSS. (first deposit), Parcel 2, Folder C (i), item 9, 'The Control of Unemployment', n.d.

the right of the workman to leave a job or employer or district that he disliked was a liberty which ought not to be curtailed by insurance, but should be regarded as 'part of the general risk of unemployment'—particularly as the workman himself would be the chief contributor to the fund. Investigation of such cases would involve the exchanges in adjudication of personal quarrels between masters and workmen, and 'once we get into these jungles we are lost. . . . I would rather reduce the benefits to cover greater risks than plunge into the system of inquisitorial checks.' Churchill argued that the limitation of benefits to contributions would gradually expel the habitual malingerer and truly superfluous workman from the scheme; and he suggested that 'it would pay the insurance fund to keep always available a certain proportion of tame jobs as testers, which could be offered in doubtful cases, refusal to accept which would disqualify for benefit and thus relieve the fund.'[1] Churchill's view was, nevertheless, overridden, either by his permanent officials or more probably by the Cabinet committee. Workmen who had been discharged for misconduct or who had abandoned their job 'without just cause' were excluded from benefit for the first six weeks of unemployment;[2] and over 50,000 claims for benefit were disallowed for these reasons during the first two years of the unemployment insurance scheme.[3]

Throughout the spring and summer of 1909 administrative adjustments were made to perfect the machinery of the scheme. A system of 'no-claim bonuses' was arranged for workmen who rarely came upon the fund;[4] and Beveridge

---

[1] Beveridge MSS. (first deposit), Parcel 2, Folder C (i), item 11, W. S. Churchill, 'Note on Malingering', 6 June 1909.

[2] Section 87 (2) of the National Insurance Act, 1911.

[3] Cd. 6965/1913, *First Report on the Proceedings of the Board of Trade under Part II of the National Insurance Act, 1911*, p. 31; Beveridge MSS., D. 030, 'Labour Exchanges and Unemployment Insurance. Report of the Proceedings of the Board of Trade', p. 81, Table LXXXVIII.

Examples of misconduct leading to dismissal on account of which claims were disallowed were absence from work without leave, 'seeing workmates off to the war', refusal to carry out foreman's instructions, drunkenness, insolence, sleeping during working hours, smoking, 'larking', bad timekeeping (*National Insurance Acts, 1911 to 1915. Unemployment Insurance. Decisions given by the Umpire respecting Claims to Benefit*, ii. 403–9).

[4] Beveridge MSS. (first deposit), Parcel 2, Folder C (i), item 1, 'Return of Premiums in Case of No Claim', by W. H. Beveridge, 17 Aug. 1909.

and Llewellyn Smith considered and rejected proposals that contributions should be graduated according to the age or earnings of individual workmen insured.[1] They also considered various ways of limiting each workman's claim upon the Insurance Fund, and thus excluding workmen who were virtually unemployable. Beveridge suggested that, rather than limiting benefits to a certain number per year, this could be done by the adoption of Sidney Webb's plan for limiting a workman's claim to one benefit for every five contributions. This would enable young and regular workmen to use the Insurance Fund as a provision against unemployment in later life; and it also 'gives the scheme a better appearance *morally* by treating men's lives as a whole. This is in accord with Charity Organisation Society principles and may therefore secure valuable expert support.'[2] Llewellyn Smith agreed that insurance should be regarded as a long-term individual investment, and that benefits should be proportionate to contributions paid over the whole of an insured person's working life. But he objected that this would be politically embarrassing if the whole scheme proved a failure and had to be abandoned; and it was therefore arranged that workmen who entered the scheme from the beginning should be gratuitously credited with several weeks prior contributions, so that they would have no cause for complaint if the scheme eventually folded up.[3]

It was agreed that trade unions in insured trades which already paid an out-of-work donation equivalent to that proposed by the State could treat their benefits and contributions as transactions under the national insurance scheme, and would be entitled to reclaim a part of their expenditure from the State.[4] It was expected, however, that most unions in insured trades would prefer to keep their private schemes as an addition to rather than as a substitute for national insurance, in order to retain a marginal advantage over un-

[1] Ibid., item 2, 'Differentiation of Contributions by Earnings', n.d.
[2] Ibid., item 6, 'Proportioning of Benefits to Contributions', by W. H. Beveridge, 10 June 1909.
[3] Ibid., item 7, 'Notes on first Memorandum', by H. Llewellyn Smith, n.d.; item 8, 'Further Memorandum', by W. H. Beveridge, 1 Sept. 1909.
[4] Ibid., item 13, 'Arrangements with Trade Unions', by W. H. Beveridge, n.d., with notes by H. Llewellyn Smith.

organized workmen.[1] In the case of unions outside the insured trades the State's contribution would be limited to three-quarters. In both cases the subsidy would be retrospective, and it was hoped that this would deter unions from being over-generous with public funds. All competent workmen in receipt of benefit would be entitled to refuse work under conditions inferior to those generally prevailing in their district; sub-standard workmen could refuse work under conditions inferior to their previous situation; and trade unions that managed their own funds under the national insurance scheme would be allowed to insist on full recognition of the standard rate.[2]

The Board of Trade's scheme for unemployment insurance was clearly ready for the first stages of legislation early in the summer of 1909. But, as Churchill told the House of Commons on 19 May, there were five reasons why its introduction had to be postponed. First of all, labour exchanges had to be set up. Further negotiations had to be conducted with unions and employers. Parliamentary time could not be set aside in an already overcrowded session. The scheme had to be integrated with Lloyd George's plan for invalidity insurance. And money had to be raised—which meant that the controversial 1909 Budget had to be passed before any progress could be made with the introduction of insurance.[3]

Churchill started negotiations with interested groups on unemployment insurance in the summer of 1909. In August he tried to persuade the engineering and shipbuilding employers that they could transfer their contributions to the workmen, and that in any case insurance would be cheaper than charitable subscriptions to the unemployed.[4] He assured representatives of the skilled trade unions that insurance would give them additional protection by gradually ex-

[1] Beveridge MSS. (first deposit), Parcel 2, Folder C (ii), item 14, 'Arrangements with Trade Unions. Safeguards against Extravagance', by H. Llewellyn Smith (after consultation with D. C. Cummings and I. Mitchell, 25 Jan. 1909).

[2] Beveridge MSS. (first deposit), Parcel 2, Folder C (i), item 13, 'Arrangements with Trade Unions', by W. H. Beveridge, n.d., with notes by H. Llewellyn Smith.

[3] *Hansard*, 5th series, vol. 5, cols. 510–11.

[4] LAB 2/211/LE. 500, Typescript report of a Conference with the Engineering Employers' Association and Shipbuilding Employers' Federation, 18 Aug. 1909, pp. 29–30.

pelling inferior and cut-price workmen from the labour market.[1]

Nevertheless, opposition among organized workmen to the contributory principle threatened to be formidable. At the Ipswich conference of the T.U.C. in September 1909 the Parliamentary Committee urged trade unionists to support the state scheme; but the conference nevertheless condemned compulsory insurance unless managed by organized labour.[2] Trade-union opposition to the scheme was, however, very diverse in its motives. J. M. Jack of the Scottish Ironfounders 'personally believe[d] that men should insure themselves individually with present insurance offices'.[3] John Clynes, the Secretary of the Gasworkers and General Labourers' Union, objected to the flat rate of contribution on the ground that 'it would not be fair that the state should receive as much from the lowly-paid workman as from the highly-paid workman for the benefits received';[4] George Barnes, the Labour M.P. for Glasgow, thought that a much lower proportion of the cost should be borne by the workers;[5] and T. E. Naylor of the London Compositors thought that workmen should pay nothing at all towards insuring themselves against a hazard which was the responsibility of the employer and the State.[4]

On 1 March 1910 the Parliamentary Committee of the T.U.C. was interviewed by the new President of the Board of Trade, Sydney Buxton. The Committee proposed that the unemployment insurance scheme should be confined exclusively to trade unionists, leaving non-unionists to be protected by the Trade Boards Act. 'Otherwise you will have men to support who never have been and never will be self-supporting. They are at present parasites on their more industrious fellows and will be the first to avail themselves of the funds

[1] LAB 2/211/LE. 500, Typescript report of a Conference with the Federation of Engineering and Shipbuilding Trades, 18 June 1909.

[2] *Report of the Ipswich T.U.C.*, 1909, pp. 56, 188–9.

[3] Beveridge MSS. (first deposit), Parcel 2, Folder A, 'Criticisms of Workmen's Insurance by members of the Executive Council of the Shipbuilders and Engineering Trade Federation', compiled by D. C. Cummings, 29 May 1909.

[4] *Report of the Ipswich T.U.C.*, 1909, p. 108.

[5] Beveridge MSS. (first deposit), Parcel 2, Folder A, 'Unemployment Insurance. Criticisms', Comment by G. N. Barnes, n.d.

the Bill provides.'[1] Buxton replied that it would be unreason-able to exclude non-unionists from insurance; but he pro-mised that trade unions would have 'full representation' on advisory committees, and would be allowed to manage their share of the national insurance funds.[2]

By far the most serious obstacle to the progress of the un-employment insurance scheme, however, was the preoccupa-tion of Lloyd George with his Budget proposals and the even-tual rejection of these proposals by the Lords in November 1909. Without the Budget there could be no new social expenditure, in spite of the continuous rise in the yield of direct taxation during the depression of 1907–9.[3] Moreover, the Budget distracted Lloyd George from invalidity insur-ance; and Churchill in May 1909 had realized that provision against sickness was not merely logically but practically com-plementary to an unemployment scheme.[4] Otherwise the unemployment fund would be constantly besieged with claims from workmen who were unemployed through sick-ness rather than through the state of the labour market. This danger could be partially forestalled by stipulating that no workman should be eligible for benefit who was physically unable to accept any suitable job that was offered him; but clearly the most effective way of protecting the unemploy-ment fund from the claims of the unfit was by the provision of insurance against ill-health.

The rejection of the Budget, however, brought in its train a constitutional crisis and two General Elections, which

---

[1] 5th Quarterly Report of the Parliamentary Committee of the T.U.C., Mar. 1910, p. 41.

[2] Ibid., pp. 42–5.   [3] Above, p. 270.

[4] Beveridge MSS. (first deposit), Parcel 2, Folder C (i), item 11. 'Note on Malingering', by W. S. Churchill, 6 June 1909: 'If . . . a man loses his employment and becomes unemployed through sickness, he is clearly not a subject for the Unemployment Insurance fund, which is limited to dealing with the evil of a man able and willing to work but without work. Suppose a man loses his job through illness of a temporary but recurring character . . . suppose he is ill for a week and out of work for three months in consequence of losing his situation, how does he stand ? In logic he clearly receives no unemployed benefit while unfit to work. The moment he becomes fit to work, he is an unemployed workman duly qualified for the Insurance fund and benefits should begin. Can you in practice, in the possible absence of an universal invalidity insurance scheme, refuse a man who loses his employment through sickness, the benefit which he needs most urgently during the month of sickness, and begin to pay it to him the moment he is physically fit.'

effectively diverted the Chancellor of the Exchequer from health insurance for over a year. The Budget itself became a factor in the unemployment debate, since the opponents of graduated taxation and the proposed tax on undeveloped site values argued that the Budget would drive capital abroad and thereby deprive the working classes of employment.[1] Liberal apologists on the other hand argued that the development of unused land and increased public expenditure of £14,000,000 a year on social reform and shipbuilding would create additional employment and increase consumer demand.[2] Nevertheless, the question of unemployment was only a secondary issue in both the elections of 1910; and Liberal agents in the south of England reported that the electorate seemed to know little or nothing of the Government's promise of national insurance.[3] Political discussion of unemployment still centred on the question of tariff reform, and Unionist leaders had to restrain their agents and party officials from making rash promises about the abolition of unemployment under a system of protection.[4]

In the political turmoil of 1910 little progress was made with either of the proposed schemes of insurance. Early in March Sydney Buxton re-circulated to the Cabinet the draft Bill and memorandum by Llewellyn Smith that had been considered in April of the previous year, with the suggestion that it should be introduced into Parliament under the 'ten minutes Rule . . . so that whatever happens our proposal may

[1] *Liberal Magazine*, Oct. 1909, pp. 542–4, Speech by Lord Rosebery at Glasgow, 10 Sept. 1909. J. Calvert Spensley, 'Urban Housing Problems', *Journal of the Royal Statistical Society*, 81 (Mar. 1918), 197, suggested that the Budget intensified the building depression and made 'investors distrust and their advisers discourage investment in house property and mortgages'.

[2] Sir Francis Mowatt, *The Budget and Unemployment*, Liberal Publication Department, Leaflet No. 2246, 7 Oct. 1909; A. C. Pigou (ed.), *Memorials of Alfred Marshall*, p. 464; Alfred Marshall to Lord Reay, 12 Nov. 1909.

[3] Philip Whitwell Wilson, 'A Workman's Charter', *Daily News*, 3 Feb. 1910.

[4] Liberals claimed that during the election of Jan. 1910 the Central Conservative Association had sent out vans proclaiming 'Fiscal Reform means work for all' (*Hansard*, 5th series, vol. 14, col. 417). In the subsequent Parliament Balfour repudiated this view, stating that protection might reduce but not abolish unemployment (ibid., cols. 400–8). Nevertheless, this promise was clearly made not merely by party Whips and irresponsible backbenchers but by the leaders of the Unionist party (Sir Charles Petrie, *Life and Letters of Austen Chamberlain*, pp. 238–9).

be on record'.[1] His intention was presumably to make sure, in the fluid political situation that prevailed, that the Liberals should have the credit of being the first party to introduce a bill for National Insurance.[2] But the Bill again got no further than Cabinet discussion. Throughout the summer officials of the labour-exchanges branch were absorbed in putting the new system of exchanges into operation. In August Lloyd George proposed unsuccessfully that unemployment along with other social questions should be dealt with by a coalition of the two major parties;[3] and Beveridge publicly prophesied that it would take two years to implement a scheme of unemployment insurance.[4] In September Llewellyn Smith outlined to the British Association the main principles of the Board of Trade's scheme; and he announced that collective action against the 'irregularity of working-class incomes so far as affected by irregular demand for labour' had become a question of 'high policy'.[5] But it was not until December 1910 that Lloyd George turned his serious attention to insurance, and appointed William Braithwaite of the Inland Revenue and John Bradbury, a Treasury official who had been Asquith's private secretary, to draft a scheme for invalidity.[6] The Attorney-General, Sir Rufus Isaacs, was as-

[1] CAB 37/102/8, 'Unemployment Insurance Bill', circulated by Sydney Buxton, 8 March 1910.

[2] A week earlier Buxton had received a letter from H. B. Lees-Smith, the new Liberal M.P. for Northampton, to this effect: 'May I express the hope that you will use your influence to see that the scheme for *Unemployed Insurance* is not abandoned. It ought to be non-contentious, but it would be of great assistance to us in the next election if we have it to our credit. Moreover, the Labour Exchanges are not likely to be very effective till Unemployed Insurance is coupled with them. It would be a great pity if we allowed any possibility whatever that the Conservatives should, for years to come, be able to claim that they were the authors of what undoubtedly would be an experiment infinitely more popular than the Labour Exchanges' (Buxton MSS., unsorted, H. B. Lees-Smith to Sydney Buxton, 23 Feb. 1910).

[3] Lucy Masterman, op. cit., p. 164; Sir Charles Petrie, *Life and Letters of Austen Chamberlain*, p. 384; B. Semmel, *Imperialism and Social Reform*, pp. 242–6.

[4] *Westminster Gazette*, 10 Aug. 1910.

[5] Beveridge MSS., Coll. B, vol. iv, item 3, Copy of Sir Hubert Llewellyn Smith's 'Presidential Address to the Economic and Statistical Section of the British Association for the Advancement of Science', p. 17.

[6] Lucy Masterman, op. cit., p. 179. William Braithwaite (1875–1938), Assistant Secretary, Board of Inland Revenue 1910–12; chief civil service architect of Part One of the National Insurance Act. In 1912 Braithwaite was disappointed in his hope that Lloyd George would appoint him to the Chairmanship of the English National Insurance Commission, the post being given to Sir Robert Morant.

signed the task of helping the Treasury to prepare the scheme for legislation; and the Solicitor-General, Sir John Simon, was commissioned to assist the Board of Trade.

Consultations between the two departments began early in 1911. Beveridge attended several Treasury conferences on invalidity;[1] and on 11 January Llewellyn Smith reported to Buxton that he 'had just had a long talk with the Chancellor of the Exchequer about Insurance. He is modifying his scheme very greatly but I foresee further difficulties. However, he is concentrating his mind on the subject and facing the problem.'[2] From the start, however, the process of integrating health and unemployment insurance was fraught with interdepartmental rivalry and personal conflict. In January 1911 the health scheme had no objective existence outside the mind of the Chancellor; even there it consisted of little more than a series of disconnected propositions. 'Rufus Isaacs compares [Invalidity Insurance] to his ancestor's task of making bricks without straw' Llewellyn Smith commented maliciously to Sydney Buxton on 24 January.[3] Unemployment insurance, on the other hand, had been on the legislative production line for more than two years. Its financial and statistical foundations had been meticulously calculated, administrative machinery had been established in the form of labour exchanges, and arrangements had been made to involve both capital and labour in the collection of contributions and the administration of benefits. The original deci-

Braithwaite became Secretary of the National Insurance Joint Committee, 1912, and Commissioner for Special Purposes of Income Tax, 1913.

John Bradbury (1872–1950), Asquith's private secretary 1905–8; a principal clerk in the Treasury 1909 and National Insurance Commissioner 1912; Permanent Secretary to the Treasury with responsibility for finance 1913–19; Baron Bradbury of Winsford 1925.

[1] Braithwaite MSS., Ia, Diary, 9 Jan. and 13 Jan. 1911.
[2] Buxton MSS., unsorted, H. Llewellyn Smith to Sydney Buxton, 11 Jan. 1911. According to Llewellyn Smith, Lloyd George's 'present idea is that it would be a good thing to get Unemployment Insurance and Invalidity Insurance read a second time before Easter, and sent to one (or two) *Select* Committees which could hear actuarial, trade union and other evidence, not of course with Counsel. There is much to be said for and against this procedure: unless you are certain of an autumn session it would probably prevent either bill passing this year. Then the composition of the Select Committees would need the most careful consideration . . .'. Nothing, however, came of this idea.
[3] Ibid., H. Llewellyn Smith to Sydney Buxton, 24 Jan. 1911.

sion to separate unemployment insurance from labour-exchange legislation had been made at the instigation of Lloyd George in December 1908; and the scheme had been further postponed for two years largely because of his failure to produce a plan for health insurance and to make the necessary provision for finance.

At the beginning of 1911 Llewellyn Smith and Beveridge were therefore inclined to look upon the Treasury's hypothetical health scheme with ill-disguised contempt. 'Lloyd George talks as though both schemes were equally matured and could be fitted together somehow,' reported Llewellyn Smith to Buxton on 24 January. 'Everyone else knows of course that as yet there is no invalidity scheme in any real sense . . . the only thing to urge on the Treasury people is to press on with a draft of their part of the Bill.'[1] In fact the unemployment scheme was less complete than Board of Trade officials liked to suggest; but Llewellyn Smith took the view 'that Health Insurance must be modified to suit Unemployment and give way to it'.[2] And Beveridge even went so far as to supply the Treasury with an unsolicited memorandum on 'sickness, invalidity and allied risks', which proposed that the cost of health insurance should be borne entirely by the State and the workers, in order to reconcile employers to the burdens of unemployment insurance and workmen's compensation.[3]

Treasury officials, on the other hand, resented the Board of Trade's 'attitude of insolent superiority to our ill-thought-out proposals';[4] and they objected to the assumption of Beveridge and Llewellyn Smith that the introduction of unemployment insurance was a virtual *fait accompli*. It soon became apparent, moreover, that far from tailoring health to unemployment insurance, Lloyd George was trying to take over the Board of Trade's scheme in order to merge it with his own.[5] This was quite contrary to the Board of Trade's original conception of

---

[1] Buxton MSS., unsorted, H. Llewellyn Smith to Sydney Buxton, 24 Jan. 1911.

[2] H. N. Bunbury (ed.), *Lloyd George's Ambulance Wagon. Being the Memoirs of William J. Braithwaite* (1911–12), p. 149.

[3] Braithwaite MSS., II, item 8, Memorandum on Sickness and Invalidity Insurance, by W. H. Beveridge, Jan. 1911.

[4] H. N. Bunbury, op. cit., p. 141.

[5] R. Churchill, op. cit., p. 306.

national insurance, since although Churchill had recognized that health and unemployment insurance were complementary neither he nor his officials had envisaged any structural amalgamation between the two schemes. Beveridge had written in June 1909 that

the best chance for the Unemployment Insurance scheme is to prevent it from being in any way knit up with other schemes or involved in their fortunes. The arguments in favour are much stronger and the obstacles altogether less in the case of unemployment insurance than in regard to any other form of social insurance, so that the former is strongest when standing alone.[1]

Lloyd George, however, took the opposite view. He regarded himself as the originator of the idea of compulsory unemployment insurance and thought that both schemes would ultimately be dovetailed into a single national insurance plan.[2] Moreover, there were clearly advantages to be gained from attaching his own cumbersome and controversial scheme to the relatively simple and problem-free measure that had been devised by the Board of Trade. He therefore urged the Liberal Chief Whip, Alexander Murray, to persuade Asquith that he should be given control of the entire insurance scheme, leaving the Board of Trade in charge only of administrative details.[3] This tactic failed; but the Cabinet agreed that the two measures should be introduced as one bill rather than two—thereby annoying the Board of Trade officials, who surmised that the 'Cabinet decision' was really a personal decision taken by Lloyd George.[4]

Having failed to appropriate unemployment insurance for his own department, Lloyd George proceeded to launch an attack upon the financial and administrative structure of the Board of Trade's scheme.[5] At a conference between Treasury

---

[1] Beveridge MSS. (first deposit), Parcel 2, Folder C (ii), item 25, 'Accident Insurance' by W. H. Beveridge, 8 June 1909.
[2] A. C. Murray, *Master and Brother. Murrays of Elibank*, p. 88. But see also W. H. Beveridge, *Power and Influence*, p. 81.
[3] A. C. Murray, op. cit., pp. 88–9.
[4] Buxton MSS., unsorted, H. Llewellyn Smith to Sydney Buxton, 24 Jan. 1911.
[5] A. C. Murray, op. cit., p. 89. Lloyd George told the Liberal Chief Whip, Alexander Murray that 'he had always thought Winston's Bill was fairly watertight, and both Winston, and subsequently Buxton, had told him that it *was* so. But when a short time previously he had examined the Bill closely he had found

and Board of Trade representatives on 23 January 1911 he was 'much down on Parliamentary aspects of insuring certain trades only against unemployment'.[1] He declared that the proposed state contribution of one-third was politically indefensible, particularly as the State was to pay for only a quarter of the cost of the health scheme; and he predicted that unless the State's contribution was reduced, the unemployment scheme would be '"stampeded" by representatives of excluded trades wanting to come in'. Llewellyn Smith denied that this was at all likely, and thought that in any case 'the administrative difficulties of all trades coming in would far outweigh the financial'.[2] The Board of Trade acquiesced, however, in the reduction of the state contribution to one-quarter, and agreed that employers' and workmen's contributions to the unemployment fund should each be raised to $2\frac{1}{2}d.$ a week.[3]

Llewellyn Smith's arrogance and Lloyd George's jealousy nevertheless continued to generate friction between the two departments. On 23 March Braithwaite recorded that in a conference with Buxton and Llewellyn Smith 'L. G. absolutely refus[ed] to give them their full terms or to give more proportionately to Unemployment than to Sickness'.[4] At the end of March Whitwell Wilson, the patron of the Labour party's Right to Work Bill, infuriated the Chancellor by suggesting in the *Daily News* that unemployment insurance 'was being blocked and held back' by the invalidity scheme. Wilson ascribed his story to 'Lobby gossip, Horatio Bottomley etc., etc.'; but Lloyd George told Braithwaite that 'he would smash to atoms Sydney Buxton or Llewellyn Smith or anyone else who had taken a hand in this game.'[5]

it to be quite the contrary. As far as he could make out, Winston had barely read the Bill before it was circulated to the Cabinet, and, instead of spending months as he ought to have done in thinking out ways and means of overcoming obstacles and solving the problem on the most approved lines, he had confined himself to a very eloquent speech in the House, setting forth a number of views which he (Lloyd George) had given him on return from the Continent.'

[1] Braithwaite MSS., Ia, Diary, 23 Jan. 1911.
[2] Buxton MSS., unsorted, H. Llewellyn Smith to Sydney Buxton, 24 Jan. 1911.
[3] These were the figures in the draft Bill circulated to the Cabinet on 3 Apr. 1911 (CAB 37/106/46, 'Unemployed Insurance Bill', by S. Buxton).
[4] Braithwaite MSS., Ia, Diary, 23 Mar. 1911.
[5] H. N. Bunbury, op. cit., p. 134.

The new scale of contributions was incorporated in the draft Bill which was circulated to the Cabinet on 3 April 1911.[1] Benefits were fixed at 7s. for five weeks and 6s. for ten weeks, but Buxton remarked that he hoped it would be possible to alter this to a uniform rate of 7s. The proposed fund for private insurers had been dropped, and the subvention to unions outside the insured trades had been reduced to one-sixth. Contributions would be remitted to firms that adopted 'systematic short time due to depression of trade'; and compulsory technical instruction would be provided by the State for workmen who through sheer incompetence constantly made claims upon the fund. This last clause was strongly criticized by Sir Horace Monro, the new Permanent Secretary of the Local Government Board, who argued that it put a premium on incompetence and that, unless new sources of employment were provided, it would merely train unemployed workmen to displace those already employed.[2] Llewellyn Smith replied that, while not essential to the structure of an insurance scheme, technical training would 'be useful for the purpose both of testing and increasing the competence of unemployed workmen claiming to follow a particular trade. It [would] not be open to the workmen to *claim* to be sent to a technical school at the public expense!'[3]

Both the Board of Trade and the Treasury schemes were exhaustively discussed by the Cabinet early in April. 'Warm and unanimous approval was given to the main principles of the [health] scheme'; but the unemployment scheme was referred to a Cabinet committee for further discussion of the cost of administration.[4] Lloyd George was jubilant at being

---

[1] CAB 37/106/46, 'Unemployed Insurance Bill', by Sydney Buxton, 3 Apr. 1911.

[2] PIN 3/3, f. 104, 'Unemployment Insurance Bill', Typescript memorandum by Sir Horace Monro, 8 Apr. 1911; ibid., f. 106, Sir Horace Monro to H. Llewellyn Smith, 8 Apr. 1911.

[3] PIN 3/3, ff. 110–11, H. Llewellyn Smith to H. C. Monro, 10 Apr. 1911. This clause (section 100 of the final draft of the Bill) was welcomed as 'the most daring proposal' of the whole National Insurance Act (*The Economist*, 23 Dec. 1911, p. 1312). But it was not in fact put into operation until 1925 (W. H. Beveridge, *Unemployment: A Problem of Industry* (1930 ed.), p. 308). The Chairman of a Court of Referees in 'a large Midland industrial area' later recorded that during the first few years of the administration of the Act he came across only one case in which retraining seemed to be necessary (Sir Frank Tillyard and F. N. Ball, *Unemployment Insurance in Great Britain 1911–48*, p. 35).

[4] CAB 41/33/9, H. H. Asquith to the King, 5 Apr. 1911.

'able to retaliate on Buxton by regretting that Health Insurance must now unfortunately wait until the Unemployment Bill is ready'.[1]

Sydney Buxton—'a gentleman and a Radical, a man whom everyone liked and trusted'[2]—was singularly ill fitted for the kind of personal vendetta in political affairs that was relished by Lloyd George; and for a time it seemed as though the Board of Trade would entirely capitulate to the Chancellor's demands. The estimated cost of the administration of unemployment insurance was reduced from £400,000 to £235,000 a year; but this was still unacceptable to the Cabinet, and a further committee, consisting of Lloyd George, Buxton, Burns, Pease, and Churchill, was appointed to consider 'some difficult points of detail'.[3] Braithwaite noted in his diary that 'the Cabinet will not hear of the Unemployment bill in its present shape and the Chancellor is going to try to reconstitute it'.[4] On 19 April Lloyd George argued that the State 'cannot guarantee (unemployment) benefits on the basis of the present evidence', and suggested that smaller benefits should be paid to insured trades with an abnormally high rate of unemployment.[5] The Board of Trade therefore reduced the proposed level of benefit in the building trades to 6s. a week.[6]

Charles Hobhouse, the Financial Secretary to the Treasury, then informed Buxton that the Chancellor intended to reduce the State's contribution to the unemployment fund to 1d. a week.[7] Buxton and Llewellyn Smith dug in their heels at this point. Buxton composed a long personal letter to Lloyd George, at once indignant and conciliatory, pointing out that several ministers had already publicly promised that the

[1] H. N. Bunbury, op. cit., p. 141.
[2] W. H. Beveridge, *Power and Influence*, p. 87.
[3] CAB 41/33/10, H. H. Asquith to the King, 11 Apr. 1911. This committee was also attended by Llewellyn Smith, Bradbury, Braithwaite, and the actuary, Thomas Ackland. The account in Braithwaite's diary suggests that they were not merely witnesses but full members (H. N. Bunbury, op. cit., pp. 146–9).
[4] H. N. Bunbury, op. cit., p. 145.   [5] Ibid., p. 147.
[6] Draft of a Bill to Provide for Insurance against Loss of Health and for the Prevention of Sickness and for Insurance against Unemployment and for Purposes Incidental thereto. Printed 4 May 1911.
[7] Buxton MSS., unsorted, draft of a letter from Sydney Buxton to D. Lloyd George, n.d., ? Apr. 1908. The following account of Buxton's reaction to Lloyd George's proposal is summarized from this letter.

Government's contribution to unemployment insurance would be £1¼ million a year. This figure had been passed without protest by the Cabinet a year previously and it had formed the basis of the Board of Trade's negotiations with trade unions. At the request of the Treasury the Board of Trade had already reduced the over-all estimate for unemployment insurance to £1,100,000 per year. Any further reduction would alienate the unions; and 'I am afraid', concluded Buxton, 'that I cannot go any further without endangering the scheme.' Lloyd George was forced to climb down[1] and the State's contribution was restored to one-third of the weekly sum paid by employers and workmen in the draft bill published on 4 May.[2]

Strong criticism of the unemployment scheme also came from the Treasury establishments department, although this was probably a reflex of normal Treasury control rather than a result of direct interference from the Chancellor's private office. Sir George Murray, the Permanent Secretary, resisted the idea that monetary incentives should be given to social organization, and in January 1911 he protested against the Board of Trade's proposal to give refunds to employers who engaged workmen through a labour exchange or on a long-term basis and to workmen who made few claims upon the fund.[3] These issues were raised at an interdepartmental conference on 16 March,[4] and reluctantly conceded by the Treasury in April 1911.[5] In July Sir Robert Chalmers,[6] Murray's successor as Permanent Secretary, criticized the

[1] Ibid., Note by S. Buxton, attached to previous letter: '. . . the Chancellor of the Exchequer gave way and the proportion was fixed finally. S. B.' It is probable that the matter was settled in a personal interview and that Buxton's letter of protest, quoted above, was never actually dispatched.

[2] These proportions of 2½d., 2⅓d., and 1⅔d., were eventually incorporated in the final version of the National Insurance Bill passed in Dec. 1911.

[3] LAB 2/1483/LE. 1006, Sir George Murray to the Secretary of the Board of Trade, 20 Jan. 1911.

[4] Buxton MSS., unsorted, typescript memorandum on 'Unemployment Insurance', suggested points for discussion at the Conference on Thursday, 16 Mar., 3.50 p.m.

[5] LAB 2/2483/LE. 1006, Sir George Murray to Sir Hubert Llewellyn Smith, 22 Apr. 1911.

[6] Robert Chalmers (1858–1938), upper-division Treasury clerk 1882; Assistant Secretary 1907; Chairman of the Board of Inland Revenue 1908–11; Permanent Secretary to the Treasury 1911–13; Governor of Ceylon 1913–16; Master of Peterhouse 1919–31; 1st Baron Chalmers of Northiam 1919.

proposal that the state subsidy to unemployment insurance should be based on contributions levied, rather than, as in the case of health insurance, on benefits paid. It was, Chalmers protested, an 'objectionable characteristic . . . from the point of view of public finance' that money should be raised not only 'for expenditure to be incurred within the year but . . . to meet contingencies which have not yet arisen'.[1] Chalmers reluctantly accepted Llewellyn Smith's argument that such an alteration in the method of payment by the State would be financially unsound, because it would be impossible to make an accurate estimate of a subsidy that varied with annual fluctuations of unemployment.[2] Llewellyn Smith proved in fact to be more than a match for both Chalmers and Lloyd George;[3] but throughout 1911 the Board of Trade scheme was doubly threatened, by Treasury parsimony on the one hand and by the personal hostility of Lloyd George on the other. The one was a normal hazard of social legislation; but the other was incalculable, particularly since the Chancellor was quite prepared to jettison unemployment insurance entirely, if he could thereby secure the passage of his own scheme.[4]

The health and unemployment schemes were introduced into the House of Commons by Lloyd George as Parts One and Two of the National Insurance Bill on 4 May 1911. The Chancellor stated that continental experience had shown that unemployment insurance must be compulsory, subsidized, and organized on a trade basis. The scheme would encourage employers to undertake decasualization, but it was primarily designed to protect workmen against trade fluctuations, which were to a certain extent predictable by the State but

[1] LAB 2/1483/LE. 9203, Sir Robert Chalmers to the Secretary of the Board of Trade, 26 July 1911.

[2] LAB 2/1483/LE. 9203, Copy of letter and memorandum by W. H. Beveridge to the Secretary of the Treasury, corrected and signed by H. Llewellyn Smith, 31 July 1911; Sir Robert Chalmers to the Secretary of the Board of Trade, 4 Aug. 1911.

[3] 'Sir H. Ll. Smith won't admit that he has been worsted, and indeed has shown himself a very strong man', recorded Braithwaite on 21 Apr. 1911 (H. Bunbury, op. cit., p. 149). See also G. Askwith, op. cit., p. 352.

[4] H. N. Bunbury, op. cit., p. 195, quoting Braithwaite's diary, 28 July 1911: '[Lloyd George] knows the terms of the Unionist party, and can satisfy them by dropping Part II (Unemployment) and a few minor alterations.'

entirely outside the control of the individual workman.[1] The Bill was welcomed on behalf of the Unionists by Austen Chamberlain, who criticized the flat rate of contribution but endorsed the principle of compulsion. He was sceptical about the actuarial basis of Part Two of the Bill, but agreed that 'so long as our trade conditions practically require a certain amount of unemployment in connexion with the organisation of industry you cannot treat the people so unemployed as not having some claim upon the industry and expect individuals to bear the whole burden'.[2] Ramsay Macdonald also welcomed the Bill, provided that the non-corporate status of unions was not going to be compromised; and he urged that the unemployment scheme should be reinforced by a system of technical training.[3] Only a small handful of dissident back-benchers complained that Part Two of the Bill would positively encourage unemployment or that its actuarial basis was conjectural and unsound.[4]

By contrast with Part One of the Bill, Part Two had a relatively easy passage through Parliament. This was largely because, as Beveridge pointed out in 1908, there were no professional or profit-making groups whose livelihood was threatened by a state-controlled system of unemployment insurance.[5] Hence there was no parliamentary opposition from representatives of organized commercial interests. It is true that the consensus of approval for unemployment insurance was less universal than its Board of Trade promoters liked to imply, and Part Two was included in general criticisms of the 'national insurance' principle that were levied most directly at Part One of the Bill. Left-wing M.P.s attacked the flat rate of contributions as inequitable and regressive and 'an income-tax double-weighted upon the poor'.[6]

[1] *The Times*, 5 May 1911. [2] *Hansard*, 5th series, vol. 25, col. 651.
[3] Ibid., cols. 658–61. [4] Ibid., cols. 657–77.
[5] Beveridge MSS. (first deposit), Parcel 2, Folder A, Questions and Answers (submitted to W. H. Beveridge by H. Llewellyn Smith), Oct. 1908, Answer 8. On the reaction of commercial interests to Part I of the Bill, see B. G. Gilbert, 'The British National Insurance Act of 1911 and the Commercial Insurance Lobby' *Journal of British Studies*, 4, no. 2 (May 1965), 127–48.

[6] *Edinburgh Evening News*, 9 Oct. 1911, report of a speech by George Lansbury (newscutting, Lansbury MSS., vol. 29, f. 226). This aspect of the scheme was virtually ignored by the Treasury until more than a year after the passage of the Act (T. 171/47, Revenue Bill Papers 1913, 'Some Notes on the Incidence of Taxa-

Representatives of organized charity objected that subsidized thrift without an internalized obligation on the part of the individual was merely a disguised form of public relief.[1] And the National Committee for the Prevention of Destitution, organized by the Webbs and the Fabian Society, protested against the expenditure of nearly £20,000,000 a year on the relief of sickness and unemployment with no provision for their cure.[2]

Nevertheless, these theoretical controversies made very little impact upon the discussion of unemployment insurance by the Board of Trade. In so far as they impinged upon the Government their brunt was borne by Lloyd George, and during the summer of 1911 Board of Trade negotiations concerning the Bill were almost exclusively confined to the amendment and refinement of individual clauses. In June Sydney Buxton received a deputation from the Shipbuilding Employers' Federation which accused the Government of 'pandering to the trade unions to the detriment of employers' in order to capture the working-class vote. Frederick Henderson, the president of the Federation, objected to the clause of the Bill that allowed trade unions to share in the management of public funds; and he feared that the long-term engagements encouraged by the Bill would give workmen unnatural security and would incite them to abandon their jobs 'just to spite . . . their employers'. He predicted that the imposition of such a prohibitive tax on industry would cripple Britain's

tion on the Working-Class Family' by F. W. Kolthammer of the Rata Tata Foundation. On the results of Kolthammer's inquiry see Appendix B, Table 9, p. 380).

[1] *The Times*, 2 Nov. 1911, C. S. Loch to the Editor.

[2] B. Webb's Diary, 13 May 1911, quoted in *Our Partnership*, p. 473. Even so, the Webbs preferred Part II to Part I of the Bill, since they envisaged that unemployment insurance might 'bring inadvertently the compulsory use of the labour exchange, and the standardisation of the conditions of employment' (ibid., p. 468). The Webbs' subsequent attitude to insurance was ambiguous. Shortly before the slump of 1929–32 they described it as 'the most nearly effective' method of relieving unemployment (S. and B. Webb, *English Poor Law History*, II. ii. 663). But in 1931 Beatrice Webb told the Royal Commission on Unemployment Insurance that unemployment under capitalism was an 'uninsurable risk' and that ' "Out of Work Pay" cannot—at least in the present organisation of society— safely be made *large enough to maintain either the unemployed men or their families in health*' (Fabian Society MSS., Box 20, 'Royal Commission on Unemployment Insurance, Memorandum by Mrs. Sidney Webb', 30 Nov. 1931, pp. 1–2).

competitive capacity in world markets, and could only result in wage-cuts or industrial ruin. Buxton pointed out, however, that many of the burdens, both financial and administrative, that the Act imposed on employers had been borne by their foremost continental rivals for the previous twenty years.[1]

To trade-union and labour representatives the Board of Trade was more conciliatory. On 15 August a deputation from the parliamentary Labour party proposed that the Government should pay half instead of a third of the sum contributed by workmen and employers; and they asked that the rebate to employers should be abandoned, that workmen should be allowed to insist upon 'standard' and not merely 'current' wages and that the rate of benefit should be the same in all insured trades.[2] Llewellyn Smith was not prepared to abandon the employer's rebate or to raise the State's contribution; and he insisted that disqualification from benefit on account of strikes and lockouts should apply to all workmen in a workshop where a dispute was in progress, whether or not they were antagonists in the dispute. He agreed, however, that insured workmen should in most cases be allowed to insist upon the standard rate,

provided that it be made clear that . . . where the workman through incompetence is incapable of earning that rate, the individual test shall prevail. It would I think be regarded as very unjust that an incompetent waster who had never earned and never expected to earn the standard rate, should nevertheless be able to make that rate an excuse for idleness while battening on the fund.[3]

Llewellyn Smith was, moreover, willing to concede the Labour party's demand that benefits should initially be equalized for all insured trades.[4] This had indeed been his original intention, and the lower scale of benefits for workmen in the building trade had been imposed on the scheme by the

[1] LAB 2/1483/LE. (1) 7150, Labour Exchanges Central Office, Report of the Proceedings at a meeting with a Deputation from the Shipbuilding Employers' Federation, 11 June 1911.
[2] LAB 2/1483/LE. (1) 9169, 'Unemployment Insurance; Report by W. H. Beveridge on a Conference with the Labour Party', 15 Aug. 1911.
[3] Ibid., H. Llewellyn Smith to J. Ramsay Macdonald, 13 Oct. 1911.
[4] Ibid., H. Llewellyn Smith to J. Ramsay Macdonald, 12 Oct. 1911.

Cabinet and Treasury.[1] This demand was supported by an emergency committee set up by the shipbuilding and engineering unions, which urged that unemployment benefit in all trades should be raised to a minimum of 7s. a week. Beveridge argued on their behalf that the 'power to vary rates and in particular to reduce a number of weeks benefit to zero really gives ample safeguard against bankruptcy'.[2] On 13 October he reported to Llewellyn Smith that the Chancellor had agreed to the equalization of benefits 'as far as the actuarial data allows'—which meant increasing the weekly levy on one or more contributors—but that Chalmers and Bradbury were 'adamant' against any increase of the contribution from the State.[3] At the end of October, however, Llewellyn Smith persuaded the Treasury that by relinquishing the Exchequer's share of the bonus refunded in respect of regular employers, it would be possible to make available an extra £66,000 a year.[4] The uniform benefit of 7s. a week for up to fifteen weeks of unemployment was introduced into the schedules of the Bill when it was discussed by a Grand Committee of the Commons early in November 1911.[5]

The piloting of Part Two through the Grand Committee was carefully stage-managed by the Board of Trade officials, who used the inarticulate Sydney Buxton to wrap up controversial items in 'decent obscurity' and the eloquent Solicitor-General to make popular clauses 'crystal clear'.[6] Consequently, Part Two of the Bill passed through the committee stage 'very easily with few amendments and quite unexpected rapidity, without any use of the closure'.[7] Most of the substantive changes made by the Grand Committee emanated

---

[1] Above, pp. 326.

[2] LAB 2/1483/LE. 11614, W. H. Beveridge to H. Llewellyn Smith, 25 Sept. 1911.

[3] Ibid., W. H. Beveridge to H. Llewellyn Smith, 13 Oct. 1911.

[4] Ibid., H. Llewellyn Smith to the Secretary of the Treasury, 16 Oct. 1911; Sir Robert Chalmers to H. Llewellyn Smith, 26 Oct. 1911. It was originally intended under this provision (Section 94 of the final draft of the Bill) that a proportion of both the employers' and the State's contribution should be remitted in respect of continuous employment.

[5] National Insurance Bill (as amended in Committee), 21 Nov. 1911, Seventh Schedule, p. 121.

[6] W. H. Beveridge, *Power and Influence*, p. 88.

[7] Passfield MSS., ii, 4, e, item 49, Clifford Sharp to B. Webb, 29 Nov. 1911.

from the Board of Trade;[1] and the Bill received its third reading and passed into law early in December 1911.[2] Apart from the simple insurance of workmen in scheduled trades Part Two of the Act contained five clauses providing for special arrangements between either workmen or employers and the Board of Trade. Section 94 allowed rebates to regular employers. Section 96 allowed refunds to employers who arranged for short-time working. Section 99 provided that where employers agreed to engage workmen solely through a labour exchange, their administrative responsibilities under both parts of the Act could be transferred to the exchange. Section 105 prescribed that benefits to organized workmen could be paid through their trade unions rather than through a labour exchange. And Section 106 implemented the 'Ghent' system by allowing the Board of Trade to refund up to one-sixth of out-of-work benefits paid by unions inside or outside the insured trades, providing that the rate of benefit did not exceed 12s. per workman per week.[3]

As soon as the Act was passed Board of Trade officials began to adapt the machinery of labour exchanges to the management of insurance. Responsibility for the central direction of unemployment insurance was given to the Labour Exchanges Central Office; and a new Labour Exchanges and Unemployment Insurance department, under an 'Assistant Secretary and Director' was established in 1913.[4] An extra 169 labour exchanges and 1,066 'local agencies' were

---

[1] These were as follows: It was laid down that 10 per cent of the income from unemployment insurance contributions should be set aside for the cost of administration. Insured workmen were to have no claim to benefit during the first six months of the scheme. The Umpire was empowered not merely to settle disputed claims but to determine which groups of workmen were insured under Part Two of the National Insurance Act; and this was to be decided with reference to the job of the workman rather than the trade or profession of his employer. Rebates of one-third were to be allowed to employers who had paid forty-five contributions in respect of a workman during the course of a year (National Insurance Bill, as amended in committee, 21 Nov. 1911).

[2] *Hansard*, 5th series, vol. 32, col. 1525.

[3] Sections 94 and 96 were both repealed in 1920 (F. Tillyard and F. N. Ball, op. cit., pp. 33–4).

[4] J. A. M. Caldwell, 'The Genesis of the Ministry of Labour', loc. cit., pp. 386, 391. Beveridge MSS., L.ii.218d, H. Llewellyn Smith to W. H. Beveridge, 8 May 1913. Beveridge and Llewellyn Smith attached great importance to the retention of the title 'director', presumably because of its business connotation.

established throughout the United Kingdom.[1] Eighty advisory panels or courts of referees, composed of employers nominated by the Board of Trade and workmen elected by ballot at local exchanges, were established to adjudicate in disputed cases of benefit.[2] Regulations were issued prescribing the procedure of applying for benefit and appealing against refusal; and employers who made an arrangement under Section 99 of the Act to transfer their administrative responsibility for insurance to a labour exchange were required to deposit three months' contributions with the Board of Trade.[3] Over two and a half million insurance cards were issued in respect of about two and a quarter million workmen in insured trades.[4] In spite of protests from Divisional Officers that the date should be deferred, the first contributions were paid into the insurance fund on 1 July 1912;[5] and the first unemployed workmen became eligible for benefit on 1 January 1913.[6]

UNEMPLOYMENT AND NATIONAL DEVELOPMENT

The idea that the State should promote useful but unprofitable public works and services had a long and ambiguous history in the discussion both of unemployment and of general

[1] Cd. 6965/1913, *First Report on the Proceedings of the Board of Trade under Part II of the National Insurance Act, 1911*, p. 42.

[2] H. of C. 527/1913, *Courts of Referees*, Return setting forth the Statutory Provisions relating to the Constitution of Courts of Referees etc., by H. Llewellyn Smith, 14 Feb. 1913, pp. 4–7. 1,145 employers' representatives had been appointed by the Board of Trade by 14 Feb. 1913. Voting for workmen's representatives took place at labour exchanges on 16 Nov. 1912. Insured workmen were divided into 'wards' in order to secure fair representation of different trades and localities. Of 1,499 places allotted to workmen, 295 were filled without a contest; for 92 no valid nomination was received; the rest were contested by 2,866 candidates. 174,669 insured workmen cast their votes. Of the successful candidates 36 per cent were in the building and construction industries, 49 per cent in shipbuilding, engineering, or ironfounding, 6 per cent in vehicle-construction, 9 per cent in other trades.

[3] LAB 2/1484/LE. 22633/2, Central Office Unemployment Insurance Regulations, signed by H. Llewellyn Smith, 6 May 1912.

[4] Cd. 6965/1913, *First Report of the Proceedings of the Board of Trade under Part II of the National Insurance Act, 1911*, p. iii. The difference between the number of books issued and the number of workmen insured at any given time arose from the fact that a large margin of workmen constantly moved in and out of insured trades.

[5] S. Tallents, *Man and Boy*, pp. 189–90.

[6] Cd. 6965/1913, para. 54.

economic policy. In theory public works as a means of increasing employment had been prohibited for most of the nineteenth century by the belief that public investment merely depressed wages and withdrew capital from private enterprise; but such theoretical constraints had often been thrust aside in times of commercial crisis.[1] Furthermore, it was recognized that there were certain kinds of development project which, though unprofitable to the individual, were necessary or desirable to the community as a whole; and since 1817 the central government had assisted local authorities to execute public works of social utility by granting long-term Exchequer loans.[2]

From the 1870s onwards local authorities had greatly increased their expenditure on public works;[3] and during the 1880s representatives of the unemployed began to demand that part at least of this expenditure should be arranged to coincide with depressions of trade.[4] During the 1890s the unorthodox economists, Hobson and Robertson, urged that additional public works should be financed out of progressive taxation, with the short-term aim of reducing unemployment and the long-term aim of increasing consumer demand;[5] and in 1895 the I.L.P. put forward a programme for the abolition of unemployment, which included redistribution of income, universal education, public control of industry, and the development by the Government of vacant agricultural land.[6] At the same time members of the business community began to press for increased 'state aid' to industry and commerce, in the form of public investment in national communications,

---

[1] Above, p. 3.

[2] M. W. Flinn, 'The Poor Employment Act of 1817', *Economic History Review*, 2nd series, 14, no. 1 (1961), 82–92. This Act established commissioners who were empowered to make loans of up to £1,750,000 during periods of three years, for assisting public works and reducing unemployment. The Act continued in force for twenty-five years. In 1842 the commissioners were converted into the Public Works Loans commissioners; the system of loans-authorization remained virtually unchanged, but the new body had no specific responsibility for promoting employment.

[3] B. R. Mitchell and P. Deane, *Abstract of British Historical Statistics*, pp. 373 and 416–17.

[4] Above, p. 73–5.

[5] J. M. Robertson, *The Fallacy of Saving* (1892), p. 138; J. A. Hobson, *The Problem of the Unemployed. An Inquiry and an Economic Policy*, pp. 99–103, 147–60.

[6] Tom Mann, *The Programme of the I.L.P. and the Unemployed*, Clarion Tract 6.

technical education, and scientific research.[1] And in the early 1900s a group of radical journalists, headed by Charles Masterman, proposed that the next Liberal government should finance works of national development, and that the expenditure not merely of local authorities but of private firms and of the central government should be scheduled to counteract fluctuations of trade.[2]

These arguments increasingly permeated the discussion of unemployment in both major political parties between 1886 and 1905. During the unemployment crisis of the 1880s both Liberal and Conservative backbenchers had called for counter-depressive public investment; and Salisbury and Balfour had tentatively endorsed the demand for public works.[3] The issue of the Chamberlain circular, though practically ineffectual, was a tacit acknowledgement of the principle that the expenditure of local authorities could be used to counteract unemployment caused by seasonal or cyclical depression of trade.[4] In 1893 Gladstone had vetoed the proposal that this principle should be extended to the expenditure of the central government;[5] but a plan for increasing employment by national investment in 'useful' and 'reproductive' works was secretly approved by several members of the Liberal shadow-cabinet during the winter of 1904–5.[6]

Nevertheless, the Liberal government which came to power at the end of 1905 had only a very vague commitment to a policy of public works; and the domestic programme of the Campbell-Bannerman administration during its first two years of office was one of financial retrenchment rather than social reform.[7] Between 1907 and 1909, however, a series of official inquiries endorsed the demand of radical backbenchers for additional public investment in scientific research and education, national communications, and 'reproductive' public works. In 1908 a departmental committee of the Board of Agriculture advised the Government to subsidize experimental farming and forestry in order to increase agri-

---

[1] Add. MS. 48661, f. 105, Sir Edward Hamilton's Diary, 28 Oct. 1893. Above, pp. 216–18.
[2] *Towards a Social Policy: or Suggestions for Constructive Reform*, pp. 68–71.
[3] Above, p. 74; *Hansard*, 3rd series, vol. 303, col. 672.
[4] Above, pp. 76.   [5] Above, p. 89.
[6] Above, pp. 223–4.   [7] Above, p. 233.

cultural employment and to maintain 'the physical standard of the race'.[1] In 1909 the Consultative Committee of the Board of Education suggested that unemployment could be reduced and 'economic efficiency' increased by additional public expenditure on technical education;[2] and in the same year a Royal Commission under Lord Shuttleworth proposed that the Government should promote domestic commerce by taking over and developing the national system of canals.[3]

The inquiry that examined most carefully the relationship between public investment and the absorption of surplus labour was the Royal Commission on Coast Erosion and Afforestation appointed under the chairmanship of Ivor Guest in 1906.[4] The terms of reference of this Commission were initially limited to the legal, technical, and administrative aspects of land reclamation.[5] In the autumn of 1907, however, a sub-committee under Rider Haggard inquired into the possibility 'of using the labour of unemployed persons upon such work of reclamation';[6] and in March 1908 the commissioners were asked to consider 'whether, in connection with reclaimed land or otherwise, it is desirable to make an experiment in afforestation as a means of increasing employment during periods of depression.'[5]

The Commission reported early in 1909 that for many reasons afforestation was peculiarly suitable for public invest-

---

[1] Cd. 4206/1908, *Departmental Committee on Agricultural Education in England and Wales, Report*, paras. 127–8.

[2] Cd. 4757/1909, *Board of Education. Report of the Consultative Committee on Attendance, Compulsory or Otherwise, at Continuation Schools*, pp. 176–7.

[3] Cd. 4979/1909, *RC on the Canals and Inland Navigations of the United Kingdom, Final Report*, p. 188. This commission did not consider the impact of canal construction on the demand for labour; but representatives of organized labour claimed canal building would help to provide work for the unemployed (*7th Quarterly Report of the Parliamentary Committee of the T.U.C.*, Dec. 1910, pp. 3, 11).

[4] Ivor Churchill Guest (1873–1939) was a cousin of Winston Churchill, with whom he crossed the floor of the Commons on the 'free food' question in 1906.

[5] Cd. 4460/1909, *RC on Coast Erosion and Afforestation, Second Report*, p. v.

[6] Cd. 4461/1909, *RC on Coast Erosion and Afforestation, Minutes of Evidence and Appendices thereto accompanying the Second Report*, Appendix No. XXXII, p. 6. The evidence received by this committee was highly contradictory. Salvation Army officials claimed that many of the unemployed would be almost as efficient as normal workmen (ibid., QQ. 29–30, 102, 178); whereas business-men who considered the question from a 'commercial stand-point' were convinced that the unemployed lacked the moral and physical stamina for arduous work (ibid., QQ. 322–9, 336–78).

ment that was designed to absorb unemployed workmen and to counteract cyclical and seasonal depression.[1] It was unattractive to private investors, because it required large-scale capital expenditure with no expectation of profit for periods of up to eighty years; and this initial expenditure could easily be adjusted within a ten-year margin to coincide with cyclical depression of trade. Moreover, forestry involved many different kinds of work, both skilled and unskilled, which could be adapted to the capacities of different unemployed workmen; and its busiest season coincided with the slack season in many other trades. The Commission admitted that the 'unemployed' were unlikely to be as efficient as experienced foresters hired through the normal labour market; but it claimed that there were two very different kinds of unemployed workmen. There were those who were permanently unfit or 'so morally deteriorated . . . so lacking in determination, as to be incapable of persisting in any useful employment'; and there were those who were merely temporarily inferior through under-nourishment and lack of physical exercise. The commissioners thought that the latter class 'with proper food and a certain period of training . . . can in time be rendered fit to perform manual labour of the character of tree-planting'. Moreover, the additional cost would be socially justified by the reduction in migration from the countryside, the prevention of moral and physical deterioration, and the saving of public expenditure on unproductive forms of relief. It was recommended, however, that separate accounts should be kept for ordinary labour and for unemployed workmen, so as not to confuse business with relief work; and, finally, the Commission concluded that, whether or not it gave additional employment, 'afforestation reacts so advantageously upon the social and economic conditions of the country, as to justify it being undertaken upon its own merits'.[2]

Public investment in land-reclamation and forestry was also recommended by the Minority Report of the Royal Commission on the Poor Laws. But the Webbs' view of the relationship between public works and the reduction of un-

[1] Cd. 4460/1909, *RC on Coast Erosion, etc., Second Report*, para. 80.
[2] Ibid., paras. 32–61.

employment was fundamentally different from that of the Guest Commission. They thought that it was 'vital . . . that there should be no attempt to employ the unemployed as such'; and that afforestation and reclamation should be undertaken 'not as Relief Works . . . but as public enterprises of national importance'.[1] Such works should, however, be concentrated into years when the demand for labour was low, so as to stabilize employment and to take advantage of cheap rates of borrowing. For these reasons, certain local authorities had already begun to reserve extraordinary public expenditure for periods of depression; but the Webbs claimed that the development of such a policy within the central government had been frustrated by the Treasury's rigid insistence on annual public accounting. They suggested that, if counter-depressive public works were planned over ten-year periods, employment would be not only stabilized but increased, since the employment of otherwise idle labour and capital would enable both workmen and investors to increase their personal consumption.[2]

By 1909 there was clearly a considerable body of expert opinion in favour of public investment in certain kinds of project, industrial, agricultural, and educational, which would assist and not compete with private enterprise. Moreover, it was believed that some at least of these projects could relieve unemployment, either directly by absorbing unemployed workmen or indirectly by raising the level of general labour demand. In the autumn of 1908 several Liberal ministers publicly recommended that government expenditure should be regulated according to the state of the labour market;[3] and on 26 December Winston Churchill informed Asquith that he had just read Ivor Guest's draft report on Afforestation, which was a 'first-class document' and would 'serve as an admirable basis for action'.[4] A few days later he wrote again to Asquith, urging that 'special expansive state industries', particularly forestry and road-building, should be included in a new comprehensive policy

---

[1] Cd. 4499/1909, RC on the Poor Laws, Minority Report, p. 1197.
[2] Ibid., p. 1198.   [3] Above, p. 274.
[4] Asquith MSS., vol. 11, f. 243, W. S. Churchill to H. H. Asquith, 26 Dec. 1908.

of social organization.[1] A similar programme was advocated by the parliamentary Labour party, which early in 1909 called for policies of afforestation, reclamation, road-construction, 'the development of neglected national resources', and 'the creation of a better-organised social state' through the 're-establishment of our rural and village population'.[2]

It was, however, the Chancellor of the Exchequer who in the spring of 1909 put forward a proposal for the creation of two special funds for public investment in national development. Churchill was initially inclined to disparage the idea of a 'development fund', possibly feeling that Lloyd George had stolen a march on him. It was contrary to orthodox financial precedent, Churchill protested, to create a fund in advance of the specific objects for which it was required; and he objected to the fact that the Treasury, as controller and arbiter of such a fund, would become in effect one of the 'spending departments'.[3]

Churchill's criticisms were disregarded, however, and Lloyd George's plans for development were outlined in his Budget speech on 29 April 1909. He stated that Britain was spending less on the development of 'national resources' than any other industrialized country; and suggested that the State should promote such development 'by instruction, by experiment, by organisation, by direction, and even, in certain cases which are outside the legitimate sphere of individual enterprise, by incurring direct responsibility.'[4] He therefore proposed to create two new central public authorities. Firstly, a board would be established to control traffic and to build motor-roads, which the advent of the motor-car had made 'part of the essential development of the country'.[5] And, secondly, the Chancellor planned to appoint a Development Commission, which would be responsible for the promotion of scientific research and education, experimental farming and schools of forestry, and the organization of co-operative marketing and rural transport. The Commission would also

[1] Asquith MSS., vol. 11, ff. 249–53, W. S. Churchill to H. H. Asquith, 29 Dec. 1908.
[2] *Report of the 9th Annual Conference of the Labour Party, Portsmouth*, 27–9 Jan. 1909, Appendix I on 'Unemployment', p. 93.
[3] R. Churchill, *Winston S. Churchill*, ii, Companion Part 2 (1907–11), pp. 885–6, W. S. Churchill to D. Lloyd George, Apr. 1909.
[4] *Hansard*, 5th series, vol. 4, col. 493.   [5] Ibid., cols. 495–8.

give assistance to land-reclamation, the creation of small-holdings and other measures for attracting labourers back to the land. It would take over the management of existing government grants for 'development' purposes, and would receive an additional income of £200,000 a year, plus any surplus revenue that accrued to the Exchequer.[1] This plan of national development would not, Lloyd George admitted, eliminate the fluctuations of the trade cycle;[2] but he claimed that 'every acre of land brought into cultivation . . . means more labour of a healthy and productive character . . . cheaper and better food for the people.'[3] He conceded also that not all the works authorized by the Commission would be immediately useful or profitable; but he argued that their value should be assessed in social and human as well as economic and financial terms.[4]

Lloyd George's plan for national economic development did not meet with the unanimous approval of his Cabinet colleagues. It increased the friction between himself and John Burns, who condemned the proposal to finance development out of the Sinking Fund and 'determinedly opposed' the Road Board 'on score of economy, efficiency and probity'.[5] Throughout the summer and autumn of 1909 he pressed for the control of both funds to be vested in his own department, which had the necessary supervisory staff and half a century's experience in administering public works,[6] but his

---

[1] *Hansard*, 5th series, vol. 4, cols. 493–5. Earlier statutes which empowered the central government to make grants or loans for national development were listed in H. of C. 278/1909, *Memorandum on Existing Powers as to Making Grants for Various Purposes*. These included grants for technical instruction, fisheries and agricultural research and loans for harbour construction. The Congested Districts Boards of Ireland and Scotland had power to acquire land and promote rural industries under the Purchase of Land (Ireland) Act, 1891, and the Congested Districts (Scotland) Act, 1897.

[2] *Hansard*, 5th series, vol. 4, col. 488.   [3] Ibid., col. 494.

[4] Ibid., col. 490: 'A State can and ought to take a longer view and a wider view of its investments than individuals. The resettlement of deserted and impoverished parts of its own territories may not bring to its coffers a direct return which would reimburse it fully for its expenditure: but . . . a State keeps many ledgers, not all in ink, and when we wish to judge of the advantage derived by a country from a costly experiment we must examine all those books before we venture to pronounce judgment.'

[5] Add. MS. 46327, Burns Diary, 9 May 1909 and 19 Aug. 1909.

[6] Ibid., 5 May 1909 and 8 Oct. 1909; H.L.G. 29/95, vol. 87, ff. 248–54, Memorandum on the Development and Road Improvement Bill, Part II, Road Improvement, para. 12.

protests were overridden by Asquith, Churchill, and Lloyd George.[1]

The development scheme was also strongly criticized in a memorandum by Edwin Montagu, the Prime Minister's parliamentary private secretary, on 14 May 1909. Montagu was a keen advocate of national development and was anxious to keep the Budget proposals to the forefront of the Chancellor's mind.[2] But he anticipated that the creation of a Development Commission might meet with opposition from other government departments, particularly from the Board of Agriculture which would be virtually stripped of all its functions if it lost control of smallholdings, afforestation, reclamation, and agricultural research. It would be possible, Montagu suggested, to entrust all the measures of development proposed by the Chancellor to existing government departments, in which case the Board of Agriculture would expand into a ministry of the first rank—a status not inappropriate to the political importance of the agricultural interest and the urgency of agricultural problems. He conceded, however, that a proposal merely to expand existing departments lacked 'the boldness of outline and the lurid attractiveness of the Chancellor of the Exchequer's scheme'. Moreover 'it also leaves no machinery for ensuring perspective and orderly distribution of means among desirable but conflicting ends, and this is of course a far more discomforting objection'. Montagu therefore proposed that a body should be set up whose responsibilities were primarily financial, and that 'this Board, Commission or Treasury Committee' should be represented in Parliament 'preferably by the Financial Secretary to the Treasury'. It should also have power to suggest development schemes and to determine priorities among the proposals of other departments; but the actual planning and management of such schemes should be left to the Board of Agriculture, the Local Government Board, and the Board of Trade.[3]

---

[1] Add. MS. 46327, Burns Diary, 9 May 1909 and 19 Aug. 1909.

[2] S. D. Waley, Edwin Montagu, *A Memoir and an Account of his Visits to India*, p. 33. Edwin Montagu (1879–1924), Liberal M.P. for Chesterton 1906–22; private secretary to Asquith 1906–10; Secretary of State for India 1917–22.

[3] Asquith MSS., vol. 22, f. 196, 'The Development Commission', by E. S. Montagu, 14 May 1909.

There is in fact no evidence to suggest that the Chancellor had any intention of creating a new executive department; and a memorandum by R. G. Hawtrey[1] in July 1909 made it clear that the Commission was to be little more than an advisory committee of the Treasury, responsible merely for making recommendations about the allotments of grants and loans. The Treasury would also nominate the Road Board, which was to administer the 'road improvement fund' derived from new taxes on motor-cars and petrol; and both authorities would be permitted to make provision for unemployed labour only in the case of work 'which has already been approved under the Act'.[2]

The Development and Road Improvement Funds Bill was introduced into Parliament by the Chancellor of the Exchequer on 26 August 1909.[3] The Conservatives took objection to the fact that Lloyd George had published a memorandum explaining the objects of the Bill in the *Westminster Gazette* before it was circulated to the House of Commons; but opposition to the Bill was largely stifled by the direct bounty given to agriculture, which was 'the apotheosis and final realisation of what the Unionist party . . . have been demanding for two generations'.[4] The support of the Nationalists had been won by Lloyd George's assurance that part of the funds would be spent in Ireland;[5] and the Bill was welcomed by Keir Hardie on behalf of the Labour party as the most 'revolutionary' measure ever introduced by a government and an implicit recognition of the 'right to work'.[6]

Criticism of the Bill on its second reading was therefore

[1] Ralph Hawtrey (1879–), Treasury official 1904–45. Hawtrey was at this time a private secretary to Lloyd George; he subsequently became Director of Financial Inquiries 1919–45. Author of many economic works, including *Good and Bad Trade* (1913).

[2] T. 170/4, Notes on the 1909 Finance Bill: Memorandum on Development and Road Improvement Funds Bill, July 1909. This memorandum also increased the proposed income of the Development Commission to a minimum of £500,000 p.a. for five years—the sum which was laid down by the Development Act.

[3] *Hansard*, 5th series, vol. 9, col. 2313.

[4] Ibid., vol. 10, col. 926.

[5] Ibid., cols. 944–7. Conservative critics cited this promise as evidence of corruption; but T. P. O'Connor claimed that it was merely an endorsement of a promise made by Arthur Balfour in 1888.

[6] *The Times*, 15 Sept. 1909.

confined to a small handful of 'diehard' Conservatives. Viscount Morpeth prophesied that the Bill would encourage 'corruption' among politicians who sought the favour of supposedly distressed constituencies;[1] and Henry Chaplin protested that Lloyd George had stolen the idea of national 'development' from a report of the Tariff Commission published two years before.[2] Lord Robert Cecil condemned the discretionary power given to the Treasury to promote development schemes by 'administrative orders'; and he claimed that many of the projects envisaged by the Bill would compete with private enterprise and therefore be tantamount to 'state trading'.[3] When the establishment of the Development Fund was put to a vote, however, only six M.P.s opposed the abandonment of a theoretical principle that had governed orthodox financial policy for nearly a hundred years.[4]

The Bill was further attacked at the Committee stage on the ground that the funds should be administered, not by commissioners nominated by the Treasury, but by the L.G.B., the Board of Agriculture, or the Board of Trade.[5] Lloyd George refused, however, to relinquish direct control of the new funds; and he argued that, far from turning the Treasury into a 'spending department', the Development Commission would assist the Treasury to distinguish and discriminate against unreasonable financial demands. He denied, moreover, that the Bill was a concession to the 'right to work' and insisted that its primary purpose was to promote economic development and only incidentally to relieve the unemployed.[6]

The Development Act became law early in December 1909. The Development Commission established under Part One of the Act was empowered to give financial assistance to agriculture and 'rural industries'; to forestry, land-reclamation, and road transport; and to the 'development and

---

[1] *Hansard*, 5th series, vol. 10, cols. 919–21.

[2] Ibid., col. 936.   [3] Ibid., col. 911.

[4] Ibid., cols. 1063–4. Twenty years later Lloyd George's advisors claimed that the Conservatives had 'strenuously opposed' the Development Act; but this was clearly incorrect (*We Can Conquer Unemployment* (1929), p. 11).

[5] *The Times*, 16, 17, 24 Sept. 1909.

[6] *The Times*, 17 Sept. 1909. An amendment to the effect that consideration of the level of employment should be entirely excluded from development schemes was, however, overruled by 22 votes to 15 (*The Times*, 1 Oct. 1909).

improvement' of fisheries and the 'construction and improvement' of harbours and inland navigations. It was also endowed with powers of compulsory purchase, limited by the proviso that an 'undue or inconvenient quantity of land' should not be taken from any one person, and that the commissioners should 'so far as practicable avoid displacing any considerable number of agricultural labourers or others employed on or about the land'. The Road Board established under Part Two was empowered to construct and maintain new roads and to give financial assistance to local highway authorities. In both cases the allocation of each grant or loan was to be separately approved by the Treasury; and under Section 18 of the Act both the Development Commission and the Road Board were obliged when considering schemes of work to have regard 'so far as is reasonably practicable to the general state and prospects of employment'.

The Development Commission was not established for some months after the passage of the Act, largely because Lloyd George was undecided about the final form of the Commission and who should be its permanent head.[1] At first he hoped that some kind of informal 'arrangement' would suffice, but Sir George Murray pointed out that this would not be binding on other ministers and future Chancellors;[2] and Lloyd George then apparently came temporarily round to the view that the development fund should be administered by the Board of Agriculture and not by a specially constituted Development Commission.[3] Eight commissioners were, however, eventually nominated in May 1910.[4] Two paid commissioners, Lord Richard Cavendish and Sir Francis

[1] The post of 'vice-chairman' was initially offered to Sir Robert Morant (Runciman MSS., Sir Robert Morant to W. Runciman, n.d., c. May 1910). B. M. Allen, *Sir Robert Morant. A Great Public Servant*, pp. 248–9, states that Morant rejected the post because it offered fewer 'opportunities for good work' than his existing post as Permanent Secretary to the Board of Education. But Morant's letter to Runciman suggests that Morant accepted the new post and actually held it for several days, and that Lloyd George subsequently changed his mind. The position was probably then offered to and refused by Sir Thomas Elliott, the Permanent Secretary of the Board of Agriculture (Runciman MSS., Sir Robert Morant to W. Runciman, pencilled postcard, n.d., c. May 1910).

[2] Runciman MSS., Sir Robert Morant to W. Runciman, n.d., c. Feb. 1910.

[3] Ibid., Sir Robert Morant to W. Runciman, n.d., c. May 1910.

[4] *1st Report of the Development Commissioners*, 1910–11, pp. 1–5.

Hopwood, were appointed as chairman and vice-chairman;[1] and the unpaid commissioners included Daniel Hall, the Director of the agricultural research station at Rothamstead, and Sidney Webb. At the same time a Road Board, together with an advisory committee of consulting engineers and surveyors, was established under the chairmanship of an ex-general manager of the North Eastern Railway, Sir George Gibb.[2] 'Development' and 'road improvement' accounts were opened with the Bank of England; and the two authorities began to consider applications for grants and loans in July 1910.

Both Lloyd George and Churchill originally saw these three measures—the Labour Exchanges Act, the National Insurance Act, and the Development Act—as part of a much wider programme of social reform, which would ultimately revolutionize the relationship between the worker and the State. Lloyd George at least in private admitted that he saw insurance merely as a transitional measure, and hinted at the recognition of some kind of 'right to work'. In March 1911 he told Ralph Hawtrey that ultimately the

State will acknowledge a full responsibility in the matter of making provision for sickness, breakdown and unemployment: it really does so now through the Poor Law: but conditions under which this system had hitherto worked have been so harsh and humiliating that working-class pride revolts at accepting so degrading and doubtful a boon. Gradually the obligation of the state to find labour or sustenance will be realised and honourably interpreted.[3]

Churchill at one stage envisaged that all the operations of the labour market might be brought under centralized government control; and in June 1909 he unfolded to Lloyd George a plan whereby the information services of the Board of

---

[1] Lord Richard Cavendish (1871–1946), nephew of the 9th Duke of Devonshire; Liberal Unionist M.P. for Lonsdale 1895–1906.

Sir Francis Hopwood (1860–1947), Permanent Secretary to the Board of Trade 1901–7; Permanent Under-Secretary of State for the Colonies 1907–10; Civil Lord of the Admiralty 1912; first Lord Southborough 1917. Hopwood remained with the Commission for only a year; in 1912 his post was taken over by Vaughan Nash, another of Asquith's private secretaries, who held it until 1929.

[2] *1st Report of the Proceedings of the Road Board*, 1910–11, pp. 1–7.

[3] D. Lloyd George to R. G. Hawtrey, 7 Mar. 1911; printed in *The Window*, July 1962, p. 76.

Trade would be developed on the lines of military intelligence until the timing and geographical location of unemployment could be accurately known in advance. The public works schemes of the Development Commission, the armed forces, and all local authorities could then be co-ordinated into a national programme of counter-depression, with special provision for areas that were abnormally distressed. The whole process of predicting and preventing unemployment and poverty should, Churchill suggested, be directed by a permanent 'social policy' committee, chaired by the Chancellor of the Exchequer and modelled on the Committee of Imperial Defence.[1] None of these plans ever came to fruition, however, during the period under discussion; and after 1911 the whole question of preventing and forestalling unemployment was temporarily eclipsed by other social and industrial problems and by the booming conditions of trade.

[1] R. Churchill, *W. S. Churchill*, ii, Companion Part 2 (1907–11), pp. 895–8, W. S. Churchill to D. Lloyd George, 20 June 1909.

VII

SUMMARY AND AFTERMATH

NEARLY thirty years' experience in the relief of unemployment had generated many conflicting explanations and many contradictory remedies for the recurrent problem of the unemployed. By the end of the period under discussion, however, a majority of social reformers and public administrators, although differing on specific remedial policies, were agreed on two main points. Firstly, that whether loss of employment was caused by personal deficiencies or by an adverse industrial environment, its effects were socially destructive, economically inefficient, and possibly even politically dangerous; and the widespread fatalism that had characterized administrative attitudes to unemployment in the late-nineteenth century had therefore been replaced by a conviction that unemployment could to a certain extent be prevented and ought to be relieved. Secondly, the results of numerous experiments in the relief of unemployment had shown that voluntary and local efforts were quite inadequate to deal with a problem that fell so erratically and often unpredictably on different industries, different regions, and different types of workmen. By 1908–9 it was therefore widely accepted by politicians and administrators that, whatever kind of remedial policy was adopted, it should be national rather than local in its organization and aims. In this chapter it is proposed to examine briefly three interrelated questions arising out of the Liberal government's programme for the treatment of the unemployed. Firstly, how far were the Liberals successful in realising Churchill's vision of a 'scientific' remedy for unemployment? Secondly, what were the main ideas and principles that governed the introduction of a centralized national unemployment policy? And, thirdly, what other remedies were available to social reformers and how far were they plausible alternatives within the political and intellectual context of the day?

A CRITIQUE OF LIBERAL POLICIES

The answer to the first of these questions—how successful were the Liberals in devising a 'scientific' policy for un-employment—must largely depend on what is meant in this context by the term 'scientific'. It would clearly be inappro-priate to indict the Liberals before 1914 for failing to imple-ment policies that were only made possible by later develop-ments in fiscal and monetary theory. It is, however, legitimate to consider, firstly, to what extent Liberal unemployment policy dealt with the different aspects of the problem revealed by contemporary expert analysis; secondly, whether it was a consistent and co-ordinated policy, or merely a series of *ad hoc* measures designed to stave off a crisis situation; and, thirdly, how far this policy was successful, within the limits imposed by the policy-makers, in its aims of streamlining the labour market, reducing unemployment, and relieving dis-tress among the unemployed.

Firstly, it is clear that Liberal policy dealt with only an iceberg-tip of the unemployment problem as defined in the previous twenty years by Booth, Llewellyn Smith, Beveridge, and the Webbs. The so-called 'unemployable', who had supposedly sabotaged earlier attempts to relieve unemploy-ment, were left completely untouched by any legislation of this period.[1] The treatment of the casual-labour problem, which was regarded by many reformers as the most uneco-nomic and socially costly form of irregular employment, con-sisted of little more than a footnote to the legislation for labour exchanges and national insurance; and, far from re-lieving the most distressed class of workmen, the unemploy-ment-insurance scheme referred to a group of predominantly skilled and organized workmen of the kind most able to make provision for themselves when unemployed.[2] Moreover, very

[1] As the Webbs subsequently pointed out, the term 'unemployable' was rendered virtually meaningless by the absorption of the surplus adult labour force into war production in 1915–16 (S. and B. Webb, *English Poor Law History*, II. ii. 668).

[2] 63 per cent of the insured were skilled workmen in 1912–13 (Cd. 6965/1913, *First Report of the Proceedings of the Board of Trade under Part II of the National Insurance Act, 1911*, para. 147).

In the summer of 1911 Llewellyn Smith and Beveridge seriously considered the possibility of including dock labourers in the unemployment insurance scheme on rather less advantageous terms than other insured workmen. This proposal received

little was done to deal with the problem of 'juvenile labour' which the Royal Commission on the Poor Laws had identified as one of the chief causes of unemployment among workmen in later life. In 1910 the Education (Choice of Employment) Act empowered local authorities to provide vocational guidance for school-leavers, and in the following year special grants were made available for this purpose by the Board of Education.[1] But only a minority of local authorities took advantage of these measures,[2] which were mainly inspired by the desire of Board of Education officials to wrest control of the juvenile labour market from the Board of Trade.[3] Between 1911 and 1913 a committee under Sir Mathew Nathan secured the partial abolition of 'boy labour' in the General Post Office;[4] but there was no attempt on the part of the Government to implement the demand of educational reformers that the school-leaving age should be raised to 'fourteen plus' and that further technical instruction should be made compulsory up to the age of seventeen.[5] Finally, no provision was made for adult workmen who were technologically redundant or deficient in industrial skill, since the technical retraining schemes authorized by Section 100 of the National Insurance Act were not brought into operation until 1925.[6]

Treasury approval; and it is not clear why it was abandoned by the Board of Trade (LAB 2/1483/LE. 9203, Draft letter to the Treasury, written by Beveridge, corrected and signed by H. Llewellyn Smith, 20 July 1911; Sir Robert Chalmers to H. Llewellyn Smith, 26 July 1911).

[1] Board of Education Circular 782, 'Exercise of Powers under the Education (Choice of Employment) Act, 1910', 17 Aug. 1911.

[2] By June 1912 only 3 counties, 26 county boroughs, and parts of 12 other county areas had started 'choice of employment' schemes (Cd. 6707/1913. Report of the Board of Education, 1911–12, pp. 146–7).

[3] This controversy arose partly from Morant's conviction that the treatment of the juvenile labour problem was logically a function of the Board of Education, and partly from Beveridge's conviction that, since the problem had in practice been neglected by local education authorities, it should be dealt with by the Board of Trade. Papers relating to the dispute and its resolution are preserved in Ed. 24/246, 247, 248 and 249; and in Beveridge MSS. D. 046. See also Stephen Tallents, Man and Boy, pp. 186–7.

[4] Cd. 7556/1914, Standing Committee on Boy Labour in the Post Office, Fourth Report, pp. 3, 15.

[5] e.g. Cd. 4757/1909, Board of Education Report of the Consultative Committee on Attendance, Compulsory or Otherwise, at Continuation Schools, pp. 233–6.

[6] E. M. Burns, British Unemployment Programs, 1920–38, p. 75. These training centres never catered for more than a tiny fraction of the numbers unemployed (ibid., pp. 77, 373).

The policy of the Liberal government was in fact mainly concerned with only three aspects of the unemployment problem. It was designed through the Labour Exchanges Act to reduce frictional unemployment and to facilitate the normal movement of labour; through the National Insurance Act to relieve unemployment in certain industries caused by temporary fluctuations; and through the Development Act to ward off the incidence of cyclical depressions of trade.

Secondly, how far can Liberal unemployment policy be , seen as part of a consistent plan of social reform? Both Lloyd George and Churchill hinted that these measures were merely the first instalments of a much larger programme for dealing with unemployment, which would include reform of the Poor Law and local taxation and the long-term regulation of central and local expenditure on public works. But although they were originally conceived as part of a grand design, the planning and execution of these measures were remarkable for lack of discussion and direction at a Cabinet level. The principles and administrative structure of labour exchanges, unemployment insurance, and the Development Commission were worked out by departmental ministers in conjunction with their permanent officials; and the four Cabinet committees appointed to consider unemployment policy in October 1908, April 1909, and April 1911 confined themselves almost exclusively to criticizing points of administrative and financial detail. There is very little evidence to suggest that—at a Cabinet level—the general principles involved in labour organization, insurance against unemployment, and public expenditure on works of national development were ever seriously discussed.

Between the different departments concerned with social administration, moreover, there were many differences of opinion and often open hostility on important policy issues. These conflicts arose partly from the territorial ambitions of ministers and their permanent officials, partly from a genuine conflict of principle between the treatment of different social problems or different methods of treating the same social problem. They were usually resolved, however, not by high-level discussion of political priorities but by demonstrations of departmental strength. Asquith's administration has become

notorious for the ignorance that prevailed among ministers
about the policies of their colleagues and for the competition
that raged among departments whose administrative functions were supposedly complementary;[1] and these characteristics were just as evident in social and industrial policy as in
foreign affairs and national defence. In the sphere of industrial
organization, supporters of the Minority Report of the Poor
Law Commission proposed that this excessive departmental-
ism might be avoided by the creation of a Ministry of Labour,
which would take over the 'labour' functions of existing
ministries and also be responsible for predicting and forestalling the onset of trade depressions. A series of private
members' bills to this effect, however, failed to reach a second
reading in the Commons;[2] hence the almost total absence of
a consistent national labour policy which became apparent at
the outbreak of war in 1914.[3]

Finally, how effective were the Liberal government's
measures in streamlining the normal market for labour and
in reducing unemployment and relieving the unemployed?
The answer to these questions is to a certain extent conjectural, because the First World War and the subsequent era of
depression fundamentally altered the balance of prosperity
among different sections of the British economy soon after
legislation dealing with unemployment was first introduced;
and the slump of 1929–32 generated unemployment on
a scale and of a duration not anticipated by even the gloomiest
predictions of the pre-war period.

Even before the war, however, it was clear that labour
exchanges had failed to make the decisive impact on the labour
market that had been envisaged by Beveridge and Churchill
in 1909. Instead of the specially constructed multi-purpose
offices located at focal points of commerce and industry that
had been recommended by the departmental committee of
1909,[4] the Treasury and Office of Works insisted that labour
exchanges should mainly be hired and converted from exist-

---

[1] R. B. Haldane, *An Autobiography*, pp. 216–18, 225–9.

[2] H. of C. Bill 37/1913, a Bill to establish a Minister of Labour to make provision
for the prevention and treatment of unemployment; introduced by Robert Harcourt
and Leo Chiozza Money, 14 Mar. 1913.

[3] Humbert Wolfe, *Labour Supply and Regulation*, pp. 7–12.

[4] Above, p. 289.

ing premises; hence they were often dismal and inconvenient buildings, situated 'in the slummiest parts of the town'.[1] The experiment also largely failed in its aim of creating an entirely new breed of civil servants, who would bridge the gulf between Whitehall and the shop floor and constitute a 'new model army of vigilant administrators, supplanting property by organisation'.[2] The unconventional method of making labour-exchange appointments, and the preference given to men with business rather than bureaucratic experience, was strongly criticized in Parliament and by witnesses before the Royal Commission on the Civil Service as savouring of 'jobbery' and political patronage[3] and tending 'to lower public confidence in the purity of administration';[4] and from 1911 onwards senior labour-exchange officials, both local and central, were mainly recruited from inside the permanent Civil Service.[5] After 1912, when labour-exchange officials became responsible for the execution of Part Two of the National Insurance Act, their original function of matching the needs of the labour market to the needs of the individual workman became increasingly submerged beneath a formal clerical routine.[6]

The greatest shortcoming of the labour-exchange service,

[1] J. B. Adams, 'Reminiscences of a Divisional Controller', *Minlabour*, 14, no. 1 (Jan. 1960), 5.

[2] E. T. (ed.), *Keeling Letters and Reminiscences*, pp. 59–60.

[3] *Hansard*, 5th series, vol. 25, col. 831; Cd. 6535/1912–13, *RC on the Civil Service, Minutes of Evidence*, QQ. 2536–54.

[4] Cd. 6534/1912–13, *RC on the Civil Service, Second Report*, Appendix U, pp. 484–5.

[5] Beveridge MSS., L.i. 209, W. H. Beveridge to Jeannette Tawney, 16 Aug. 1912. Beveridge was highly critical of the type of official appointed under the orthodox system: 'What a muddle through it has all been and will be again! It needn't have been, if we had had the proper scales of salaries, to get proper staff (£700–900 in place of £500–750 for our chiefs of section). Of course *outside* the Civil Service one could have as good a chance of getting the right man for the £500 as for £700, but inside (and we've been limited to choosing from inside) salaries are everything, and also, everyone tends to look to security of future promotion. We can't give that because we haven't the higher posts.'

[6] The Royal Commission on Unemployment Insurance in 1931–2 found that between 40 per cent and 60 per cent of labour-exchange work was performed by unestablished officials, temporarily engaged during periods of crisis. The Commission strongly recommended that the labour-exchange service be placed on a more permanent basis and that specialist officials be employed on the different functions of insurance and organization (Cmd. 4185/1932, *Royal Commission on Unemployment Insurance*, Final Report, paras. 594–6).

however, was its limited success in winning and keeping the confidence of trade unionists and employers of labour. It has been shown that Churchill went to considerable lengths to conciliate these groups, using often incompatible arguments to convince both sides of industry of the advantages of an organized labour market.[1] Nevertheless, the new service was bitterly attacked by representatives of both skilled and un-skilled workmen at the Sheffield Conference of the T.U.C. in 1910. It was asserted that exchanges had greatly increased black-legging and that labour-exchange officials had been recruiting women and children from workhouses in order to boost the number of registrations and to undermine the standard rate. Ben Tillet strongly criticized the Government for not setting up joint advisory committees until eight months after the opening of the first exchanges. This was, he claimed, 'the deadliest attack on unskilled and manual labour that the most malicious body of employers could have conceived' and he prophesied that henceforward the unions would have to 'fight the police, the employer and the Government at the same time'.[2] A resolution condemning the management of labour exchanges and instructing the Parliamentary Committee to insist on the observance of trade-union conditions was passed by a majority of 1,147,000 votes to 272,000.[3] This kind of antagonism underlined Churchill's argument that some kind of incentive was necessary to persuade the better class of workmen to register at an exchange; and it was not until the payment of benefits came into operation in 1913 that the prejudice of organized workmen, at least in the insured industries, was to a certain extent broken down.[4] It proved even more difficult to attract the custom of employers of labour, many of whom were not so much hostile as indifferent to the facilities afforded by an exchange.[5] When the exchanges first opened, officials spent much of their time

---

[1] Above, pp. 290–2.
[2] *Report of the Sheffield T.U.C.*, 1910, pp. 160–2.　　　　　[3] Ibid., p. 164.
[4] The registration of members of insured trades increased from 642,547 in 1912 to 1,433,700 in 1913 (Beveridge MSS., D. 030, 'Labour Exchanges and Unemployment Insurance. Report of the Proceedings of the Board of Trade . . . to July 1915', p. 40, Table VII).
[5] Sir Frank Tillyard and F. N. Ball, *Unemployment Insurance in Great Britain 1911–48*, p. 33.

touring local factories and canvassing reluctant employers;[1] but this kind of publicity necessarily came to an end when the pressure of work increased.[2] The majority of employers continued to recruit their labour through their own private exchanges or directly at the factory gate;[3] and it was estimated that only a third of vacant situations in insured trades were filled by the Board of Trade's labour exchanges by 1913.[4]

The effectiveness of labour exchanges in actually reducing frictional unemployment is difficult to calculate, since although they were filling 3,000 vacancies a day in 1914 it is possible that these vacancies might have been filled just as adequately by direct application. Moreover, for every applicant who found a job between 1910 and 1914, three were turned away.[5] As Llewellyn Smith had anticipated in 1895,[6] exchanges were most conspicuously successful in periods and areas of low unemployment, least successful in periods and areas most affected by depression of trade.[7] In 1920 Beveridge argued in defence of labour exchanges that they had never been given an opportunity to realize their true potential for minimizing unemployment, because they had always been encumbered with secondary responsibilities, such as the administration of insurance or post-war demobilization.[8] But this was an unconvincing protest, since both Beveridge and Churchill had claimed that the registration of vacancies was not merely an end in itself but a prerequisite of other forms of social organization and other methods of dealing with the unemployed; and labour exchanges were in fact made re-

---

1 'News from all Quarters 1910', *Minlabour*, 14, No. 1 (Jan. 1960), 9, Note by D. W. Richards.

2 Beveridge MSS., D. 030, 'Labour Exchanges and Unemployment Insurance. Report of the Proceedings of the Board of Trade . . . to July 1914', para. 231.

3 J. B. Seymour, *The British Employment Exchange* (1928), p. 99, ascribed this partly to the fact that 'the worse class of employers' were reluctant to publicize their low rates of pay.

4 Beveridge MSS., D. 030, 'Labour Exchanges and Unemployment Insurance. Report of the Proceedings of the Board of Trade . . . to July 1914', para. 294.

5 Ibid., p. 37, Table V.

6 H. of C. 365/1895, *SC on Distress from Want of Employment, Minutes of Evidence*, QQ. 4863–72.

7 Beveridge MSS., D. 030, 'Labour Exchanges and Unemployment Insurance. Report of the Proceedings of the Board of Trade . . . to July 1914', para. 301.

8 Evidence to the Labour Exchanges Committee of Enquiry, 1920, quoted in Beveridge, *Unemployment* (1930 ed.), p. 305, fn. 1.

sponsible for very few of the subsidiary administrative functions which had originally been claimed for them. They never became either the new kind of public service envisaged by Beveridge or the golden road to social organization envisaged by Winston Churchill in 1909;[1] and in the 1920s and 1930s this institutional embodiment of the idea of 'national efficiency' became a symbol of national despair and indifference to several millions of the unemployed.[2]

The other items of Liberal legislation were even less successful in preventing unemployment or reducing the number of unemployed. It has been shown that Part Two of the National Insurance Act was designed not merely to give unemployment benefits, but to penalize casual employment, to weed out chronically irregular workmen, and to give incentives to employers to engage workmen for long periods 6r to hire all their labour through an exchange.[3] These clauses of the Act proved almost entirely abortive in most areas, however, for three reasons. Firstly, because of the resistance of employers and workmen to schemes of decasualization.[4] Secondly, because of the rigorous control exercised by the Treasury over special arrangements with employers under Section 99.[5] And, thirdly, because successive amendments to the National Insurance Act in the 1920s relaxed the limitations on a workman's claim to benefit and therefore positively

[1] W. H. Beveridge, *Power and Influence*, p. 33; *Hansard*, 5th series, vol. 1, col. 194.

[2] e.g. Max Cohen, *I was One of the Unemployed* (1945), p. 5.

[3] Above, p. 333.

[4] Beveridge, *Unemployment* (1930 ed.), pp. 313–14. Decasualization schemes were started among dock labourers in Goole and Liverpool and among ship repairers in South Wales in 1912–13; but only the Liverpool scheme survived until the 1930s. For a detailed discussion of the arrangements made with employers in Liverpool, see R. Williams, *The First Year's Working of the Liverpool Docks Scheme*, 28 Nov. 1913.

[5] Under Section 99 the Board of Trade was empowered to make such arrangements with employers or groups of employers to transfer their administrative duties under Parts I and II of the Act to an exchange. Employers who entered into such an arrangement were not required to pay a full week's contribution in respect of each separate casual engagement—the aim of the section being to induce employers to hire casual labourers only through an exchange. The Treasury refused, however, to give the Board of Trade financial discretion to make such arrangements except with firms where at least 80 per cent of the employees were covered by Part II of the National Insurance Act; with the result that many Section 99 arrangements had to be abandoned (LAB 2/1483/LE. 2211/30, T. L, Heath to H. Llewellyn Smith, 18 Dec. 1912; and list of 'Terminated Arrangements under Section 99').

subsidized irregular employment by allowing casual workmen to draw indefinitely upon the insurance funds.[1] Moreover, in 1913 it was discovered that casual labourers who failed to obtain employment on the first day of a week were being forced to stamp their own cards with a whole week's contributions under Part One of the Act in order to persuade employers to employ them on subsequent days.[2] To remedy this situation the National Health Insurance Commission introduced an amendment to Part One of the National Insurance Act which permitted employers and workmen to pay daily rather than weekly contributions in respect of casual engagements.[3] The aim of this amendment was to prevent 'physical and moral deterioration' among 'underemployed casual labour', which was thought to be 'a very fruitful source of disease';[4] but it was strongly criticized by officials of the Board of Trade, who pointed out that any alleviation of personal hardship in the casual-labour force merely hindered the process of decasualization by subsidizing inefficient workmen instead of driving them out of the market.[5]

From the point of view of prevention, however, the most conspicuous failure of Liberal unemployment policy was the work of the Development Commission. The Development Act of 1909 had recognized a principle that had been advanced by the Minority Report of the Poor Law Commission and by supporters of the Labour Party's Right to Work Bill —the principle that public expenditure could be used not

[1] Beveridge MSS., L. ii. 218, W. H. Beveridge to W. S. Churchill, 5 Feb. 1930 (typescript copy). Churchill was inclined to blame the Lloyd George peace-time coalition of 1918–22 for the 'degradation of the scheme'; but Beveridge blamed 'the bad advice of the rather stupid Blanesborough Committee' of 1927, which 'made the insurance benefit unlimited in time and finally divorced the claim to benefit from payment of contributions'.

[2] LAB 2/1484/LE. 23987/5, Memorandum by E. F. Wise of the National Health Insurance Commission on 'A Suggested Casual Labour Amendment to the Act', 22 Apr. 1913, para. 1 (1).

[3] 3 & 4 Geo. V, c. 37, an Act to amend Parts I and III of the National Insurance Act, Section 19.

[4] LAB 2/1484/LE. 23987/5, 'A Suggested Casual Labour Amendment to the Act', para. 1 (4).

[5] LAB 2/1484/LE. 23987/27, 'Notes on the 2d Daily Stamp Scheme', by W. H. Beveridge, 16 June 1913; LAB 2/1484/LE 23987/28, W. H. Beveridge to Herbert Samuel, 18 June 1913.

merely to create specific employment but to regulate the level
of labour demand.[1] This principle was refined by the commissioners, who claimed that

it would not be proper that a Fund intended for 'economic development' should be used to employ the class of man who is generally known
as 'unemployed'. They have in view . . . the provision of work for good
and skilled workmen . . . not the quasi-charitable employment of men
not economically worth their wages.[2]

They therefore advised local authorities and other bodies
responsible for development schemes merely to concentrate
the work of skilled workmen into periods when trade was
slack, in the 'hope that any increase of skilled employment . . .
will react beneficially on the market for unskilled labour'.[3]

Nevertheless, the Commission largely failed in its task of
promoting schemes of economic development, either with
or without the intention of employing the unemployed. In the
period before 1914 this was partly because the commissioners
themselves thought that counter-cyclical expenditure was
unnecessary in the booming state of trade.[4] But the main
reason was that no adequate machinery had been established
for the planning and execution of the schemes envisaged by
the Development Act. In their first report the commissioners
emphasized that their role was purely advisory, and that they
were empowered merely to consider requests for financial
assistance from other administrative bodies.[5] They discovered, however, that it was often difficult to find suitable
authorities to carry out schemes of development, and that
many local authorities were reluctant to initiate large capital
projects with no immediate prospect of profit, particularly
since assistance from the Development Fund was initially

---

[1] Above, pp. 236–7, 339. The Development Act also prescribed many of the
schemes for counter-depressive public expenditure that were subsequently included
in Lloyd George's famous 'Yellow Book' programme of 1928–9 (*We Can Conquer
Unemployment* (Mar. 1929), pp. 9–24, 40–3, 47–9).

[2] T. 171/87, Budget Papers 1914, Copy of Memorandum by the Development
Commissioners to the L.G.B., n.d.

[3] *2nd Report of the Development Commission* (1911–12), p. 27.

[4] T. 171/87, Budget Papers 1914, H. E. Dale (Development Commissioner) to
H. P. Hamilton, 8 May 1914, enclosing copy of memorandum by Development
Commissioners to the L.G.B.

[5] *1st Report of the Development Commission* (1910–11), p. 6.

guaranteed for a period of only five years.[1] Largely for this reason the Commission spent less than 5 per cent of its income of over £12,000,000 between 1910 and 1915; and even in the 1920s its expenditure only once exceeded a third of its income in any one year.[2]

It was almost certainly these administrative factors, rather than residual prejudice against the creation or regulation of public employment, that frustrated the policy of 'national development' between 1909 and 1914. By 1914 politicians of all parties were alarmed by the escalation of public expenditure on defence, education, and social administration during the previous twenty years;[3] but they were not opposed to the regulation of such expenditure in order to provide work for the unemployed. This was apparent in March 1914 when Edmund Harvey introduced a resolution in the House of Commons proposing that the extension of unemployment insurance under the National Insurance Amendment Act should be reinforced by national and municipal regulation of the demand for labour.[4] This resolution was passed without criticism by the House of Commons, and during the debate on the resolution Herbert Samuel announced that the Treasury was appointing a committee to inquire into the possibility of eliminating depressions by expenditure on public works.[5] This committee was established under the chairmanship of Percy Alden in May 1914, but its proceedings were swal-

[1] 4th Report of the Development Commission, p. 4. It was found that local authorities most willing to undertake development projects were usually those already burdened with heavy rates (ibid., p. 1). The Commission nevertheless specifically rejected the idea that the Fund should be used to relieve local rates (3rd Report of the Development Commission, pp. 3–4).

[2] Appendix B, Table 10, p. 381. The difficulty experienced by the Road Fund in this respect was less severe, because road-construction—unlike afforestation, reclamation, etc.—was one of the traditional functions of local authorities. In the spring of 1914 the Board urged local highway authorities to prepare schemes in readiness for periods of depression (4th Annual Report of the Proceedings of the Road Board (1913–14), p. 12); with the result that over £200,000 was dispensed to distressed areas for road-construction during the brief period of high unemployment at the outbreak of war (5th Annual Report of the Proceedings of the Road Board (1914–15), p. 7). Even so, less than a quarter of the income of the Road Board during the ten years of its existence was spent on the construction of roads (Appendix B, Table 11, p. 382).

[3] Hansard, 5th series, vol. 52, cols. 264–8, 295–6; vol. 60, cols. 807–75.

[4] Ibid., vol. 60, cols. 1127–37, 1159.

[5] Ibid., col. 1154.

lowed up by the outbreak of war and there is no evidence to suggest that it ever produced a report.[1]

In the short term the most successful item of Liberal unemployment policy was the payment of benefits to unemployed workmen under Part Two of the National Insurance Act. Over two and a half million unemployment insurance books were issued during the first year of the scheme, and it was estimated that 2,325,598 of these books were current in July 1914.[2] Of the workmen insured in 1913, 63 per cent were skilled workmen;[3] but only one-fifth had previously been covered by a private out-of-work insurance scheme.[4] The introduction of the scheme coincided with a peak of trade prosperity, and a reserve of £3,211,379 had been accumulated in the unemployment fund by July 1914.[5] Nevertheless, during the first full year of benefits, which ended in July 1914, over 23 per cent of insured workmen made a claim upon the fund. £533,016 was paid in benefits in respect of 1,092,288 claims from 550,000 individuals; and of this sum 69·4 per cent was paid directly through exchanges, while 30·6 per cent was paid through trade unions that had made a special arrangement with the Board of Trade to manage their own share of the unemployment insurance fund.[6] Slightly more than half of the days of unemployment

[1] The Year-Book of Social Progress 1914–15, pp. 407–8. Beveridge, Unemployment (1930 ed.), p. 401.

[2] Cd. 6965/1913, Unemployment Insurance. First Report of the Proceedings of the Board of Trade under Part II of the National Insurance Act, 1911, para. 145: Beveridge MSS., D. 030, 'Labour Exchanges and Unemployment Insurance. Report of the Proceedings of the Board of Trade . . . to July 1914', p. 66, Table LXV.

[3] Cd. 6965/1913, para. 147.

[4] Geoffrey Drage, The State and the Poor, p. 136.

[5] Beveridge MSS., D. 030, 'Labour Exchanges and Unemployment Insurance. Report of the Proceedings of the Board of Trade . . . to July 1914', para. 642.

[6] Ibid., paras. 397, 461. The proportion of claims made through trade unions rose considerably between Jan. 1913 and Jan. 1914, largely owing to the rise of union membership in insured trades (ibid., paras. 376–7). This increase of membership may have been influenced by the fact that the Act encouraged many unions to start additional private insurance schemes (LAB 2/1483/LE. 10237/8, 'List of Unions which now pay Unemployment Benefit but which did not do so prior to the commencement of the National Insurance Act, 1911, and with which an arrangement has been made under S. 105 of the Act', 26 Aug. 1913). There is little specific evidence, however, to support Halevy's suggestion that the increase was caused by the desire of workmen to participate in the actual administration of the Act (E. Halevy, The Rule of Democracy, p. 479).

incurred by insured workmen were covered by the fund, the rest falling either in the 'waiting week' or after the exhaustion of their right to benefit; and the average duration of insured spells of unemployment was slightly more than a week.[1] These figures were well inside the actuarial predictions on which the scheme had been based; but they showed that, even at the peak of the trade cycle, nearly a quarter of the labour force in insured industries was liable to an average of more than four weeks of unemployment during the course of a year. 'The . . . picture thus presented', concluded Beveridge in his first report on Unemployment Insurance, '. . . is that of a constant irregularity of employment even when employment is at its best, a ceaseless shifting from job to job, a recurrent loss of productive power and of wages in the interval between one job and the next.'[2]

Nevertheless, the working of the insurance scheme in 1913–14 was in no sense a test of the effectiveness of insurance as a permanent means of relieving unemployed distress. The experience of a group of predominantly skilled workmen in a year of booming trade was in no way indicative of the experience of the whole of the labour force during a period of high unemployment. This fact was largely ignored, however, when for political reasons insurance was gradually extended to cover all kinds of unemployment in 1920, 1927, and 1930;[3] hence the virtual breakdown of the unemployment insurance system in 1931. The year 1913–14 demonstrated that unemployment insurance was a useful means of assisting workmen during the time-lags between engagements that in a free market economy were a necessary feature even of a period of full employment; it failed to reveal, however, that insurance offered no solution to the problem of unemployment during a prolonged depression of trade.

[1] Beveridge MSS., D. 030, 'Labour Exchanges and Unemployment Insurance. Report of the Proceedings of the Board of Trade . . . to July 1914', p. 89, Table XCVI.

[2] Cd. 6965/1913, para. 248.

[3] On the statutory extensions of the unemployment insurance system, see E. M. Burns, op. cit., pp. 35–51.

THE ORIGINS AND PRINCIPLES OF
STATE INTERVENTION

What were the political forces and administrative principles that governed the introduction of a national policy for the unemployed? It has been shown that, in spite of Churchill's dreams of social empire-building, there was little conscious reference to theory in the Liberal reforms; and in tracing the evolution of a national unemployment policy it is difficult to point to the decisive influence of any single set of reforming ideas or to discover any logical sequence of institutional change. The adoption of such a policy cannot be ascribed to the substitution of an 'economic' for a 'moral' view of the causes of unemployment,[1] since in certain respects the prevalent attitude towards victims of unemployment was more severe in the 1900s than it had been twenty years before. It cannot be directly ascribed to the impact of the new 'sociology', since although social inquiries were undoubtedly important in forming public opinion and in reinforcing political argument, many of the administrative remedies devised to deal with unemployment were based on very inadequate statistical knowledge; and to a large extent new information about the state of the labour market was not the cause but the product of experiments in social reform.[2] It cannot be seen as part of a cumulative process of governmental growth, since until 1908 the group of public officials most directly concerned with the problem were almost uniformly hostile to any extension of public responsibility for the relief of the unemployed.[3] Moreover, many of the remedies devised to

---

[1] See e.g. Calvin Woodard, 'Reality and Social Reform: the Transition from Laissex-Faire to the Welfare State', *Yale Law Journal*, 72, no. 2 (Dec. 1962), 322.

[2] A. C. Pigou, *Unemployment* (1913), p. 19.

[3] Above, p. 100. In certain respects the development of remedies for unemployment in the 1900s closely conformed to the 'model' of governmental growth suggested by Professor O. Macdonagh ('The Nineteenth Century Revolution in Government: A Reappraisal', *Historical Journal*, 1, no. 1 (1958), 52–67). Public exposure of the evils of unemployment was partially responsible for the passage of a weak and locally administered statute, the Unemployed Workmen Act. Experience gained in the administration of this Act led to further more comprehensive legislation, which established central administrative machinery and conferred wide discretionary powers on public officials. The analogy breaks down, however, in certain important particulars. Firstly, there was no administrative continuity between the Unemployed Workmen Act and subsequent legislation; and, secondly,

meet unemployment proved to be administrative cul-de-sacs which supplied only negative evidence about methods of treating the unemployed. Finally, it cannot be explained by an advance of theories of 'collectivism',[1] since state intervention was brought about less by ideological factors than by the proven inadequacy of voluntary and local forms of unemployment relief. In so far as they had a discernible ideological content, the policies introduced by the central government were mainly designed to combat the threat of more extreme measures of state intervention, whether in the shape of the 'right to work' or 'tariff reform'; and in so far as they were an expression of collectivism, it was primarily administrative collectivism of a kind that had been invoked to deal with many of the social consequences of industrialization during the previous eighty years.[2]

At each stage the intervention of the central government —through the Chamberlain circular, the Unemployed Workmen Act, and the Liberal measures of 1909–11—was primarily pragmatic in its motivation and aims, evoked partly by the fear of being politically outmanœuvred and partly by the practical inadequacy of existing forms of unemployment relief. The Unemployed Workmen Act aroused a certain amount of theoretical controversy about the doctrine of the right to work, but the measures passed by the Liberals were remarkable for the absence of political and economic discussion of the long-term implications of labour organization, national insurance, and counter-depressive public works.

Nevertheless, as an experiment in social reform the Liberal measures of 1909–11 were ultimately more significant in principle than successful in practice; and three underlying characteristics, which were to a certain extent mutually

pressure for centralization came, not from public officials or from a central-government inspectorate, but from reforming politicians, from members of distress committees, and from freelance experts on unemployment like Beveridge and the Webbs.

[1] For a classic expression of this view, see A. V. Dicey, *Law and Opinion in England* (2nd ed.), p. xxxviii: '. . . the National Insurance Act admits the so-called "right to work" . . .'

[2] O. R. McGregor, 'Social Research and Social Policy', *British Journal of Sociology*, 8, no. 2 (June 1957), 154.

contradictory, can be discerned in Liberal unemployment policy which are of general interest to the historical analysis of the 'welfare state'.

Firstly, an implicit concern of Liberal unemployment policy was to preserve and enhance the 'free market' situation which throughout the nineteenth century orthodox economists had regarded as most conducive to maximum industrial efficiency and the accumulation of wealth. This was apparent in the desire to promote labour mobility and working-class independence; and in the belief that public investment should refrain from competition with private commercial interests and avoid the creation of 'artificial' work. These attitudes clearly distinguished the Liberal leaders and their civil-service advisors from the left wing of the Labour party and from economic radicals like Hobson and Percy Alden, who believed that the free market was obsolete, inefficient, and personally destructive to individual workmen.

Secondly, in spite of this concern for the preservation of the free market, the administrative machinery established to deal with unemployment was capable of greatly extending the central government's control over industrial and economic affairs. This extension was potentially most far-reaching in the case of the Development Act, which established a precedent for state involvement in certain aspects of economic life; but the administrative body set up by the Act was too weak to capitalize on the innovation of principle which the Act involved. Both the Labour Exchanges Act and the National Insurance Act, however, created an administrative system that was designed to impose an entirely new kind of discipline on private industry. This was evident in the incentives given to both employers and workmen to register the movement of labour; and in the penalization of casual and irregular employment and, by inference, of casual and irregular habits of life. It was evident in the framing of qualifications for unemployment benefit, which insisted on the diligence, sobriety, and good time-keeping that private employers had been trying to impose on their workmen since the start of industrialization.[1] In the determination of right to benefit these rules were enforced by the kind of quasi-judicial statutory committees

---

[1] S. Pollard, *The Genesis of Modern Management*, pp. 181–9

that had been used since the 1840s to regulate industrial society;[1] and as Beatrice Webb had predicted, the administration of unemployment benefit made possible 'the increased control of the employer and the wage-earner by the state'.[2]

Thirdly, in spite of the opposition that was evoked from employers and organized labour, it is important not to underrate the Liberal government's desire to consolidate social institutions and to achieve a political consensus at a time of rising antagonism in the political and industrial world.[3] The Development Act made concessions to supporters of the Liberal, Conservative, Labour, and Nationalist parties. The advisory committees of the labour exchange and insurance system were specifically designed to force capital and labour into consultation and collaboration, and it was hoped that labour exchanges would 'so far demonstrate the complexity of the problem as to dispose finally of all rough-and-ready revolutionary solutions'.[4] Moreover, contributory insurance was seen as a means of achieving two apparently incompatible aims—the realization of the Treasury ideal of 'broadening the basis of taxation', without thereby irrevocably alienating the political support of the working class. 'The idea is to increase the stability of our institutions by giving the mass of industrial workers a direct interest in maintaining them', Winston Churchill told a *Daily Mail* reporter in August 1909: 'With a "stake in the country" in the form of insurances against evil days these workers will pay no attention to the vague promises of revolutionary socialism.' And he concluded these Bismarckian utterances with an account of unemployment insurance that was more reminiscent of 'Tory

[1] e.g. H. Parris, *Government and the Railways in Nineteenth Century Britain*, pp. 210–11.

[2] B. Webb's diary, 13 May 1911, quoted in *Our Partnership*, p. 475. On the subsequent influence of the British unemployment insurance system in 'establishing norms and standards in regard to a tremendously wide range of industrial practice and economic policies', see E. M. Burns, 'Unemployment Compensation and Socio-Economic Objectives', *Yale Law Journal*, 55, no. 1 (Dec. 1945), 18.

[3] On contemporary industrial unrest, see E. Phelps Brown, *The Growth of British Industrial Relations*, Ch. VI, and R. V. Sires, 'Labor Unrest in England 1910–14', *Journal of Economic History*, 15, no. 3 (Sept. 1955), 246–66.

[4] Beveridge MSS., Coll. B., vol. xiv, Item 25, 'Labour Exchanges General Memorandum', p. 12.

democracy' than of the Liberal tradition of social reform. 'It is a scheme', he stated,

with a strictly limited risk. There is no sentiment about it. It does not interfere with any natural law . . . it will help to remove the dangerous element of uncertainty from the existence of the industrial worker. It will give him an assurance that his home, got together through long years and with affectionate sacrifice, will not be broken up, sent bit by bit to the pawnshop, just because through no fault of his own maybe he falls out of work. It will make him a better citizen, a more efficient worker, a happier man.[1]

NEGLECTED REMEDIES FOR UNEMPLOYMENT

What, if any, were the alternative policies for counteracting unemployment available to social reformers before 1914? The most serious political challenge to the Liberal programme came from the protectionists, who argued that domestic employment could be increased by a general tariff. But the plausibility of this argument was called in question by evidence of high unemployment in Germany and America;[2] and the tariff-reform programme was in any case rejected by the British electorate in 1906 and 1910. A second alternative was a public guarantee of the 'right to work' and a recognition on the part of the state of an obligation to provide work for the unemployed. But acceptance of this obligation merely begged the question of how the unemployed were to be employed; and the solution to this problem advanced by theoretical socialists—namely, the public ownership and regulation of industry—was at no time politically feasible before 1914. A third alternative often advocated by proponents of 'national efficiency' was that surplus workmen should be withdrawn from the labour market by a system of naval and military conscription;[3] but it is highly unlikely that politicians would have been able to introduce in peacetime a policy that subsequently met with considerable opposition

[1] *Daily Mail*, 16 Aug. 1909, 'Insurance against Unemployment. Mr. Churchill's scheme. Full Explanation' (Press-cutting, Beveridge MSS., Parcel 2, Folder C (ii), 279, item 28).

[2] *Hansard*, 5th series, vol. 60, col. 1142.

[3] J. W. Bennington, *Unemployment and the Remedy* (1908), pp. 4–12; G. Drage, *The State and The Poor* (1914), p. 240; and 'National Decadence a Myth', reprinted in *Ephemera* (1915), pp. 395–6.

even in the midst of a world war.[1] A fourth alternative was the policy advocated by J. A. Hobson and by certain leaders of the labour movement of transferring income from the saving to the spending classes by progressive taxation and counter-depressive public works. Underconsumptionist arguments had a certain amount of influence upon Liberal policy-makers, and were used to justify both the Development Act and the Budget of 1909;[2] but Hobson's remedy for unemployment required a degree of redistributive taxation that was scarcely conceivable to non-socialists in Edwardian England, particularly since the boom in the export of both goods and capital shielded investors and manufacturers from the consequences of a deficiency in home demand.[3]

In the long run the history of the unemployment policies introduced before 1914 was a history of almost unmitigated failure. This was due partly to the changing nature of the problem and to the emergence of a new type of unemployment after 1918, caused by chronic stagnation in the heavy manufacturing industries. But it was also due to the failure of social reformers fully to comprehend the problem, even as it existed before 1914, and to the concentration of politicians and administrators on measures of relief and organization at the expense of measures for stabilizing and increasing labour demand. Both Beveridge and the Webbs subsequently admitted that they had underestimated the severity of the problem and the degree of state intervention necessary for its solution in 1909;[4] and, far from being diminished by the 'organization of the labour market', the average level of unemployment in the 1920s and 1930s was three times as

[1] T. Wilson, *The Downfall of the Liberal Party*, pp. 78–85.

[2] Above, p. 319. Many Liberal politicians were undoubtedly familiar with Hobson's views (e.g. Add. MS. 46327, Burns Diary, 2 Apr. 1909). One of Hobson's leading disciples, J. M. Robertson, actually served in the Liberal government as Parliamentary Secretary to the Board of Trade in 1911–15, although there is no evidence to suggest that he used his position to advance 'underconsumptionist' views.

[3] Exported goods rose by 70 per cent in volume and 80 per cent in value between 1900 and 1913 (*The Economist. Commercial History and Review of 1913*, 21 Feb. 1914, p. 415). Income from overseas investments rose during the same period by over 100 per cent (B. R. Mitchell and P. Deane, *Abstract of British Historical Statistics*, pp. 334–5).

[4] W. H. Beveridge, *Full Employment in a Free Society* (1944), pp. 90–2; B. Webb, *Our Partnership*, p. 484.

high as it had been in the thirty years before the First World
War.[1] The labour-exchange system, which had been designed
to maximize the mutual interests of employers and workers,
tended instead—according to one high authority—to empha-
size 'the division of classes into two camps', and to 'form a
prop to class war'.[2] The system of unemployment insurance
broke down in the face of prolonged stagnation in the staple
exporting industries, and in some quarters was seen as a
principal cause of Great Britain's insolvency in the world
monetary crisis of 1931.[3] Orthodox objections to counter-
depressive public expenditure, which by 1914 had been
explicitly or tacitly rejected by many Liberal, Conservative,
and Labour politicians, were revived in the 1920s when
retrenchment and deflation were thought to be essential to
the restoration of business confidence and of international
trade.[4] The national exchequer, which in 1909 had accepted
a share of financial responsibility for the promotion of public
employment, tried in the early 1920s to thrust this responsibility back upon local authorities; but, as in the 1880s and
1890s, the areas with the greatest need for local relief schemes
proved to be those with the heaviest existing burden of local
rates.[5] The experience of the pre-war period was thus in
many respects forgotten; and one of the main trends of forty-
five years of unemployment policy was reversed by the
National government when a large number of unemployed
workmen were restored to the care of a refurbished Poor Law
system in 1931.[6]

[1] W. H. Beveridge, op. cit., p. 72.
[2] G. Askwith, *Industrial Problems and Disputes* (1920), p. 278.
[3] R. Skidelsky, *Politicians and the Slump*, p. 365.
[4] Ibid., pp. 156–7, 227; K. Hancock, 'The Reduction of Unemployment as a Problem of Public Policy', *Economic History Review*, 2nd series, 15, no. 2 (Dec. 1962), 336–7.
[5] K. Hancock, 'The Reduction of Unemployment as a Problem of Public Policy', loc. cit., pp. 335–6.
[6] Under the Unemployment Insurance (National Economy) Orders of Oct. 1931 insurance benefit was again limited to 26 weeks in any one year. Workmen who had paid at least 30 contributions but had exhausted their claim to benefit were henceforth transferred to 'transitional payments', donated by the national exchequer but subject to a means test applied by the local 'public assistance committees' which had replaced the Poor Law guardians in 1929. At the nadir of the slump in Sept. 1932, 1,345,000 workmen were supported by unemployment insurance, 1,018,000 by transitional payments, and 140,000 by public assistance (E. M. Burns, *British Unemployment Programs, 1920–38*, pp. 111–15).

APPENDIX A

SOCIAL POLICY AND THE PROBLEM OF LOCAL TAXATION

THROUGHOUT the period covered by this book local developments in social policy were hampered by disputes about the allocation of financial responsibility for local burdens. In 1888 the traditional system of 'grants-in-aid' from the national exchequer to local authorities was replaced by a system of 'assigned revenues', initially derived from licence and probate duties. The Chancellor of the Exchequer, Goschen, originally planned to distribute the assigned revenues on the basis of indoor pauperism—this being seen as the most significant index of local poverty. But this proposal was abandoned under pressure from representatives of rural areas with low indoor and high outdoor pauperism; and assigned revenues were distributed locally in the same proportions as the old grants-in-aid—a system that was admitted by the Treasury to be increasingly 'obsolete and inequitable'.[1]

The aggregate contribution of the national exchequer to local expenditure increased considerably after Goschen's innovations;[2] but nevertheless there was a recurrent demand among social reformers for more national alleviation of local burdens, for redistribution of income between rich and poor local authorities, and for rates to be levied on personal as well as real property. These problems were exhaustively considered by a Royal Commission on Local Taxation under Lord Balfour of Burleigh in 1896–1901. The only substantial reform recommended by the majority of this Commission was that different kinds of rateable property (i.e. commercial, agricultural, residential, etc.) should be subject to a uniform system of valuation.[3] Individual commissioners, however, made more far-reaching proposals. Balfour of Burleigh himself recommended that 'assigned revenues' should be replaced by annual grants from the Consolidated Fund, to be distributed partly according to the size of local population, partly according to the level of local expenditure, and partly according to the degree of

---

[1] T. 168/34, Memorandum on 'Imperial Relief of Local Burdens and the System of Imperial and Local Taxation', by Edward Hamilton, Mar.–Apr. 1897; E. Cannan, *The History of Local Rates in England*, p. 145; S. Webb, *Grants in Aid*, pp. 28, 84–5.
[2] E. P. Hennock, 'Finance and Politics in Urban Local Government in England, 1835–1900', *Historical Journal*, 6, no. 2 (1963), 224–5.
[3] Cd. 638/1901, *RC on Local Taxation, Final Report*, pp. 62–3.

local poverty. He suggested also that additional revenue for local purposes might be raised through taxes on houses, bicycles, and advertisements.[1] The Treasury officials Sir Edward Hamilton and Sir George Murray agreed with Balfour of Burleigh's analysis of how national subsidies to local areas should be distributed. They emphasized that, contrary to a widely held belief, the 'Local Taxation question' was 'a question not of real versus personal property, but of national versus local services'; and they proposed that the central government should pay for up to one-half of all local expenditure on 'national services', such as highways and education.[2] Finally, several of the commissioners recommended that the 'site value' of property should be assessed and rated separately from the buildings erected upon it—the tax on site values being deducted from the rent paid to ground landlords as a form of income-tax. Such a tax, it was argued, would relieve tenants and owners of buildings at the expense of the ground landlords (the value of whose property was largely determined by the economic and social activity of the surrounding community). It would tend to 'rectify inequalities between one district and another'; it would head off demands for measures of property confiscation; and it would disprove the radical argument that existing systems of rating and taxation left a large potential source of revenue entirely untapped.[3]

Throughout the 1900s radical politicians were demanding reform along these lines; and successive Chancellors promised to tackle the problem of relieving and equalizing local rates.[4] Very little had been done in this direction, however, by 1914; hence the reluctance of many local authorities to assume responsibility for the expensive new local services, such as those envisaged by the Reports of the Royal Commission on the Poor Laws in 1909.[5]

[1] Cd. 638/1901, *Separate Recommendations* by Lord Balfour of Burleigh, pp. 70–2, 75–6.

[2] Cd. 638/1901, *Report* by Sir Edward Hamilton and Sir George Murray, pp. 128, 144.

[3] Cd. 638/1901, *Separate Report on Urban Rating & Site Values*, pp. 175–6.

[4] Above, pp. 233, 267–8.

[5] On the reluctance of local authorities to take on additional financial responsibilities in this period see F. Honigsbaum, 'The Struggle for the Ministry of Health', *Occasional Papers in Social Administration*, No. 37, p. 25.

APPENDIX B

UNEMPLOYMENT STATISTICS BEFORE 1914

FOR most of the period covered by this study the only continuous and nation-wide record of the level of unemployment was the index published by the Labour bureau (subsequently Labour department) of the Board of Trade. For the period prior to 1888 this was calculated retrospectively from the records of 'out-of-work' benefit paid by certain trade unions. From 1888 onwards it was based on monthly returns from benefit-paying unions of members in receipt of 'out-of-work' benefit, supplemented where available by the total number of union members unemployed. These figures were used to calculate not the volume but the trend of unemployment in certain trades. In the 1880s and 1890s the index tended to exaggerate fluctuations of employment, since it over-represented workmen in the engineering and shipbuilding industries, which were peculiarly liable to cyclical depressions of trade.[1] Stable occupations like transport and agriculture were not included; and the index under-represented the building trade, which was particularly vulnerable to seasonal unemployment, and the mining and textile industries, which adjusted to slack periods by working short time. Between 1894 and 1908, however, the index was gradually adjusted to increase the representation of stable and short-time trades.[2] Information was also obtained by the Board of Trade from employers in the coal-mining, iron-mining, iron, steel, and tinplate industries about the number of days per week worked by their employees, from which the level of unemployment in these industries could to a certain extent be inferred.[3]

These figures could be supplemented by the data collected by the Local Government Board about the number of able-bodied paupers in receipt of indoor and outdoor relief. Only a small percentage of able-bodied paupers were destitute through want of employment; and the reliability of statistics of able-bodied pauperism was in any case

[1] H. of C. 365/1895, SC on Distress from Want of Employment, Minutes of Evidence, QQ. 4562–3, Statement of Llewellyn Smith.

[2] On changes in the composition of the Board of Trade's index see N. B. Dearle, 'English Statistics of Unemployment', International Conference on Unemployment, Paris 1910, Tome 3, Report 25.

[3] Cd. 5068/1910, RC on the Poor Laws, Appendix XXI (A), 'Method of Compilation of Board of Trade's Percentages of the Unemployed and the extent to which they may be taken as a Measure of Unemployment', pp. 587–8.

questionable, because of widely varying local interpretations of the concept of 'able-bodied'.[1] But, nevertheless, the returns of able-bodied pauperism did reflect the operations of the trade cycle, usually reaching their highest point slightly more than a year after the nadir of a depression of trade.[2] In 1904 it was observed by Arthur Bowley that this time-lag was decreasing, from which he surmised that either the sources of the unemployed were diminishing or that they were less reluctant than formerly to accept outdoor relief.[3]

Several attempts were made during this period to calculate on a wider basis the extent of unemployment and the effect of fluctuations of trade. In 1912 Bowley compiled an index of unemployment for a wide cross-section of the labour force, based on the quantification of verbal descriptions of the state of the market in different industries as 'very good', 'good', 'fair', 'bad', and 'very bad'.[4] Charles Booth in 1895 suggested to the Select Committee on Distress from Want of Employment that an index of national employment trends might be constructed from 'a combination of statistics, such as those relating to revenue and traffic, banking and pauperism';[5] and in 1903 the Board of Trade began to compile quarterly reports on 'Employment and Trade', which summarized information about wage-rates, pauperism, trade-union unemployment, railway 'goods traffic' receipts, imports and exports, and bankers' clearances in London.[6] In 1899 G. H. Wood made a pioneering attempt to relate the level of unemployment to the consumption of basic commodities;[7] and in 1907 W. H. Beveridge tried to chart the 'pulse of the nation', by correlating figures relating to unemployment and pauperism, interest-rates, marriage, and crime.[8]

More detailed information about the incidence of unemployment outside the benefit-paying unions only became available, however, as the result of measures of unemployment relief; and not until the opening of labour exchanges in 1910 did the Board of Trade have

[1] B. Abel-Smith and R. Pinker, *Changes in the Use of Institutions in England and Wales between 1911 and 1951*, pp. 21–2.

[2] W. H. Beveridge, 'The Making of Paupers', *Toynbee Record*, Nov. 1904, 27–9.

[3] Arthur Bowley to the Editor, *Toynbee Record*, Dec. 1904, 48–9.

[4] Arthur Bowley, 'The Measurement of Employment: An Experiment', *Journal of the Royal Statistical Society*, 75 (July 1912), 791–829.

[5] H. of C. 365/1895, *SC on Distress from Want of Employment, Minutes of Evidence*, Q. 10520.

[6] e.g. T. 168/60, Financial Papers 1903–4, Memoranda by H. Llewellyn Smith on 'Employment and Trade', 30 Mar. 1903; and 'Employment and Trade in the First Two Months of 1904'.

[7] G. H. Wood, 'Some Statistics Relating to Working Class Progress since 1860', *Journal of the Royal Statistical Society*, 62 (Dec. 1899), 639–75.

[8] W. H. Beveridge, 'The Pulse of the Nation', *Albany Review*, 2 (Nov. 1907), 160–70.

access to more reliable information about unemployment among unemployed workmen. The following tables summarize some of the information that is available about the incidence of unemployment before 1914; and some of the results of administrative measures for the relief of the unemployed.

TABLE I

Estimate of able-bodied males relieved by the Poor Law in England and Wales on account of destitution arising from unemployment on 1 January 1886–1912

	Relieved inside workhouse	Percentage of total able-bodied indoor pauperism	Relieved outside workhouse	Percentage of total able-bodied outdoor pauperism
1886			4,420	5·3
1887			3,506	4·2
1888			3,554	4·2
1889			2,021	2·6
1890			738	1·0
1891	7,338	26·4	1,514	2·2
1892	6,499	20·3	722	1·1
1893	8,023	23·1	2,951	4·1
1894	9,912	25·5	2,981	3·8
1895	10,583	26·2	2,831	3·8
1896	10,517	25·4	1,082	1·5
1897	10,205	25·3	789	1·2
1898	9,502	23·3	851	1·3
1899	7,807	20·0	520	0·8
1900	7,728	20·0	297	0·5
1901	6,570	17·5	422	0·7
1902	7,555	19·0	581	0·9
1903	8,338	19·6	931	1·5
1904	9,598	20·8	1,585	2·4
1905	11,470	22·6	7,872	9·6
1906	12,153	23·0	4,224	5·7
1907	11,508	22·2	2,235	3·2
1908	11,413	22·2	2,732	4·0
1909			6,374	8·0
1910			3,252	4·3
1911			2,676	3·6
1912			1,204	1·7

SOURCES: Based on Cd. 5077/1911, pp. 22–3, 27, and 44; and *L.G.B. Annual Reports*, 1908–9 to 1912–13. The category of 'able-bodied' was abandoned in L.G.B. records in 1912.

TABLE 2

Trade Union unemployment recorded by the Board of Trade 1870–1912

	Engineers, shipbuilders, and metalworkers per cent		Builders per cent	Woodworkers and furnishers per cent		Printers and bookbinders per cent		All unions making returns per cent
1870	4·4		3·7	4·8		3·5		3·75
1871	1·3		2·5	3·5		2·0		1·65
1872	0·9		1·2	2·4		1·5		0·95
1873	1·4		0·9	1·8		1·3		1·15
1874	2·3		0·8	2·1		1·6		1·60
1875	3·5		0·6	2·0		1·6		2·20
1876	5·2		0·7	2·4		2·4		3·40
1877	6·3		1·2	3·5		2·6		4·40
1878	9·0		3·5	4·4		3·2		6·25
1879	15·3		8·2	8·3		4·0		10·70
1880	6·7		6·1	3·2		3·2		5·25
1881	3·8		5·2	2·7		2·8		3·55
1882	2·3		3·5	2·5		2·4		2·35
1883	2·7		3·6	2·5		2·2		2·60
1884	10·8		4·7	3·0		2·1		7·15
1885	12·9		7·1	4·1		2·5		8·55
1886	13·5		8·2	4·7		2·6		9·55
1887	10·4		6·5	3·6		2·2		7·15
1888	5·5	6·0	5·7	3·6	3·1	2·5	2·4	4·15
1889	2·0	2·3	3·0	2·6	2·4	2·1	2·5	2·05
1890	2·4	2·2	2·2	1·5	2·5	1·9	2·2	2·10
1891	4·4	4·1	1·9	1·7	2·1	2·9	4·0	3·40
1892	8·2	7·7	3·1	2·4	3·8	3·6	4·3	6·20
1893		11·4	3·1		4·1		4·1	7·70
1894		11·2	4·3		4·4		5·7	7·20
1895		8·2	4·4		3·6		4·9	6·00
1896		4·2	1·3		2·0		4·3	3·35
1897		4·8	1·2		2·2		3·9	3·45
1898		4·0	0·9		2·3		3·7	2·95
1899		2·4	1·2		2·1		3·9	2·05
1900		2·6	2·6		2·8		4·2	2·45
1901		3·8	3·9		3·7		4·5	3·35
1902		5·5	4·0		4·1		4·6	4·20
1903		6·6	4·4		4·7		4·4	5·00
1904		8·4	7·3		6·8		4·7	6·40
1905		6·6	8·0		5·8		5·1	5·25
1906		4·1	6·9		4·8		4·5	3·70
1907		4·9	7·3		4·6		4·3	3·95
1908		12·5	11·6		8·3		5·5	8·65
1909		13·0	11·7		7·6		5·6	8·70
1910		6·8	8·3		5·4		4·9	5·10
1911		3·4	4·2		3·3		5·1	3·05
1912		3·6	3·7		3·1		5·2	3·15

SOURCE: W. H. Beveridge, *Unemployment. A Problem of Industry* (1930 ed.), pp. 39 and 432; and Cd. 5068/1910, *RC on the Poor Laws*, Appendix XII (B), p. 597.

TABLE 3

Charity Organisation Society: cases of unemployment relieved by London District Committees 1886–1906

	Persons assisted to find employment*	Persons who received help in emigration†
1886	1,536	184
1887	623	229
1888	821	204
1889	753	138
1890	683	84
1891	810	84
1892	721	111
1893	722	115
1894	699	98
1895	685	67
1896	676	52
1897	577	24
1898	267**	13
1899	543	24
1900	516	23
1901	546	20
1902	509	45
1903	618	130
1904	624	131
1905	633	238
1906	318***	423

*   includes provision of tools, migration, and situations for children.
** six months only—returns not available for May–Oct.
*** Jan.–June only—series thereafter abandoned.
†   not including dependants.

SOURCE: *Charity Organisation Review.*

TABLE 4

Salvation Army: work for the unemployed in the U.K. 1903–1912

Year ending 30 Sept.	Number of applications to Labour bureaux	Number received into factories	Number for whom employment (permanent or temporary) found
1903	12,863	4,872	14,062
1904	12,765	4,016	15,531
1905	18,135	2,475	23,057
1906	27,260	3,019	13,913
1907	28,000	4,410	38,515
1908	39,864	4,177	42,493
1909	17,935	6,425	22,194
1910	13,009	6,754	20,210
1911	6,237	8,645	12,234
1912	5,455	12,956	11,827

SOURCE: *Reports on Social Service in the Salvation Army*, 1903, 1904, 1906, 1908, 1909, 1910, 1912.

TABLE 5

Numbers relieved by Distress Committees under the Unemployed Workmen Act 1905/6 to 1913/14

	Number of Distress Committees receiving applications		Numbers relieved (including dependants)		Numbers relieved per 1,000 of population in areas covered by Distress Committees
	London	*Provinces*	*London*	*Provinces*	
1905–6	29	85	92,876	180,906	1·7
1906–7	29	76	60,416	152,801	1·4
1907–8	29	69	54,613	150,971	1·4
1908–9	29	95	136,589	376,043	3·1
1909–10	29	87	82,349	236,094	2·0
1910–11	29	65	51,828	149,087	1·5
1911–12	29	45	37,643	105,819	1·2
1912–13	29	43	30,662	87,912	1·0
1913–14	29	30	16,349	47,318	0·7

SOURCE: *Annual Distress Committee Returns.*

TABLE 6

Income and expenditure under the Unemployed Workmen Act

	Central Unemployed Body and London Distress Committees		Provincial Distress Committees	
	Receipts	Expenditure	Receipts	Expenditure
	£	£	£	£
1905–6	46,757	32,718	80,775	55,996
1906–7	138,121	133,682	106,723	94,063
1907–8	186,350	138,098	91,108	87,589
1908–9	127,363	155,586	189,009	171,893
1909–10	146,176	144,728	110,638	128,943
1910–11	87,832	101,999	74,293	81,586
1911–12	109,572	106,975	55,990	58,835
1912–13	103,654	98,317	54,559	59,032
1913–14	62,483	71,516	37,631	41,437

SOURCE: *Annual Distress Committee Returns.*

TABLE 7

Labour exchange registrations 1911–1920

	Number of registrations	Number of individuals registered	Number of vacancies notified	Number of vacancies filled	Number of individuals found work	Vacancies filled as a percentage of registrations
1911	1,965,991	—	769,661	608,475	—	30·95
1912	2,362,225	—	1,033,780	809,553	—	34·27
1913	2,836,366	1,783,951	1,183,356	895,273	632,666	31·56
1914	3,305,056	2,076,187	1,436,663	1,086,738	792,034	32·88
1915	3,047,025	2,232,390	1,761,090	1,279,918	1,037,689	42·00
1916	3,436,405	2,660,425	2,017,363	1,534,928	1,334,896	44·67
1917	3,432,154	2,729,285	1,974,383	1,536,383	1,359,704	44·76
1918	3,594,383	2,924,471	2,039,931	1,495,774	1,312,579	41·61
1919	5,928,947	4,774,011	1,909,489	1,258,965	1,111,847	21·23
1920	1,597,971	1,234,241	753,168	517,667	445,806	32·40

SOURCE: Based on Cmd. 1054/1920, *Report of the Committee of Enquiry into the Work of Employment Exchanges*, p. 29, Appendix III.

TABLE 8

National Insurance Act 1911

Unemployed Insurance Fund Account: select items of income and expenditure 1912–1921 (to nearest £)

Year ending July	(a) Contributions from employers and workmen £	(b) Contribution from the State £	(c) Unemployment benefit paid direct to workmen £	(d) Payments to trade unions under Section 105 £	(e) Refunds to employers in respect of regular workmen £	(f) Refunds to employers for working short time £	(g) Refunds to workmen at age 60 £
1912–13	1,622,038	378,000	183,193	25,124	—	112	—
1913–14	1,802,940	602,000	364,555	166,038	113,107	204	—
1914–15	1,649,641	546,666	249,533	169,158	120,476	2,578	245
1915–16	1,694,115	538,863	39,973	38,998	94,035	410	532
1916–17	2,699,932	746,372	24,133	10,175	107,405	—	773
1917–18	3,277,123	1,007,541	75,128	11,025	117,035	—	3,082
1918–19	2,871,640	994,402	148,882	3,839	137,242	—	11,253
1919–20	3,043,252	912,701	869,424	139,702	117,391	—	27,531
1920–1	11,303,175	2,168,239	30,111,070	4,005,125	150,384	—	40,516

SOURCE: *Unemployment Insurance Fund Accounts.*

TABLE 9

National Insurance Act 1911

Relation between Contributions and Income in Working-Class Families
1912

Total family income per week	Percentage of income paid in taxes on food, tobacco, and alcohol	Percentage of income paid in Part I contributions	Percentage of income paid in Part II contributions	Percentage of income paid in indirect taxation and insurance contributions by families covered by Parts I and II of the National Insurance Act
18s.	7·10	2·0	1·15	10·25
21s.	6·10	1·72	0·99	8·81
25s.	5·12	1·40	0·83	7·39
30s.	4·27	1·20	0·69	6·16
35s.	3·65	1·03	0·59	5·27

SOURCE: Based on T. 171/47, Revenue Bill Papers 1913, 'Some Notes on the Incidence of Taxation on the Working Class Family' by F. W. Kolthammer.

TABLE 10

Income and expenditure of the Development Fund 1910–1931

	(a)			(b)			(c)
	Income			Grants and loans			Column (b) as a percentage of column (a)
	£	s.	d.	£	s.	d.	
1910–11	900,000	0	0	8,543	12	3	0·95
1911–12	1,987,989	6	3	70,945	0	0	3·57
1912–13	7,227,349	13	0	136,055	5	6	1·88
1913–14	5,281,907	5	11	294,808	13	2	5·58
1914–15	4,886,761	7	10	349,979	13	5	7·16
1915–16	4,138,858	14	0	283,377	7	5	6·85
1916–17	3,944,430	4	1	201,865	2	5	5·12
1917–18	3,670,420	7	1	384,771	7	10	10·48
1918–19	3,289,309	3	6	220,050	0	6	6·69
1919–20	5,663,924	4	3	368,360	16	4	6·50
1920–1	4,773,844	6	1	568,852	5	9	11·92
1921–2	2,954,724	15	0	385,184	9	3	13·04
1922–3	4,653,411	13	3	378,840	13	8	8·14
1923–4	2,261,159	8	0	321,965	0	6	14·24
1924–5	2,043,064	1	2	455,368	8	3*	22·29
1925–6	1,702,799	11	11	474,200	15	4	27·85
1926–7	1,414,365	4	2	514,480	9	7	36·38
1927–8	1,284,972	10	4	373,679	14	3	29·08
1928–9	1,699,861	5	6	353,982	12	4	20·82
1929–30	1,887,368	1	7	435,592	11	11	23·08
1930–1	2,462,427	4	5	580,508	9	10	23·57

* includes £78,772. 11s. 10d. paid to Irish Free State in respect of allocation of capital of Development Fund.

SOURCE: *Development Fund Accounts*, 1910–11 to 1930–1.

TABLE II

Income and expenditure of the Road Fund 1910–1921

	(a) Income			(b) Grants and loans	(c) Column (b) as a percentage of column (a)
	£	s.	d.	£	
1910–11	1,426,640	15	9	8,420	0·59
1911–12	3,570,220	13	2	270,661	7·58
1912–13	5,056,741	0	2	459,929	10·00
1913–14	3,573,552	10	2	936,605	26·21
1914–15	4,044,071	10	5	1,369,153	33·86
1915–16	4,305,995	5	2	598,480	13·90
1916–17	2,375,183	12	6	428,760	18·05
1917–18	2,720,537	8	2	285,636	10·50
1918–19	4,001,937	8	4	256,796	6·42
1919–20	12,594,804	9	2	3,605,456	28·63
1920–1	41,411,636	13	7	4,706,101	11·36

SOURCE: *Annual Reports of the Proceedings of the Road Board*, 1910–21.

SOURCES AND SELECT BIBLIOGRAPHY

THESE are arranged as follows:

(1) Manuscript Sources.
(2) Official Publications.
(3) Collections of Reports, Pamphlets, and other Printed Material published between 1886 and 1914.
(4) Reports by Local Authorities and Other Institutions.
(5) Bibliographies and Works of Reference.
(6) Newspapers and Periodicals.
(7) Works with special reference to Unemployment.
(8) Political, Social, and Economic Background.
(9) Biographies and Memoirs.

The place of publication of printed works is London unless otherwise stated.

PART ONE: MANUSCRIPT SOURCES

(*a*) PUBLIC RECORD OFFICE

Board of Education Papers: Ed. 24.
Board of Trade Papers: B.T. 13; LAB 2; PIN 3 and 7.
Cabinet Papers: CAB 1; CAB 37; CAB 41.
Development Commission Papers: D. 2.
Home Office Papers: H.O. 45.
Local Government Board Papers: H.L.G. 29; M.H. 19.
Treasury Papers: T. 1; T. 168; T. 170; T. 172.
Gerald Balfour Papers: P.R.O. 30.

(*b*) BRITISH LIBRARY OF POLITICAL SCIENCE

William Beveridge Papers*
W. J. Braithwaite Papers.
Robert Giffen Papers.
J. Ramsay Macdonald Papers.
George Lansbury Papers.
Passfield Papers.

(*c*) BRITISH MUSEUM ADDITIONAL MANUSCRIPTS

John Burns Papers: Add. MS. 46282–302, 46323–35.
H. Campbell-Bannerman Papers: Add. MS. 41210–22.

* These papers consist of two collections; one deposited by Beveridge in the B.L.P.S., the other originally deposited at University College, Oxford and transferred to the B.L.P.S. in 1965. The former collection is referred to in my references as 'first deposit'.

W. E. Gladstone Papers: Add. MS. 44515–18.
Herbert Gladstone Papers: Add. MS. 45986–95, 46017–19, 46021, 46118.
Edward Hamilton Papers: Add. MS. 48642–83.
Charles Dilke Papers: Add. MS. 43877 and 43893.

(*d*) BODLEIAN LIBRARY

James Bryce Papers.
H. H. Asquith Papers.

(*e*) OTHER COLLECTIONS

Samuel and Henrietta Barnett Papers (Library of Greater London Council).
Sydney Buxton Papers (in the possession of Mrs. Elizabeth Clay).
R. C. K. Ensor Papers (Corpus Christi College, Oxford).
Fabian Society Papers (Nuffield College, Oxford).
Hicks Beach Papers (in the possession of Lord St. Aldwyn).
C. S. Loch's Diary (Goldsmiths' Library, London University).
Lancelot Phelps Papers (Oriel College, Oxford).
Salisbury Papers (Christ Church, Oxford).

(*f*) OTHER UNPUBLISHED SOURCES

Board of Trade establishment lists (Board of Trade Library).
'Notes on the Trade Union Congress 1890', by Theodore Llewellyn Davies (Nuffield College).
Runciman Papers (transcripts by Dr. C. Hazlehurst).

PART TWO: OFFICIAL PUBLICATIONS

(*a*) COMMAND PAPERS

(1) *Annual*

Annual Reports of the Local Government Board, 1885/6 to 1913/14.
Annual Reports of the Board of Education, 1908/9 to 1912/13.

(2) *Occasional*

Royal Commission on the Housing of the Working Classes, First Report, C. 4402/1884/5, xxx, p. 1.
Royal Commission on the Depression of Trade and Industry, Final Report, C. 4893/1886, xxiii, p. 507.
Tabulation of the Statements Made by Men Living in Certain Districts of London in March 1887, C. 5228/1887, lxxi, p. 303.
Royal Commission on Gold and Silver: Second Report, C. 5284/1888, xxvi, p. 381; *Third Report*, C. 5512/1888, xlv, p. 285.
Royal Commission on Labour: Minutes, Precis, Index and Appendices to the Evidence taken before the Royal Commission on Labour (Sitting as a Whole), C. 7063–I; C. 7063–II; C. 7063–III; C. 7063–IIIA/1893–4,

xxxix, Pt. I, pp. 5, 629, and 805. *Fifth and Final Report*, C. 7421/1894, xxxv, p. 9.

Board of Trade, Agencies and Methods for Dealing with the Unemployed, C. 7182/1893–4, lxxxii, p. 377.

Royal Commission on the Aged Poor: Report, C. 7684/1895, xiv, p. 1. *Minutes of Evidence*, C. 7684–I/1895, xiv, p. 123, C. 7684–II/1895, xv, p. 1.

Foreign Office. Miscellaneous Series, No. 399. Reports on Subjects of General and Commercial Interest. Report on the Society for Insurance against Want of Employment in Winter and the General Labour Registry at Cologne, C. 7920–20/1896, lxxxiv, unpaginated.

Royal Commission on Local Taxation: Final Report (England and Wales), Cd. 638/1901, xxiv, p. 413.

Royal Commission on Alien Immigration: Report, Cd. 1741/1903, ix, p. 1; *Appendix*, Cd. 1741–I/1903, ix, p. 935; *Minutes of Evidence*, Cd. 1742/1903, ix, p. 61; *Index*, Cd. 1743/1903, ix, p. 1041.

Board of Trade. Memoranda, Statistical Tables, and Charts prepared in the Board of Trade with Reference to various matters bearing on British and Foreign Trade and Industrial Conditions, Cd. 1761/1903, lxvii, p. 253; Cd. 2337/1905, lxxxiv, p. 1; Cd. 2669/1905, lxxxiv, p. 669.

Committee Appointed to Consider the Position and Duties of the Board of Trade and the Local Government Board, Report, Cd. 2121/1904, lxxxviii, p. 439.

Interdepartmental Committee on Physical Deterioration: Report, Cd. 2175/1904, xxxii, p. 1.

Report to the President of the Local Government Board on Methods of dealing with Vagrancy in Switzerland, by H. Preston Thomas, Cd. 2235/1904, lxxxii, p. 593.

Report to the Board of Trade on Agencies and Methods for dealing with the Unemployed in Certain Foreign Countries, Cd. 2304/1905, lxxiii, p. 471.

London Unemployed Fund. Preliminary Statement, Cd. 2561/1905, lxxiii, p. 415.

Report on the Salvation Army Colonies in the United States, and at Hadleigh, England; with a Scheme of National Land Settlement, by H. Rider Haggard, Cd. 2562/1905, liii, p. 359.

Departmental Committee on Vagrancy: Report, Cd. 2852/1906, ciii, p. 1; *Minutes of Evidence*, Cd. 2891/1906, ciii, p. 131.

Report by the Departmental Committee appointed to consider Mr. Rider Haggard's Report on Agricultural Settlements in British Colonies, Cd. 2978/1906, lxxvi, p. 533.

Report to the President of the L.G.B. on the Poplar Union, by J. S. Davy, Cd. 3240/1906, civ, p. 1.

Transcript of the Shorthand Notes taken at the Poplar Inquiry, by J. S. Davy etc., Cd. 3274/1906, civ, p. 97.

Royal Commission on Coast Erosion, Reclamation of Tidal Lands, and Afforestation in the United Kingdom: First Report, Cd. 3883/1907, xxxiv, p. 1; *Second Report*, Cd. 4460/1909, xiv, p. 125; *Minutes of Evidence and Appendices accompanying the Second Report*, Cd. 4461/1909, xiv, p. 185.

Departmental Committee on Agricultural Education in England and Wales: Report, Cd. 4206/1908, xxi, p. 363.

Report on Dock Labour in Relation to Poor Relief, by the Hon. Gerald Walsh, Cd. 4391/1908, xcii, p. 483.

Royal Commission on the Poor Laws and Relief of Distress: Report, Cd. 4499/1909, xxxvii, p. 1; *Appendix,* Vol. VIII, *Unemployment,* Cd. 5066/1910, xlviii, p. 1; Vol. IX, *Unemployment,* Cd. 5068/1910, xlix, p. 1; Vol. XI, *Miscellaneous Papers,* Cd. 5072/1910, li, p. 1; Vol. XVI, *Reports on the Relation of Industrial and Sanitary Conditions to Pauperism,* by A. Steel Maitland and R. Squire, Cd. 4690/1909, xliii, p. 433; Vol. XIX, *Report on the Effects of Employment or Assistance given to the Unemployed since 1886 as a Means of Relieving Distress outside the Poor Law,* by Cyril Jackson and J. C. Pringle, Cd. 4795/1909, xliv, p. 1; Vol. XXVII, *Replies by Distress Committees to Questions circulated on the Subject of the Unemployed Workmen Act,* 1905, Cd. 4944/1909, xlv, p. 43.

Board of Education. Report of the Consultative Committee on Attendance, Compulsory or Otherwise, at Continuation Schools: Report and Appendices, Cd. 4757/1909, xvii, p. 6; *Summaries of Evidence,* Cd. 4758/1909, xvii, p. 353.

Royal Commission Appointed to Enquire into and Report on the Canals and Inland Navigations of the United Kingdom: Fourth Final Report, Cd. 4979/1910, xii, p. 1.

Tables showing the Rules and Expenditure of Trade Unions in Respect of Unemployed Benefits and also showing Earnings in the Insured Trades, Cd. 5703/1911, lxxiii, p. 479.

Royal Commission on the Civil Service: Second Report, Cd. 6534/1912–13, xv, p. 255; *Minutes of Evidence,* Cd. 6535/1912–13, xv, p. 259.

First Report of the Proceedings of the Board of Trade Under Part II of the National Insurance Act, Cd. 6965/1913, xxxvi, p. 677.

Departmental Committee on Local Taxation, Final Report, Cd. 7315/1914, xl, p. 537.

Royal Commission on Unemployment Insurance: First Report, Cmd. 3872/1930–1, xvii, p. 885; *Final Report,* Cmd. 4185/1931–2, xii, p. 393.

(*b*) HOUSE OF COMMONS PAPERS

(1) *Annual*

Returns as to the Proceedings of Distress Committees, 1905/6 to 1913/14.
Reports on the Proceedings of the Development Commission, 1911 to 1914/16.
Development Fund Accounts, 1910/11 to 1929/30.
Reports on the Proceedings of the Road Board, 1911 to 1921.
Unemployment Fund Accounts, 1912/13 to 1920/21.

(2) *Occasional*

Select Committee on Smallholdings, Report, H. of C. 223/1890, xvii, p. 183.
Select Committee on the Charity Commissioners, Report, H. of C. 221/1894, xi, p. 1.
Select Committee on Distress from Want of Employment (1895): *First Report and Minutes of Evidence,* H. of C. 111/1895, viii, p. 1; *Second Report and Appendices,* H. of C. 253/1895, viii, p. 215; *Third Report, Evidence and Appendices,* H. of C. 365/1895, ix, p. 1.

Select Committee on Distress from Want of Employment (1896): *Report and Minutes of Evidence*, H. of C. 321/1896, ix, p. 301.

Report on Labour Bureaux to the President of the L.G.B. by Arthur Lowry, H. of C. 86/1906, cii, p. 363.

Select Committee on the Income Tax: Report, H. of C. 365/1906, ix, p. 659.

Select Committee on the Housing of the Working Classes Acts Amendment Bill. Report, Special Report, Proceedings, Minutes of Evidence and Index, H. of C. 376/1906, ix, p. 1.

Select Committee on Homework. Report, Proceedings and Minutes of Evidence, H. of C. 290 Ind./1907, vi, p. 55; H. of C. 246 Ind./1908, viii, p. 1.

Memorandum on Existing Powers as to Making Grants for Various Purposes, H. of C. 278/1909, lxxx, p. 47.

(*c*) OTHER OFFICIAL PUBLICATIONS

Hansard.
Board of Trade. Eight Volumes on Employment and Trade 1905–09.
National Insurance Acts, 1911 to 1915. Unemployment Insurance. Decisions given by the Umpire respecting claims, 2 vols.
Board of Trade. Labour Gazette.

PART THREE: COLLECTIONS OF REPORTS, PAMPHLETS AND OTHER PRINTED MATERIAL, PUBLISHED BETWEEN 1886 AND 1914

Beveridge Coll. B. Unemployment, Central and Local Government Reports and Pamphlets 1886–1914, Vols. I–XX [British Library of Political Science].

Burns Collection of Reports and Pamphlets on the Eight Hours Movement, 5 vols. [Training College Library of the T.U.C.].

Burns Collection of Reports and Pamphlets on Unemployment [Training College Library of the T.U.C.].

Family Welfare Association Collection. Annual Reports, Special Reports and Occasional Papers of the Charity Organisation Society [Goldsmiths' Library, University of London].

PART FOUR: REPORTS BY LOCAL AUTHORITIES AND OTHER INSTITUTIONS*

Amalgamated Society of Engineers, *Annual Reports.*
Bimetallic League, *Proceedings of the Bimetallic Conference held at Manchester*, 4 and 5 April 1898.
Central (Unemployed) Body: *Annual Reports; Minutes; Miscellaneous Papers* [Coll. B].

* Some of the principal items in this section and in section six are to be found in one or more of the collections listed under section three. They are indicated by [Coll. B], [T.U.C.], or [G.L.]. Items in the general section of London University Library are indicated by [L.U.].

Christian Social Service Union. *The Problem of the Unemployed. Notice of a Conference on 'Labour and Training Colonies'*, 5 June 1905 [T.U.C.].

Distress Committees under the Unemployed Workmen Act, 1905. *Annual Reports* [Coll. B].

Egham Free Registry, *Annual Reports* [Coll. B].

Eugenics Education Society, 'Report of the Committee Appointed to Consider the Eugenic Aspect of Poor Law Reform', *Eugenics Review*, 2, no. 3 (Nov. 1910), 167–203.

Fabian Society, *The Government Organisation of Unemployed Labour*, 1886 [T.U.C.].

General Federation of Trade Unions: *10th Annual Report, 22nd Quarterly Report* [T.U.C.].

Independent Labour Party, *Conference Reports*, 1905–10.

International Conference on Unemployment, Paris 1910. Reports of proceedings, 3 vols.

Labour Representation Committee and Labour Party, *Annual Reports*, 1900–9.

The Land. The Report of the Land Enquiry Committee, 2 vols., 1912–13.

London County Council. *Minutes of a Conference on Lack of Employment in London*, 13 Feb. and 3 Apr. 1903 [Coll. B].

London Unemployed Fund Report of the Central Executive Committee 1904–5.

Report of the Mansion House Committee, appointed March 1885 to Inquire into the Causes of Permanent Distress in London and the Best Means of Remedying the Same, presented to the Lord Mayor, Dec. 1885 [T.U.C.].

Report of Mansion House Conference on the Condition of the Unemployed, 1887–8 [Coll. B].

Report of the Mansion House Conference on the Condition of the Unemployed, 1892–3 [L.U.].

Report of the Executive Committee of the Mansion House Conference on the Unemployed, 1893–4 [T.U.C.].

Mansion House Committee on the Unemployed, 1903–4. *Report of the Executive Committee* [L.U.].

National Conference on the Prevention of Destitution, Report of Proceedings, 1911.

National Conference on the Unemployment of Women Dependent on their Own Earnings, Report, 1907 [Coll. B].

Poplar Labour Colony, Report, 1904 [L.U.].

Salvation Army, *Annual Reports on Social Work*, 1902/3 to 1915/16.

PART FIVE: BIBLIOGRAPHIES AND WORKS OF REFERENCE

Annual Charities Register and Digest. Annual Register.

Dictionary of National Biography and Supplements.

Dod, *Parliamentary Companion.*

Encyclopaedia of Social Reform (1909 ed.), 2 vols.

P. and G. Ford (1957), *A Breviate of Parliamentary Papers 1900–16. The Foundation of the Welfare State.*

P. C. Lyon, *Report on the Correspondence of the Revd. L. R. Phelps, D.C.L. from 1877 to 1936.*

I.L.O. (1926), *Bibliography of Unemployment.*

W. Poole (1891), *Index to Periodical Literature*, and supplements by W. Fletcher.

Readers' Guide to Periodical Literature.

F. I. Taylor (1909), *A Bibliography of Unemployment and the Unemployed.*

The Times, House of Commons Lists of Members.

Year Book of Social Progress 1912–15.

PART SIX: NEWSPAPERS AND PERIODICALS

(*a*) Continuously: *Charity Organisation Review, Liberal Magazine.*

(*b*) Selected periods: *Daily News, Fabian News, Pall Mall Gazette, Nation, The Times, Toynbee Record, West Ham Herald.*

(*c*) Collections of press-cuttings in Beveridge MSS.; Buxton MSS.; H.O. 45/9861/13077; and H.L.G., vol. 77.

(*d*) *Minlabour*, Labour Exchanges Jubilee edition, vol. 14, no. 1, Jan. 1960 (publ. by the Ministry of Labour).

(*e*) *The Window*, National Insurance Jubilee edition, July 1962 (publ. by the Ministry of Pensions and National Insurance).

PART SEVEN: WORKS WITH SPECIAL REFERENCE TO UNEMPLOYMENT

ADLER, N., and TAWNEY, R. H., *Boy and Girl Labour* (1909).

ALDEN, P., *The Unemployed. A National Question* (1905).

—— and HAYWARD, P. E., *The Unemployable and the Unemployed* (1908).

ANON. *The Unemployed and the Proposed New Poor Law* (1843).

ANON. (H. H. Champion), *The Facts about the Unemployed.* An Appeal and a Warning by 'one of the middle class' (1886) [T.U.C.].

BALDWINSON, P., *Unemployment: its causes and suggestions for cure* (1908).

BARNES, G. N., *The Unemployed Problem*, I.L.P. pamphlet (1909).

—— and HENDERSON, A., *Unemployment in Germany*, I.L.P. pamphlet (1908).

BARNETT, S., 'A scheme for the Unemployed', *Nineteenth Century*, 24 (Nov. 1888), 753–63.

BARTLEY, GEORGE C. T., *London and the Unemployed Problem*, COS Occasional Paper No. 5, 4th series, Jan. 1905.

BEALES, H., and LAMBERT, R. S., *Memoirs of the Unemployed* (1934).

BEARDSLEY, CHARLES, 'The Effect of an Eight Hours Day on Wages and the Unemployed', *Quarterly Journal of Economics*, 9 (1894–5), 450–9.

BENNINGTON, J. W., *Unemployment and the Remedy* (Southsea, 1908).

BEVERIDGE, W. H., 'The Vagrant and the Unemployable', *Toynbee Record*, Apr. 1904, 97–105.

BEVERIDGE, W. H., 'Unemployment in London'; I. *Toynbee Record*, Oct. 1904, 9–15; III. 'The Making of Paupers', *Toynbee Record*, Nov. 1904, 27–9. IV. 'The Preservation of Efficiency', *Toynbee Record*, Dec. 1904, 43–7; V. 'The Question of Disfranchisement', *Toynbee Record*, Mar. 1905, 100–2.

—— 'Emergency Funds for the Relief of the Unemployed: A Note on their Historical Development', *Clare Market Review*, 1, no. 3 (May 1906), 73–8.

—— 'The Problem of the Unemployed', *Sociological Papers*, 3 (1906), 323–41.

—— 'Labour Bureaux', *Economic Journal*, 16, no. 63 (Sept. 1906), pp. 436–9.

—— 'Labour Exchanges and the Unemployed', *Economic Journal*, 17, no. 65 (Mar. 1907), 66–81.

—— *Metropolitan Employment Exchanges of the Central (Unemployed) Body* (1907).

—— 'The Pulse of the Nation', *Albany Review*, 2 (Nov. 1907), 160–70.

—— 'The Unemployed Workmen Act in 1906–7', *Sociological Review*, 1 (1908), 79–83.

—— 'Public Labour Exchanges in Germany', *Economic Journal*, 18, no. 69 (Mar. 1908), 1–18.

—— 'Unemployment and its Cure', *Contemporary Review*, 93 (Apr. 1908), 385–98.

—— *Unemployment. A Problem of Industry* (1909) (1910 and 1930 eds.)

—— 'A Seventeenth Century Labour Exchange', *Economic Journal*, 24 (Sept. 1914), 371–6; (Dec. 1914), 635–6.

—— 'The Past and Present of Unemployment Insurance', *Barnett House Papers*, No. 13, 1930.

—— *Causes and Cures of Unemployment* (1931).

—— *Full Employment in a Free Society* (1944).

BOOTH, CHARLES, 'Inaugural Address as President of the Royal Statistical Society', *Journal of the Royal Statistical Society*, 60 (Dec. 1892), 521–57.

BOOTH, WILLIAM, *The Vagrant and the 'Unemployable'* (1904) [Coll. B].

—— *Emigration—Colonisation* (1905) [Coll. B].

BOSANQUET, H., *Past Experience in Relief Works*, COS pamphlet (1903).

BOURNE, H. C., 'The Unemployed', *Macmillan's Magazine*, 67 (Dec. 1892), 81–90.

BOUSFIELD, W., 'The Unemployed', *Contemporary Review*, 70 (Dec. 1896), 835–52.

BOWLEY, A., 'The Measurement of Unemployment: An Experiment', *Journal of the Royal Statistical Society*, 75 (July 1912), 791–822.

BROOKS, J. G., 'The Unemployed in German Cities', *Quarterly Journal of Economics*, 7 (1892–3), 353–8.

—— 'Insurance of the Unemployed', *Quarterly Journal of Economics*, 10 (1895–6), 341–8.

BURLEIGH, BENNET, 'The Unemployed', *Contemporary Review*, 52 (Dec. 1887), 770–80.

BURNS, E. M., *The British Unemployment Programmes 1920–38* (1941).

Burns, John, *The Unemployed*, Fabian Tract No. 47 (1893).

Caldwell, J. A. M., 'Social Policy and Public Administration 1909–11', Unpublished Nottingham Ph.D. thesis, 1956.

Campbell, D., *The Unemployed Problem—The Socialist Solution*, S.D.F. pamphlet (1892) [T.U.C.].

Chapman, S. J., and Hallsworth, H. M., *Unemployment: the results of an investigation made in Lancashire* (Manchester University Economic Series No. 12, 1909).

Chegwidden, T. S., and Myrddin-Evans, G., *The Employment Exchange Service of Great Britain* (foreword by W. S. Churchill), Industrial Relations Counsellors Inc. (New York, 1934).

Christian Social Service Union, *Labour and Training Colonies* (1905).

Cohen, J. L., *Insurance Against Unemployment With Special Reference to British and American Conditions* (1921).

Corry, B., 'Theory of the Economic Effects of Government Expenditure in English Classical Political Economy', *Economica*, n.s. 25 (Feb. 1958), 34–48.

Crabb, J., *Bad Times—Their Cause and Cure*, by a Radical (publ. by the National Fair Trade League, 1885).

Dallas, Duncan, *How to Solve the Unemployed Problem by Cooperative Organisation of the Unemployed with State Control* (1895).

Davidson, John, *Unemployment and the Wage-Fund* (publ. by National Labour Press, n.d. ? 1909).

Davison, R. C., *The Unemployed. Old Policies and New* (1929).

Dearle, N. B., *Problems of Unemployment in the London Building Trades* (1908).

Dessauer, Marie, 'Unemployment Records 1848–59', *Economic History Review*, 1st series, 10, no. 1 (Feb. 1940), 38–43.

Dodd, J. T., *The Winter's Distress: How to Provide for the Unemployed. To Boards of Guardians in Rural Districts* (1894) [T.U.C.].

Drage, G., *The Unemployed* (1894).

Emmerson, H. C., and Lascelles, E. C. P., *Guide to the Unemployment Insurance Acts* (1930).

Fels, J., and Orr, J., *The Remedy for Unemployment* (Land Valuation Publication Department, Glasgow, 1908).

Flinn, M. W., 'The Poor Employment Act of 1817', *Economic History Review*, 14, no. 1 (Aug. 1961), 82–92.

Foxwell, H. S., *Irregularity of Employment and Fluctuations of Prices* (Edinburgh, 1886).

George, Henry, *Progress and Poverty* (1883).

—— *Social Problems* (1884).

Gibb, Revd. Spencer J., *The Irregular Employment of Boys* (C.S.U. pamphlet, 1903).

Gibbon, I. G., *Unemployment Insurance. A Study of Schemes of Assisted Insurance* (1911).

Gilbert, B. B., 'Winston Churchill versus the Webbs: The Origins of British Unemployment Insurance', *American Historical Review*, 71, no. 3 (Apr. 1966), 846–62.

GILSON, M. B., *Unemployment Insurance in Great Britain. The National System and Additional Benefit Plans* (New York, 1931).

HANCOCK, K. J., 'The Reduction of Unemployment as a Problem of Public Policy 1920–29', *Economic History Review*, 15, no. 2 (Dec. 1962), 328–43.

HARDIE, KEIR, *The Unemployed Problem with Some Suggestions for Solving It*, I.L.P. pamphlet (1904) [Coll. B].

—— *John Bull and His Unemployed. A Plain Statement on the Law of England as it affects the Unemployed*, I.L.P. pamphlet (1905) [Coll. B].

HATCH, E. F. G., *Reproach to Civilisation: A Treatise on the Problem of the Unemployed* (1906).

HAWTREY, R. G., *Good and Bad Trade, An Inquiry into the Causes of Trade Fluctuations* (1913).

HENDERSON, W. O., *The Lancashire Cotton Famine 1861–1865* (Manchester University Press, 1934).

HIGGS, M., *How to Deal with the Unemployed* (1904).

HILL, ALSAGER HAY, *Our Unemployed: an Attempt to Point out Some of the Best Means of Providing Occupation for Distressed Labourers: with Suggestions on a National System of Registration; and Other Matters Affecting the Well-Being of the Poor* (1868) [T.U.C.].

—— 'Unemployed Labour. What Means are Practicable for Checking the Aggregation and Deterioration of Unemployed Labour in Large Towns?', *Transactions of the National Association for the Promotion of Social Science*, 1875, 656–73.

HOBSON, J. A., *Cooperative Labour upon the Land (And Other Papers)* (1894).

—— 'The Meaning and Measure of "Unemployment"', *Contemporary Review*, 67 (Mar. 1895), 415–32.

—— 'The Economic Cause of Unemployment', *Contemporary Review*, 67 (May 1895), 744–60.

—— *The Problem of the Unemployed. An Enquiry and an Economic Policy* (1896) (1908 ed.).

HOWELL, GEORGE, *Waste Land and Prison Labour* (1877) [T.U.C.].

HUNT, W., *Labour Colonies: What They Are and What They Can Be Made To Do* (1905–6) [Coll. B].

HYNDMAN, H. M., *Commercial Crises of the Nineteenth Century* (1892).

KEELING, F., *The Labour Exchange in Relation to Boy and Girl Labour* (1910).

LASCELLES, E. C. P., and BULLOCK, S. S., *Dock Labour and Decasualisation* (1924).

LAYTON, WALTER, 'The Government and Unemployment', *St. George Utopian Papers*, July 1909 [Coll. B].

LAZARD, MAX, *Le Chômage et la profession* (Paris, 1909).

LEWIS, F. M., *State Insurance. A Social and Industrial Need* (1909).

LOCH, C. S., 'Manufacturing a New Pauperism', *Nineteenth Century*, 37 (Apr. 1895), 697–708.

—— *Employment Relief*, COS Occasional Paper No. 23, 4th series, Nov. 1905.

—— 'La Lutte pour le travail et les inemployés'. Paper read at the Congress of the Institut Internationale de Sociologie, London, 3–6 July 1906.

MACDONALD, J. RAMSAY, *The New Unemployed Bill of the Labour Party*, I.L.P. Pamphlet (1907).

MACKAY, D. I., FORSYTH, D. J. C., and KELLY, D. M., 'The Discussion of Public Works Programmes, 1917–1935: Some Remarks on the Labour Movement's Contribution', *International Review of Social History*, 11 (1966), 8–17.

MAJOR, M. B., *Unemployment and the Gold Reserve* (1907) [L.U.].

MANN, TOM, *The Programme of the I.L.P. and the Unemployed* (1895).

MARTIN, A. J., *The Remedy for Unemployment*, Liberty and Property Defence League pamphlet (1895) [T.U.C.].

MAVOR, JAMES, *et al.*, *Report on Labour Colonies to the Glasgow Labour Centres Committee* (1892).

—— 'Setting the Poor on Work', *Nineteenth Century*, 34, Oct. 1893, pp.523–32.

—— 'Labour Colonies and the Unemployed', *Journal of Political Economy*, 2 (Dec. 1893), 26–53.

MILLS, H. V., *Poverty and the State* (1886).

—— *Home Colonisation and Work for the Unemployed*, publ. by the Home Colonisation Society (Liverpool, n.d.).

MOORE, H., 'The Unemployed and the Land', *Contemporary Review*, 63 (Mar. 1893), 423–38.

MORLEY, F., *Unemployment Relief in Great Britain: A Study in State Socialism* (1924).

NICHOLSON, J. SHIELD, *The Tariff Question with Special Reference to Wages and Employment* (1903).

OLIVER, W. H., 'The Labour Exchange Phase of the Co-operative Movement', *Oxford Economic Papers*, N.S. 10 (1958), 355–67.

PATON, J. B., *The Unemployable and the Unemployed*, Inner Mission pamphlet (1905) [Coll. B].

PIGOU, A. C., *Unemployment* (1913).

—— *The Theory of Unemployment* (1933).

RATHBONE, E., and WOOD, G. H., *Report of an Inquiry on the Conditions of Labour at the Liverpool Docks* (1904).

ROWNTREE, B. S., and LASKER, B., *Unemployment: A Social Study* (1911).

—— *The Way to Industrial Peace and the Problem of Unemployment* (1914).

SAMUELS, H. B., *What's to be Done? The Unemployed Question Considered* (1892) [T.U.C.].

SCHLOSS, D. F., *Insurance Against Unemployment* (1909).

SEYMOUR, J. B., *The British Employment Exchange* (1928).

SMART, H. RUSSELL, *The Right to Work* (1895) [T.U.C.].

Social Democratic Federation, *State Organisation of Unemployed Labour. As an Alternative to the Harmful Scheme of State-Aided Emigration* (1883) [T.U.C.]

SUTTER, J., *A Colony of Mercy* (1893).

THORESBY, FREDERICK, 'How to Deal with the Unemployed', *Westminster Review*, 165 (Jan. 1906), 36–40.

TILLYARD, F., 'Three Birmingham Relief Funds 1885, 1886 and 1905', *Economic Journal*, 15 (Dec. 1905), 505–20.

TILLYARD, F., and BALL, F. N., *Unemployment Insurance in Great Britain 1911–48* (Thames Bank Publ. Co., Leigh on Sea, Essex, 1949).

TOYNBEE, H. V., 'A Winter's Experiment', *Macmillan's Magazine*, 69 (Nov. 1893), 54–8.

—— 'The Problem of the Unemployed', *Economic Review*, xv (July 1905), pp. 291–305.

WALLACE, ALFRED RUSSELL, *Suggestions for Solving the Problem of the Unemployed*, Land Nationalisation Society, Tract No. 64 (1895) [T.U.C.].

WEBB, S., *How the Government Can Prevent Unemployment*, Pamphlet publ. by the National Committee for the Prevention of Destitution (1912).

WHETHAM, W. C. D., *Eugenics and Unemployment*, Reprint of a lecture given in Cambridge (1910).

WILLIAMS, R., *The Liverpool Docks Problem* (Liverpool Economic and Statistical Society, 1912).

—— *The First Year's Working of the Liverpool Docks Scheme* (Liverpool Economic and Statistical Society, 1914).

WILLINK, H. G., *The Report of the Departmental Committee on Vagrancy, with especial regard to Labour Colonies* (c. 1906) [L.U.].

WOODWORTH, A. V., *Report of an Inquiry into the Condition of the Unemployed*, conducted under the Toynbee Trust, winter 1895–6 (1897) [Coll. B].

WOYTINSKY, WLADIMIR, *Three Sources of Unemployment. The Combined Action of Population Changes, Technical Progress and Economic Development*, I.L.O. (Geneva, 1935).

WRIGHT, A. L., 'The Genesis of the Multiplier Theory', *Oxford Economic Papers*, N.S. 8 (1956), 181–93.

Yale Law Journal, 55 (1945–6), Special edition on Unemployment Insurance.

PART EIGHT: POLITICAL, SOCIAL, AND ECONOMIC BACKGROUND

ABEL-SMITH, B., and PINKER, R., *Changes in the Use of Institutions in England and Wales between 1911 and 1951* (Manchester Statistical Society, 1960).

ALDCROFT, D. H. (ed.) *The Development of British Industry and Foreign Competition 1875–1914* (1968).

Amalgamated Society of Engineers, *The Eight Hours Day. A Ton of Practice* (A.S.E. pamphlet, reprinted from the *Daily News*, 20 July 1897) [T.U.C.].

ARMYTAGE, W. H. G., *Heavens Below. Utopian Experiments in England 1560–1960* (1961).

ASHTON, T. S., 'The Relation of Economic History to Economic Theory', *Economica*, n.s. 13 (Feb. 1946), 81–96.

ASKWITH, G. R., *Industrial Problems and Disputes* (1920).

ASPINALL, A., *The Early English Trade Unions* (1949).

AUSUBEL, HERMAN, 'General Booth's Scheme of Social Salvation', *American Historical Review*, 56, no. 3 (Apr. 1951), 519–25.

BARNETT, H. and S., *Towards Social Reform* (1909).

BARRY, MALTMAN, *The Labour Day* (Aberdeen Trades Council pamphlet, 1890) [T.U.C.].

BEALES, H., '"The Great Depression" in Industry and Trade', *Economic History Review*, 1st series, 5, no. 1 (Oct. 1934), 65–75.

BEALEY, FRANK, 'The Electoral Arrangement between the Labour Representation Committee and the Liberal Party', *Journal of Modern History*, 28, no. 4 (Dec. 1956), 353–73.

—— 'Keir Hardie and the Labour Group', *Parliamentary Affairs*, 10 (1956–7), 81–93 and 220–33.

—— and PELLING, H., *Labour and Politics 1900–1906. A History of the Labour Representation Committee* (1958).

BELL, F., LADY, *At the Works. A Study of a Manufacturing Town* (1907).

BEVERIDGE, W. H., *Insurance for all and everything* (1924).

—— *Voluntary Action* (1948).

BLEWETT, NEAL, 'The Franchise in the United Kingdom 1885–1918', *Past and Present*, 32 (Dec. 1965), 27–56.

BOOTH, CHARLES, *Life and Labour of the People of London*, vols. i, ii, iv, v, vii, viii, ix (1892–7).

BOOTH, W., *In Darkest England and the Way Out* (1890).

BOSANQUET, H., *Rich and Poor* (1896).

BOUTMY, ÉMILE, *The English People. A Study of their Political Psychology* (transl. from French by E. English, 1904).

BOWLEY, A. L., *Changes in the Distribution of National Income 1880–1913* (1920).

—— *Wages and Income in the United Kingdom since 1860* (1937).

—— and HURST, A. R. BURNETT, *Livelihood and Poverty* (1915).

BRABROOK, E., *Provident Societies and Industrial Welfare* (1898).

BRADLAUGH, CHARLES, *The Eight Hours Movement* (pamphlet reprinted from the *New Review*, 1889).

BRIGGS, ASA, 'The Welfare State in Historical Perspective', *Archives européennes de sociologie*, 2 (1961), 221–58.

BROWN, E. PHELPS, *The Growth of British Industrial Relations. A Study from the Standpoint of 1906–14* (1959).

BROWN, J. C., 'Ideas concerning social policy and their influence on legislation in Great Britain 1902–11' (London Ph.D. thesis 1964).

BROWNE, SIR BENJAMIN, *Selected Papers on Social and Economic Questions* (ed. by his daughters E. M. B. and H. M. B.) (Cambridge, 1918).

BRUCE, M., *The Coming of the Welfare State* (1961).

BURNS, JOHN, *The Eight Hour Day. Facts to remember*, reprinted from the *Daily Chronicle* (n.d. ? 1897–8) [T.U.C.].

—— *A Straight Tip to Workmen. Brains Better than Bets or Beer*, Clarion Pamphlet No. 36 (1902) [T.U.C.].

—— *Labour and Drink* (1904) [T.U.C.].

BUXTON, SYDNEY, *Finance and Politics* (1888).

—— *A Handbook to Political Questions of the Day* (1880) (11th ed., 1903).

CAIRNCROSS, A., *Home and Foreign Investment 1870–1913. Studies in Capital Accumulation* (Cambridge, 1953).

CALDWELL, J. A. M., 'The Genesis of the Ministry of Labour', *Public Administration*, 37 (Winter 1959), 367–91.

CAMPBELL-BANNERMAN, H., *Speeches by Sir Henry Campbell-Bannerman. Selected and Reprinted from the Times* (1908).

CANNAN, E., *The History of Local Rates in England in Relation to the Proper Distribution of the Burden of Taxation* (1898) (1927 ed.).

CHAMBERLAIN, J., *The Radical Programme*, reprinted from the *Fortnightly Review*, 1884 (1885).

—— *Imperial Union and Tariff Reform*, Speeches delivered from May 15 to Nov. 4, 1903 (1903).

Charity Organisation Society, *How to Help Cases of Distress* (1894 and 1895 eds.).

CHECKLAND, S. G., 'The Propagation of Ricardian Economics in England', *Economica*, N.S. 16 (Feb. 1949), 40–52.

CLEGG, H. A., FOX, ALAN, and THOMPSON, A. F., *A History of British Trade Unions since 1889, Vol. I, 1889–1910* (Oxford, Clarendon Press, 1964).

CLEMENTS, R. V., 'British Trade Unions and Popular Political Economy 1850–1875', *Economic History Review*, 2nd series, 14, no. 1 (Aug. 1961), 93–104.

COLE, M., *The Story of Fabian Socialism* (1961) (1963 ed.).

CROMWELL, VALERIE, 'Interpretations of Nineteenth Century Administration: An Analysis', *Victorian Studies*, 9, no. 3 (Mar. 1966), 245–55.

CUNNISON, JAMES, 'Some Factors Affecting the Incidence of the National Insurance Contributions', *Economic Journal*, 23 (1913), 367–78.

DANGERFIELD, G., *The Strange Death of Liberal England* (1936).

DONALD, A. K., *The Eight Hours Working Day. Its Advantages and How to Obtain it* (1890) [T.U.C.].

DUFFY, A. E. P., 'New Unionism in Britain 1889–1890: A Reappraisal', *Economic History Review*, 2nd series, 14, no. 2 (Dec. 1961), 306–19.

Economic Club, *Family Budgets. Being the Income and Expenditure of Twenty-Eight British Households 1891–1894* (1896).

ENSOR, R. C. K., *England 1870–1914* (Oxford, Clarendon Press, 1936).

Fabian Society, Tract No. 9, *An Eight Hours Bill in the Form of an Amendment to the Factory Acts, with further provisions for the Improvement of the Conditions of Labour* (1890).

—— Tract No. 16, *A Plea for an Eight Hours Bill* (1890).

—— Tract No. 23, *The Case for an Eight Hours Bill* (1891).

GARDINER, A. G. (ed.), *To Colonise England* (1907).

GARTNER, LLOYD P., *The Jewish Immigrant in England 1870–1914* (1960).

GEORGE, D. LLOYD, *Better Times* (1910).

GIBBON, GWILYM, and BELL, REGINALD, *History of the London County Council 1889–1939* (1939).

GIFFEN, R., *The Case Against Bimetallism* (1892).

—— 'Notes on Imports *versus* Home Production, and Home *versus* Foreign Investments', *Economic Journal*, 15 (1905), 483–93.

GILBERT, B. B., *The Evolution of National Insurance in Great Britain. The Origins of the Welfare State* (1966).

GINSBERG, M. (ed.), *Law and Opinion in England in the 20th Century* (1959).

GOLDTHORPE, JOHN, 'The Development of Social Policy in England 1800–1914. Notes on a Sociological approach to a problem in historical explanation.' *Transactions of the 5th World Congress of Sociology* (Washington), 4 (1962), 41–56.

HADFIELD, R. A., and GIBBINS, H. DE B., *A Shorter Working Day* (1892).

HAGGARD, H. RIDER, *Rural England* (1902).

—— *Regeneration* (1910).

HALEVY, E., *History of the English People in the Nineteenth Century* (1932), vol. v, Imperialism and the Rise of Labour 1895–1905; vol. vi, The Rule of Democracy 1905–1914.

HART, JENNIFER, 'Nineteenth-Century Social Reform: A Tory Interpretation of History', *Past and Present*, 31 (July 1965), 39–61.

HAWKINS, C. B., *Norwich. A Social Study* (1910).

HEWITT, MARGARET, *Wives and Mothers in Victorian Industry* (1958).

HIGGS, HENRY, 'Workmen's Budgets', *Journal of the Royal Statistical Society*, 56 (June 1893), 255–94.

HIGGS, MARY, *Glimpses into the Abyss* (1906).

HOBHOUSE, L. T., *Liberalism* (1911).

HOBSBAWM, E. J., *Labouring Men. Studies in the History of Labour* (1964).

—— 'The Fabian Society 1884–1913', Cambridge Ph.D. thesis.

HOBSON, C. K., *The Export of Capital* (1914) (1963 ed.).

HOBSON, J. A., *The Evolution of Modern Capitalism. A Study of Machine Production* (1894).

—— 'The Social Philosophy of Charity Organisation', *Contemporary Review*, 70 (Nov. 1896), 710–27.

HOFFMAN, R. J. S., *Great Britain and German Trade Rivalry* (Philadelphia, 1933).

HOFFMAN, W., *British Industry 1700–1950* (Oxford, 1955).

HOMER, S., *A History of Interest Rates* (New Jersey, 1963).

HONIGSBAUM, F., 'The Struggle for the Ministry of Health', *Occasional Papers in Social Administration*, No. 37, 1970.

HOWARTH, E. G., and WILSON, M. (compilers), *West Ham. A Study in Social and Industrial Problems* (1907).

HOWE, ELLIC, and WAITE, H., *The London Society of Compositors: A Centenary History* (1948).

HUTCHINSON, T. W., *A Review of Economic Doctrines 1870–1929* (Oxford, 1953) (1962 ed.).

HYNDMAN, H. M., *Gladstone and the Eight Hours Law* (n.d.) [T.U.C.].

—— *The Eight Hours Movement* (1890) [T.U.C.].

—— and BRADLAUGH, C., *Report of a Debate on the Eight Hours Movement* (1890) [T.U.C.].

INGLIS, K. S., *Churches and the Working Classes in Victorian England* (1963).

JEANS, J. STEPHEN, *The Eight Hours Day in the British Engineering Industry. An Examination and Criticism of Recent Experiments* (1894).

JEFFERYS, J. N., *The Story of the Engineers 1800–1945* (1946).

JEVONS, W. S., *The State in Relation to Labour* (1882).

—— *Methods of Social Reform and Other Papers* (1883).

KING, A. S., 'Some Aspects of the History of the Liberal Party 1906–14', Oxford D.Phil. thesis 1962–3.

LANSBURY, G., *The Principles of the English Poor Law* (1897) [T.U.C.].

LEATHAM, JAMES, *An Eight Hours Day with Ten Hours Pay: How to get it and How to Keep it* (Aberdeen, 1890).

LEE, H. W., and ARCHBOLD, E., *Social-Democracy in Britain, Fifty Years of the Socialist Movement* (1933).

LETWIN, S. R., *The Pursuit of Certainty* (Cambridge University Press, 1965).

LIPMAN, V. D., *Social History of the Jews in England 1850–1950* (1954).

—— *A Century of Social Service 1859–1959. The Jewish Board of Guardians* (1959).

LOCH, C. S., *The Elberfield System*, COS Occasional Paper, No. 20, 3rd series, Dec. 1903.

—— (ed.) *Methods of Social Advance* (1904).

LUCAS, B. KEITH, 'Poplarism', *Public Law*, Spring 1962, pp. 52–80.

LYND, H. M., *England in the Eighteen Eighties. Towards a Basis for Social Freedom* (1945).

MACBRIAR, A. M., *Fabian Socialism and English Politics 1884–1918* (Cambridge University Press, 1962).

McCORMICK, B., and Williams, J. E., 'The Miners and the Eight-Hour Day, 1863–1910', *Economic History Review*, 2nd series, 12, no. 2 (Dec. 1959), 222–37.

MACDONAGH, O., 'The Nineteenth Century Revolution in Government: A Reappraisal', *Historical Journal*, 1, no. 1 (1958) 52–67.

McGREGOR, O. R., 'Social Research and Social Policy in the Nineteenth Century', *British Journal of Sociology*, 8 (1957), 146–57.

MACKAY, T., *The English Poor. A Sketch of their Social and Economic History* (1889).

MALLET, BERNARD, *British Budgets 1887–1913* (1913).

MANN, P. H., 'Life in an Agricultural Village in England', *Sociological Papers*, 1 (1904), 163–93.

MANN, TOM, *What a Compulsory Eight Hour Working Day Means to the Workers* (1886).

—— *The Eight Hours Movement* (1889) [T.U.C.].

—— *The Eight Hour Day. How to get it by Trade and Local Option* (1892) [T.U.C.].

MARSH, D. C., *Changing Social Structure of England and Wales 1871–1951* (1958).

MARSHALL, ALFRED, 'The Housing of the London Poor', *Contemporary Review*, 45 (Feb. 1884), 224–31.

—— *Elements of the Economics of Industry* (1892).

—— (ed. J. M. Keynes) *Official Papers* (1926).

MARSHALL, T. H., 'The Welfare State: A Sociological Interpretation', *Archives européennes de sociologie*, 2 (1961), 284–300.

—— *Social Policy in the Twentieth Century* (1965) (1967 ed.).

MARX, KARL, *Capital*, vol. i, transl. by S. Moore and E. Aveling and ed. by F. Engels 1886 (Foreign Languages Publishing House, Moscow, 1958).

MASTERMAN, C. F. G., *et al.*, *The Heart of the Empire* (1901).

—— *Condition of England* (1909).

—— *The New Liberalism* (1920).

MASTERMAN, N. (ed.), *Chalmers on Charity* (1900).

MATHER, W., *Trade Unions and the Hours of Labour* (1892).

MAYHEW, HENRY, *London Labour and London Poor*, vols. i and iii (1851–62).

Miners (Eight Hours) Bill. Reports of Deputations from Representatives of Miners of the United Kingdom to Henry Mathew Q.C. (Home Secretary), Lord Dunraven, Lord Randolph Churchill and W. E. Gladstone, 17 and 18 Feb. 1890 [T.U.C.].

MITCHELL, B. R., and DEANE, PHYLLIS, *Abstract of British Historical Statistics* (Cambridge, 1962).

MONEY, L. G. CHIOZZA, *Riches and Poverty* (1905) (3rd ed. 1906).

MOSES, R., *The Civil Service of Great Britain* (1913).

MOWAT, C. L., *The Charity Organisation Society 1869–1913. Its Ideas and Work* (1961).

OWEN, DAVID, *English Philanthropy 1660–1960* (1965).

PEACOCK, A. T., and WISEMAN, J., *The Growth of Public Expenditure in the United Kingdom* (1961).

PEASE, E., *History of the Fabian Society* (1913).

PELLING, H., *The Origins of the Labour Party 1880–1900* (Oxford, Clarendon Press, 1954).

—— *A History of British Trade Unionism* (1963).

—— *Social Geography of British Elections 1885–1910* (1967).

—— *Popular Politics and Society in Late Victorian Britain* (1969).

PERCY, C. M., *The Miners and the Eight Hours Movement*, Cobden Club pamphlet (1891) [T.U.C.].

PICHT, WERNER, *Toynbee Hall and the English Settlement Movement* (transl. from German by Lilian Cowle, 1914).

PIGOU, A. C., *Economics of Welfare* (1920) (4th ed., 1932).

PIMLOTT, J. A. R., *Toynbee Hall. Fifty Years of Social Progress 1884–1934* (1935).

POLLARD, S. *The Genesis of Modern Management* (1965).

PORTER, G. R., *The Progress of the Nation* (ed. and revised by F. W. Hirst, 1912).

RAE, JOHN, *Eight Hours for Work* (1894).

REEVES, M. S. PEMBER, *Round About a Pound a Week* (1914).

RICARDO, D., *Principles of Political Economy* (1817) (Vol. I of *Ricardo's Works*, ed. Piero Sraffa, 1951).

ROBBINS, L., *The Theory of Economic Policy in English Classical Political Economy* (1952).

ROBERTS, D. G., *Victorian Origins of the British Welfare State* (New Haven, Yale U.P., 1960).

ROBERTSON, D. H., *A Study of Industrial Fluctuation* (1915).

ROBERTSON, J. M., *The Fallacy of Saving* (1892).

—— *The Eight Hours Question* (1893).

RODGERS, BRIAN, 'The Medical Relief (Disqualification Removal) Act 1885', *Parliamentary Affairs*, 9 (1955–6), 188–94.

ROSEBERY, LORD, *National Policy* (printed as a pamphlet 1902).

ROWNTREE, B. S., *Poverty. A Study of Town Life* (1901).

—— and KENDALL, M. *How the Labourer Lives. A Study of the Rural Labour Problem* (1913).

SAMUEL, H., *Liberalism. Its Principles and Proposals* (1902).

SAVILLE, JOHN, *Rural Depopulation in England and Wales 1851–1951* (1957).

—— (ed.), 'Studies in the British Economy 1870–1914', *Yorkshire Bulletin of Economic and Social Research*, Special Number, 17, no. 1 (May 1965).

SAYERS, R., *Bank of England Operations 1890–1914* (1936).

SCHUMPETER, J., *Business Cycles*, Vol. I (New York, 1939).

SCHWEINITZ, KARL DE, *England's Road to Social Security* (1943) (Perpetua ed., University of Pennsylvania Press, 1961).

SEARLE, G. R., 'The Development of the Concept of "National Efficiency" and its relations to Politics and Government 1900–10', Cambridge Ph.D. thesis, 1965.

SEMMEL, B., *Imperialism and Social Reform* (1961).

SHADWELL, ARTHUR, *Industrial Efficiency. A Comparative Study of Industrial Life in England, Germany and America* (1905).

SHAW, G. B., et al., *Fabian Essays in Socialism* (1889).

SHEHAB, F., *Progressive Taxation. A Study of the Development of the Progressive Principle in the British Income Tax* (Oxford, Clarendon Press, 1953).

SIDGWICK, H., *The Principles of Political Economy* (1883) (1887 ed.).

SIMEY, M. B., *Charitable Effort in Liverpool in the Nineteenth Century* (1951).

SMITH, H. Llewellyn, *Modern Changes in the Mobility of Labour* (1891).

—— *The Board of Trade* (1928).

—— and NASH, V., *The Story of the Dockers' Strike* (1889).

SMYTH, R. L. (ed.), *Essays in the Economics of Socialism and Capitalism.* Selected Papers read to Section F of the British Association for the Advancement of Science 1886–1932 (1964).

STARK, W., *Jeremy Bentham's Economic Writings*, 3 vols. (ed. 1952–4).

STIRLING, J., 'Social Services Expenditure During the Last 100 years', *Proceedings of the British Association for the Advancement of Science*, 1951, 379–92.

Tariff Reform League, *Reports on Labour and Social Conditions in Germany*, Vols. I and II (1910).

THOMAS, D. S., *Social Aspects of the Business Cycle* (1925).

THOMAS, J. A., *The House of Commons 1906–11. An Analysis of its Economic and Social Character* (Cardiff, 1958).

THOMPSON, PAUL, *Socialists, Liberals and Labour. The Struggle for London 1885–1914* (1967).

TITMUSS, R. M., *Essays on the Welfare State* (1963).

TREBILCOCK, R. C., 'A "Special Relationship"—Government, Rearmament and the Cordite Firms', *Economic History Review*, 2nd series, 19, no. 2 (Aug. 1966), 364–79.

WALKER, F. A., *International Bimetallism* (1896).

WEBB, SIDNEY, 'The Reform of the Poor Law', *Contemporary Review*, 58 (July 1890), 95–120.

WEBB, S. and B., *The History of Trade Unionism* (1894).

—— *Industrial Democracy* (1897).

—— (eds.), *The Break-Up of the Poor Law: Being Part One of the Minority Report of the Poor Law Commission* (1909).

—— (eds.) *The Public Organisation of the Labour Market: Being Part Two of the Minority Report of the Poor Law Commission* (1909).

—— *English Local Government* (Vols. 8 and 9. *English Poor Law History Part II: The Last Hundred Years*, Vols. 1 and 2 (1929).

—— *English Local Government*, Vol. 10. *English Poor Law Policy* (1910) (1963 edn.).

WEBB, S., and COX, H., *The Eight Hours Day* (1891).

WILLIAMS, ERNEST, *'Made in Germany'* (1896).

WILSON, CHARLES, 'Economy and Society in Late Victorian Britain', *Economic History Review*, 2nd series, 18, no. 1 (Aug. 1965), 183–98.

WILSON, TREVOR, *The Downfall of the Liberal Party 1914–1935* (1966).

WOLFE, HUMBERT, *Labour Supply and Regulation* (1923).

WOLFE, J. N., 'Marshall and the Trade Cycle', *Oxford Economic Papers*, N.S. 8 (1956), 90–101.

WOOD, G. H., 'Some Statistics Relating to Working Class Progress since 1860', *Journal of the Royal Statistical Society*, 62 (Dec. 1899), 639–75.

—— 'Trade Union Expenditure on Unemployed Benefits since 1860', *Journal of the Royal Statistical Society*, 63 (Mar. 1900), 81–92.

—— 'Real Wages and the Standard of Comfort since 1850', *Journal of the Royal Statistical Society*, 72 (1909), 91–103.

WOODROOFE, K., *From Charity to Social Work* (1962).

WOODS, R. A., *et al.*, *The Poor in Great Cities. Their Problems and What is Being Done to Solve Them* (1896).

Workman's Times, 'The Legal Eight Hours Demonstration in London', 1 May 1891 (reprinted as a pamphlet) [T.U.C.].

YOUNG, A. F., and ASHTON, E. T., *British Social Work in the Nineteenth Century* (1956).

PART NINE: BIOGRAPHIES AND MEMOIRS

ALLEN, E. M., *Sir Robert Morant. A Great Public Servant* (1934).

ASQUITH, MARGOT, *Autobiography* (1920–22) (1962 ed.).

BARNES, G. N., *From Workshop to War Cabinet* (1924).

BARNETT, H., *Canon Barnett. His Life Work and Friends*, 2 vols. (1919).

BEGBIE, H., *Life of William Booth*, 2 vols. (1920).

BEVERIDGE, W. H., *Power and Influence* (1953).

BRIGGS, ASA, *Social Thought and Social Action: A Study of the Work of Seebohm Rowntree* (1961).

BUNBURY, H. N. (ed.), *Lloyd George's Ambulance Wagon. Being the Memoirs of William J. Braithwaite 1911–12* (1957).

CARTER, V. BONHAM, *Winston Churchill As I Knew Him* (1965) (1967 ed.).

CHAMBERLAIN, AUSTEN, *Politics from Inside. An Epistolary Chronicle 1906–1914* (1936).

CHAMPNESS, E. I., *Frank Smith M.P., Pioneer and Modern Mystic* (1943).

CHURCHILL, R. S., *Winston S. Churchill*, vol. ii (1967). Young Statesman 1901–1914 (and Companion Parts 1, 2, and 3).

COHEN, MAX, *I was One of the Unemployed* (1945).

COHEN, MORTON, *Rider Haggard. His Life and Works* (1960).

ERVINE, ST. JOHN, *God's Soldier. General William Booth*, 2 vols. (1934).

FARRER, T. C., *Some Farrer Memorials* (1923).

FELS, M., *Joseph Fels. His Life-Work* (1920).

FOWLER, E. H., *Life of Henry Fowler, 1st Viscount Wolverhampton* (1912).

FRASER, P., *Joseph Chamberlain. Radicalism and Empire 1868–1914* (1966).

GARDINER, A. G., *John Benn and the Progressive Movement* (1925).

GARVIN, J. L., and AMERY, J., *Life of Joseph Chamberlain*, 4 vols. (1932–51).

GOLLIN, A., *Proconsul in Politics* (1964).

HALDANE, R. B., *Autobiography* (1929).

HAMER, D., *John Morley. Liberal Intellectual in Politics* (1968).

HARRIS, WILSON, *J. A. Spender* (1946).

HAW, GEORGE, *The Life Story of Will Crooks, M.P. From Workhouse to Westminster* (1917).

History Group of the Communist Party, 'Tom Mann and His Times 1890–92', *Our History*, Pamphlets Nos. 26–7, Summer–Autumn 1962.

JENKINS, R., *Asquith* (1964).

JONES, T., *Lloyd George* (1951).

KENT, W., *John Burns Labour's Lost Leader* (1950).

LANSBURY, G., *My Life* (1928).

—— *Looking Backwards—and Forwards* (1935).

LAYTON, WALTER, *Dorothy* (1961).

LONG, WALTER, *Memoirs* (1923).

MACDONALD, J. RAMSAY, *Margaret Ethel Macdonald* (1912).

McKENNA, S., *Reginald McKenna 1863–1945; A Memoir* (1938).

MANN, TOM, *Memoirs* (1923).

MARCHANT, JAMES, *J. B. Paton M.A., D.D. Educational and Social Pioneer* (1909).

MASTERMAN, LUCY, *C. F. G. Masterman* (1939).

MAURICE, C. EDMUND, *Life of Octavia Hill. As Told in Her Letters* (1913).

MORLEY, JOHN, *Recollections*, vol. ii (1917).

MURRAY, ARTHUR C., *Master and Brother. Murrays of Elibanke* (1945).

OWEN, FRANK, *Tempestuous Journey. Lloyd George and his Times* (1954).

PIGOU, A. C. (ed.), *Memorials of Alfred Marshall* (1925).

POSTGATE, R., *The Life of George Lansbury* (1951).

RIDDELL, G., *More Pages from My Diary 1908–1914* (1934).

A. S. and E. M. S., *Henry Sidgwick. A Memoir* (1906).

SIMEY, T. S. and M. B., *Charles Booth, Social Scientist* (1960).

SPENDER, J. A., *The Life of the Right Hon. Sir Henry Campbell-Bannerman, G.C.B.*, 2 vols. (1923).

SQUIRE, ROSE, *Thirty Years in the Public Service. An Industrial Retrospect* (1927).

E. T., *Keeling Letters and Reminiscences* (1918).

TALLENTS, S. G., *Man and Boy* (1943).

THOMAS, HUGH PRESTON, *Work and Play of a Government Inspector* (1909).

THOMPSON, W., *Victor Grayson, His Life and Work. An Appreciation and Criticism* (1910).

THORNE, WILL, *My Life's Battles* (1925).

TILLETT, BEN, *Memories and Reflections* (1931).

Torr, Dona, *Tom Mann and His Times*, Vol. One, 1856–1890 (1956).

Tsuzuki, Chuschichi, *H. M. Hyndman and British Socialism* (1961).

Tuckwell, William, *Reminiscences of a Radical Parson* (1905).

Waley, S. D., *Edwin Montagu, A Memoir and an Account of his visits to India* (1964).

Webb, B., *My Apprenticeship* (1938) (2nd ed. 1946).

—— *Our Partnership* (ed. by Barbara Drake and Margaret Cole, 1948).

Wilson, Harold, *Beveridge Memorial Lecture* (1966).

Winterton, Earl, *Orders of the Day* (1953).

INDEX

Ackland, Thomas, 312

Acts of Parliament
  Development and Road Improvement Funds Act, 1909, 343–4, 346, 351, 357–8, 364–5, 367
  Education (choice of employment) Act, 1910, 350
  Equalization of Rates Act, 1894, 79, 84, 159
  Finance Act (1909), 1910, 268
  Inebriates Act, 1898, 133
  Labour Bureaux (London) Act, 1902, 200, 282
  Labour Exchanges Act, 1909, 288–92 passim, 351, 364
  London Government Act, 1899, 282
  National Insurance Act, 1911, 306–4 passim, 351, 353, 356–7, 360, 364, 379–80
  National Insurance Amendment Act, 1914, 359
  Smallholdings and Allotments Act, 1908, 234
  Trade Boards Act, 1908, 317

Adams, J. B., 294
Adler, N., 257
Alden, Percy, 89, 92, 151, 211, 223, 227, 230, 240–1, 265, 284, 359, 364
Alexandra, Queen of England, 166, 176
Alien Immigration, Royal Commission on, 30
Allan, William, 68
Anderson, the Revd. J., 180
Apprenticeship and skilled employment committees, 32, 201–2
Arnold, Dr. Thomas, 103
Ashley, Percy, 310
Ashley, W. T., 309
Ashton, T. S., 4
Asquith, Herbert Henry, 81–3, 220, 223–4, 231, 233, 244, 262, 265, 269–71, 274, 276–7, 306, 339, 341, 351
Aveling, Edward, 65, 80

Balfour, Arthur, 153, 164–7, 237, 262, 336
Balfour, Gerald, 162, 165, 175, 190

Barnett, Samuel, 55, 111–14, 136–7, 227, 255
Barstow, George, 310
Bartley, Sir George, 91, 163
Beaufoy, Mark, 68
Bell, Richard, 294
Benn, John Williams, 79, 91, 94, 163
Bentham, F. R., 246, 248
Bentham, Jeremy, 295
Bevan, W., 59
Beveridge, William, 11, 134, 151, 168, 184, 209, 227, 261, 276, 367
  analysis of unemployment, 21–6, 28, 31, 36, 349
  attitude to unemployed, 26, 42–3
  and Central (Unemployed) Body, 187, 189–90, 197
  on charitable funds, 179
  on decline of Poor Law, 148–9
  and labour colonies, 189
  and labour exchanges, 200–8 passim, 255–6, 284–95 passim, 352, 355–6
  and RC on Poor Laws, 206, 256, 300–3
  and unemployment insurance, 300–3, 305–34 passim, 361
  and unemployment statistics, 372
  visit to Germany, 205–6
  (see also Board of Trade; Labour Exchanges; Unemployment; Unemployment Insurance)
Blackley, Canon W., 298
Blackwood, Lord Basil, 293
Board of Agriculture, 242, 336, 342, 344–5
Board of Education, 242, 337, 350
Board of Trade, 100–1, 152, 161, 214, 218, 242, 259, 265, 293, 342–4, 350
  and labour exchanges, 200, 205, 279–95, 355
  and unemployment insurance, 300, 301, 304–34, 357, 360
  and unemployment statistics, 100–1, 371–2
  (see also Beveridge; Churchill; Labour Exchanges; Llewellyn Smith; Unemployment Insurance)

Booth, Bramwell, 129, 131–2
Booth, Charles, 11, 248
  analysis of unemployment, 15, 20–1,
    25–6, 31, 35–6, 349
  and labour colonies, 119
  and London survey, 155, 161
  and social evolution, 20, 28, 288
  and unemployment statistics, 372
  and working-class budgets, 35–6, 214
Booth, William, 124–35 passim, 142,
    207, 281
Bottomley, Horatio, 324
Bousfield, William, 91
Boutmy, Émile, 228
Bowley, Arthur, 22, 372
Brabrook, Sir Edward, 177
Bradbury, John, 320, 332
Bradford, Sir Evelyn, 83
Braithwaite, William, 320, 324
British Women's Emigration Society,
    185
Browne, Sir Benjamin, 308–9
Brunner, Sir John, 217–20, 230
Bryce, James, 1st Viscount, 223–4, 227,
    250
Burdett, H. C., 101
Burns, John, 67, 91, 192, 209, 220, 243,
    253–4, 257, 262, 312, 326, 341–2
  and eight hours movement, 69, 71
  and labour colonies, 142, 182–3, 191–
    3, 195–7
  and labour exchanges, 205, 207
  as President of LGB, 142, 167, 174,
    176, 179, 182–3, 187, 191–3, 196–7,
    238–9, 241, 266, 268, 274–6, 288
  view of social reform, 231–2, 240
  (see also Labour colonies; Local
    Government Board; Unemployed
    Workmen Act)
Buxton, Sydney, 67, 216, 266, 275
  and Unemployed Workmen Act,
    158, 163–4, 167, 226
  as President of Board of Trade, 312,
    317–32
  (see also Unemployment Insurance)

Cadman, Elijah, 128
Campbell-Bannerman, Sir Henry, 154,
    158, 164, 167, 174, 188, 238, 265,
    267, 271, 336
  and eight hours movement, 70
  and SC on Distress from Want of
    Employment, 91–5, 281

  and unemployment policy, 213–32
    passim
Canals and Inland Navigation, RC on,
    233, 337
Carrington, Lord, 216, 233–4
Cavendish, Lord Richard, 345
Central (Unemployed) Body, 167–210
    passim, 235, 295, 300
Chadwick, Sir Edwin, 74
Chalmers, Sir Robert, 327–8, 332
Chalmers, Thomas, 103
Chamberlain, Austen, 329
Chamberlain, Joseph, 75–7, 117, 213–
    14, 218, 221–2, 225
Chamberlain Circular, 75–7, 79, 96,
    101, 108, 157, 208, 219, 336, 363
Chandler, Francis, 258
Chaplin, Henry, 343
Chapman, Sidney, 301
Charity Commissioners, 104
Charity Organisation Society, 105–10,
    112–13, 118, 134, 143, 146–9, 174,
    200, 209, 250, 301, 315, 375
Charrington, Frederick, 151
Christian Social Service Union, 123,
    187–8
Church Army, 95, 185–7, 281
Churchill, Winston, 207
  and labour exchanges, 284–92, 352,
    354–5
  as President of the Board of Trade,
    207, 274–7, 284–316 passim
  and unemployment insurance, 303–
    316 passim
  views on social reform, 262–5, 271,
    339–40, 346, 351, 356, 362, 365–6
Civil Service, R.C. on, 353
Clynes, John, 317
Coast Erosion and Afforestation, RC
  on, 337–8
Cohen, C. Waley, 169
Cohen, Nathaniel, 279
Cohen, W. S., 294
Collings, Jesse, 104
Colonisation, SC on, 127
Conservative Party, 70, 74–5, 343–4, 365
Cox, Harold, 67
Crooks, Will, 164, 196, 235, 237, 240
Cunninghame-Graham, Robert, 66

Davy, J. S., 193–5, 239
Development Commission, 340–6, 351,
    357–9

Dilke, Sir Charles, 164, 276
Distress Committees, 168–77, 183–5, 202
Distress from Want of Employment, SCs on
(1895) 33, 49, 73, 86, 89–95, 132, 139, 192, 281, 372
(1896) 95–7, 101, 124, 148
Drage, Geoffrey, 21
Drummond, C. J., 309
Dunn Gardner, Rose, 189–90, 197

East End Emigration Fund, 106, 185
Edward VII, King of England, 165
Egham Free Registry, 204, 279–80
Eight Hours League, 60–1, 99
Eight Hours Movement, 58–73, 99
Emigration
as remedy for unemployment, 22, 127–8, 180, 184–7
Engels, Friedrich, 54–5, 99
English Land Colonisation Society, 123
Ensor, Robert, 255, 258
Eugenics Education Society, 46–7

Fabian Society, 12, 57, 64, 330
Fels, Joseph, 140, 155, 197–8
Fitzmaurice, Lord Edmond, 167
Foster, Sir Walter, 216
Fowler, Henry, 93, 223–4, 226
Foxwell, H. S., 9
Free Trade, 51, 214–16, 270
Friendly Societies, 295–6
Fry, George, 288
Furness, Sir Christopher, 230

Gates, Wilson, 124
General Federation of Trade Unions, 236, 239
Gibb, Sir George, 346
Giffen, Robert, 29, 72, 111, 119
Gladstone, Herbert, 213, 215–17, 229, 231
and unemployment policy, 219–27
Gladstone, W. E., 66, 69–70, 75, 100, 336
Gorst, Sir John, 130
Grayson, Victor, 273
Gretton, le Mesurier, 138
Grinling, C. H., 188, 196
Grove, Archibald, 85
Guest, Ivor, 337

Haldane, Richard, 230–1, 262, 267, 269–70
Hall, Sir Daniel, 346
Halsbury, Lord, 246
Hamilton, Lord George, 248, 250, 254–5, 261
Hancock Nunn, Thomas, 261
Hardie, James Keir, 71, 80, 82, 84–5, 93, 135–6, 163–4, 211, 236–41, 343
Harvey, T. Edmund, 359
Hawtrey, Ralph, 343
Hazell, Walter, 124, 138
Health Insurance, 306–7, 311, 316, 320–30 passim, 357
Henderson, Arthur, 162, 275
Henderson, Frederic, 292
Hill, Octavia, 248, 330
Hills, Arnold, 86–8, 113
Hirst, Francis, 232, 271
Hobhouse, Charles, 326
Hobson, J. A., 4, 10, 22–3, 125, 335, 364, 367
Holloway, H. T., 310
Home Colonisation Society, 120–3, 187
Home Office, 56, 81, 83, 100, 259, 294
Hopwood, Sir Francis, 345–6
Housing of the Working Classes, RC on, 38
Hyndman, H. M., 60–61

Income Tax, 214–15, 269–70
Independent Labour Party, 89, 223, 230, 235, 335
Irish Nationalist Party, 75, 343, 365
Isaacs, Sir Rufus, 320

Jack, J. M., 291, 317
Jackson, Cyril, 175, 252–3
James, Sir Henry, 131

Knollys, W. E., 140

Labour Colonies, 115–44 passim, 180, 187–99, 227, 237, 257, 259, 260
Labour, RC on, 68, 71–2, 200, 280–1, 300
Labour Exchanges (bureaux, registries), 84, 109, 126, 131, 180, 199–208, 242, 255–6, 260, 264, 277–95, 306, 309, 311, 351–6, 368
Labour Party, 238, 331, 340, 357, 364, 365
its Unemployed Workmen Bill, 241–4, 263

Labour Representation Committee, 157, 162, 236–7

Lamb, David, 129–30

Land Reform, 94–5, 115–44 *passim*, 216, 227, 233–4

Lansbury, George, 93, 166, 168, 186–7, 207, 237, 258–60
and labour Colonies, 139–40, 187–91, 196–8, 248
(*see also* Central (Unemployed) Body; Poplar Guardians)

Lawrence, William, 91

Lawson, H., 163

Leathes, Stanley, 294

Liberal Party, 57, 69–70, 75, 212–35 *passim*, 238, 278, 336, 362, 363, 365

Light Railways Commission, 98

Llewellyn Smith, Sir Hubert, 11, 18, 24, 29–30, 121–2, 131, 207, 283
analysis of unemployment, 12–15, 28, 33, 349
and labour exchanges, 281, 288, 292, 355
and unemployment insurance, 303–34
and unemployment statistics, 100–1, 371 n.
(*see also* Board of Trade)

Lloyd George, David, 231, 268, 271, 274, 351
and Development Act, 278, 340–7
and health insurance, 276–7, 306, 311, 316, 318–30 *passim*
and 'right to work', 226–7, 346
and unemployment insurance, 276, 322–8

Local Government Board, 75, 77, 80, 135, 142, 153, 168, 172, 219, 242, 246, 259, 342, 344, 371
and Chamberlain circular, 75–6, 81, 88, 100
and labour colonies, 188, 191–2
and labour exchanges, 203, 205
negative policies of, 78, 81–2, 85, 90, 92–4, 96, 140
not responsible for unemployed, 56, 161–2, 265, 279, 282–3
strict administration of Poor Law, 52–4, 90, 146–7
and Unemployed Workmen Act, 159–210 *passim*

Local Taxation, 159, 216, 225, 233, 267–9, 275, 351, 369–70

Loch, C. S., 41, 105, 108, 112, 118, 120, 136–7, 174–5, 248

Lockwood, Colonel, 192

Lodge, Oliver, 233

London Congregational Union, 119

London County Council, 67, 150, 202

London Trades Council, 80

London Unemployed Fund, 141, 153–8, 169, 173, 188, 195, 201, 219

Long, Walter, 140, 152–5, 158–60, 165, 176, 190, 212, 237, 240, 246–7

Lowry, Arthur, 202, 204

MacDonald, James Ramsay, 162, 169, 237, 241–2, 329

MacKay, Thomas, 92, 134

Mackenna, Reginald, 254–5, 274

Macleod, F. H., 288

Maddison, F., 244

Malthus, Thomas, 103

Mann, Harold, 143

Mann, Tom, 20, 60–1, 64–5, 71

Mansion House Committees, 38, 44–5, 104, 108, 111–15, 121, 136, 151–2, 169

Marshall, Alfred, 4, 9, 109, 119, 212

Marx, Eleanore, 65

Masterman, C. F. G., 213, 241, 265, 275, 336

Mather, William, 67, 91, 94–5

Matthews, Henry, 66

Mavor, James, 118

Mayhew, Henry, 16

Maynard, H. R., 151, 170, 255

Metropolitan Board of Works, 75, 78

Metropolitan Common Poor Fund, 79, 84, 97, 159

Metropolitan Society for Befriending Young Servants, 106

Midlands Re-afforesting Association, 233

Mills, the Revd. Herbert, 85, 119–23

Ministry of Labour, demand for, 237, 259, 282, 352

Minto, Lord, 229

Mitchell, Isaac, 162, 241

Mitchell, W. H., 294

Money, Sir Leo Chiozza, 232

Monro, Sir Horace, 325

Montagu, Edwin, 342

Moore, Harold, 131

Morley, John, 70, 223, 227, 229, 230–1, 271

Morpeth, Viscount, 344
Mowatt, Sir Francis, 220
Murray, Alexander, 323
Murray, Sir George, 327, 345

Nathan, Sir Matthew, 350
National Committee for the Prevention of Destitution, 330
'National Efficiency', 213, 231, 366–7
National Fair Trade League, 55
National Health Insurance Commission, 357
National Insurance, see Health Insurance, Unemployment Insurance
Naylor, T. E., 317

Office of Works, 352
O'Grady, J., 240
Owen, Sir Hugh, 93, 96, 135
Owen, Robert, 124

Paton, John, 123
Patten Macdougall, J., 248
Pease, Edward, 258
Pease, J., 326
Phelps, the Revd. Lancelot, 248, 250, 252, 260
Physical Deterioration, Inter-Departmental Committee on, 141
Poor Law, 1–2, 4–5, 8, 49, 51, 53–4, 81–2, 90, 92–3, 96–7, 154, 223, 278, 346
    changing attitudes towards, 147–50
    crisis in, 266
    irrelevance to unemployed, 148–50
    and labour colonies, 135–41
    and political change, 146–7
    reform of, 245–8, 261, 288, 351
    rising cost of, 145–6
    statistics, 371–3
    (See also Local Government Board; Poor Laws, RC on)
Poor Laws, RC on, 31–2, 109, 152, 164–5, 175, 178, 180–1, 188, 199, 206, 211, 233, 235, 244, 245–64, 284, 300–2, 350
    Majority Report, 26–7, 142, 260–1, 263, 265, 303
    Minority Report, 26–7, 27–8, 142, 168, 223, 243, 258–60, 263–5, 287–8, 303, 309, 338–9, 352, 357
Poplar Guardians, 139–40, 188, 192, 193–5, 266

Pigou, A. C., 9
Price Hughes, Hugh, 113–14
Pringle, the Revd. J. C., 175, 253
Provis, Sir Samuel, 160, 162, 248, 284
Public Works, 1, 3, 5, 57, 73–9, 108–9, 222–5, 227, 237, 242, 244, 259–61, 263, 334–46 passim, 351
Public Works Loan Commissioners, 78, 97

Quelch, Harry, 238

Relief Works, 86–8, 111–15, 149, 180–4
Rey, Sir Charles, 288, 293
Ricardo, David, 2
Riddell, George, 276
Rider Haggard, Henry, 130–1, 337
'Right to Work', 74, 88, 99, 120, 162–4, 227, 235–44 passim, 260, 324, 346, 363, 366
Ripon, Marquis of, 167
Ritchie, C. T., 77–8
Road Board, 341, 343, 345–6
Robertson, J. M., 10, 335
Robinson, Sir Henry, 248
Rosebery, 5th Earl of, 88, 134, 213
Ross, Bishop of, 260
Rowntree, B. S., 37, 143, 214
Russell, Thomas, 95

Salisbury, 3rd Marquess of, 66, 74, 336
Salisbury, 4th Marquess of, 160
Salvation Army, 95, 124–35 passim, 185–7, 376
Samuel, Herbert, 163, 265, 359
Scammell, E. T., 280
Schloss, David, 300
Self-Help Emigration Society, 123–4, 138, 185
Senior, Nassau, 2
Sexton, James, 162, 165
Shaw Lefevre, J., 80
Shuttleworth, Lord, 337
Sidgwick, Henry, 74
Siemens, Alexander, 291–2
Simon, Sir John, 321, 332
Smart, Bolton, 190–1
Smart, William, 260
Smith, Frank, 126, 128
Smith, Thomas, 301
Social Democratic Federation, 55, 57, 65, 80, 221

Social Policy
  continental models for, 117–18, 205–
    6, 276–8, 281, 298–300, 302, 328
  some determinants of, 266–71
  discussion of, 212–72
  underlying principles of, 362–6
Spender, J. A., 271
Squire, Rose, 36, 256
Stead, the Revd. F. H., 166
Stead, W. T., 101, 124
Steele-Maitland, Arthur, 36, 256
Strachey, St. Loe, 270

Tariff question, 57, 100, 213–17, 221–
  2, 225, 233, 270–1, 319, 363, 366
Tawney, R. H., 151, 257
Technical Education, 32–3, 217–18,
  232, 285, 300, 325, 329, 350
Tennant, May, 169, 188
Thorne, Will, 85, 87
Threlfall, James, 59
Tillett, Ben, 354
Tozer, the Revd. Wickham, 280
Trades Union Congress, 99, 152, 157,
  165, 235–6, 239, 289–91, 317, 354
Trade Unions, 57–8, 235–6
  and eight hours movement, 57–73
  and labour exchanges, 203–4, 277,
    287, 289–91, 354
  and opposition to Salvation Army,
    132–3
  and out of work funds, 52, 246, 259,
    273, 296–304
  and Unemployed Workmen Act,
    162–3
  and unemployment insurance, 315–
    18, 325, 331–3
  and unemployment statistics, 8, 371–
    2, 374
Treasury, 98–9, 173, 179, 214, 231,
  288–9, 294, 306, 320–3, 327–8, 332,
  339, 344, 352, 356, 365

Unemployed,
  attempts to organize, 55–6, 80–4, 87,
    89, 152–3, 166, 273
  characteristics of, 17–18, 77, 115–16,
    125
  composition of, 8, 28, 89, 155
  demonstrations, 55–6, 79–83, 87
  moral character of, 42–8, 114, 116,
    153, 251–2, 262, 338
  political significance of, 54, 56–7

  statistics of, 8–9, 33–4, 312, 362, 371–
    82
Unemployment
  Cabinet committees on, 180, 233,
    275, 312–14, 325–6
  and casual labour, 12–25, 47, 99, 155,
    213, 274, 277, 306, 349, 357
  and charity, 51, 54, 102–15 passim,
    157, 179–80, 209
  cyclical, 5, 12–15, 26–9, 48, 99, 223,
    304, 328, 351
  definition of, 4, 11–12
  and economic theory, 1–2, 6, 7, 9–10,
    23–4, 102–3, 105–6, 212, 334–5,
    368
  and housing, 38–9, 49
  and immigration, 29–31
  inadequate information about, 8–9,
    22, 33–4, 47, 80, 362, 371–3
  as an industrial problem, 8–33 passim
  and juvenile labour, 29, 31–2, 259–61,
    263–4, 287, 349–50
  lack of co-ordinated policy on, 351–2
  and local administration, 168–210
    passim, 216, 242–3, 274–5
  and physical disability, 40–2, 49, 125
  and Poor Law, 1–2, 5, 51, 53–4, 102–
    3, 212
  and poverty, 34–7, 49–50, 52
  seasonal, 12–13, 285, 309
  and socialism, 10, 235
  as a social problem, 8, 33–47
  and state intervention, 5–6, 88, 129–
    35, 141, 211, 216, 264, 362–6
  and underconsumption, 10, 22, 120,
    225, 235–6, 325, 367
Unemployment Insurance, 227, 240,
  259–60, 263, 274, 276–8, 285, 295–
  334 passim, 351, 356–7, 360–1, 368
'Tramp' Benefit, 296–7

Vagrancy, Departmental Committee on,
  141–2, 188

Wakefield, the Revd. Henry Russell
  and Central (Unemployed) Body,
    169, 176, 178, 183, 186, 188, 191,
    196, 207
  and RC on Poor Laws, 248, 258–9
Webb, Beatrice, 11, 62, 76, 134, 161,
  197–8, 200, 243, 248, 284
  analysis of unemployment, 8, 17–18,
    42–3, 44, 125, 349, 367

attitude to unemployed, 42–3, 103
and influence on politicians, 264–7
and RC on Poor Laws, 249–64, 338–9
and unemployment insurance, 303–4,
    330, 365
Webb, Sidney, 11, 62, 134, 161, 197–8,
    200, 207, 213, 220, 243, 284, 346
analysis of unemployment, 27–8, 42–
    3, 125, 349, 367
and eight hours day, 67, 72
and influence on politicians, 264–7
and RC on Poor Laws, 249–67, 338–9
and unemployment insurance, 303–4,
    309, 315, 330

Weiler, Adam, 59
Wemyss, Lord, 256
West Ham, unemployment problem in,
    84–9, 110, 222
White, Arnold, 142
Whitechapel, policy of guardians, 137–9
Williams, Jack, 82
Willinck, Sir Henry, 118
Wilson, Philip Whitwell, 213, 243–4,
    324
Wilson Fox, Arthur, 30
Wood, G. H., 372
Woodworth, Arthur, 45